Einstein Studies

Editors: Don Howard   John Stachel

Published under the sponsorship
of the Center for Einstein Studies,
Boston University

A.J. Kox   Jean Eisenstaedt
Editors

The Universe of General Relativity

Birkhäuser
Boston • Basel • Berlin

A.J. Kox
Universiteit van Amsterdam
Instituut voor Theoretische Fysica
Valckenierstraat 65
1018 XE Amsterdam
The Netherlands

Jean Eisenstaedt
Observatoire de Paris
SYRTE/UMR8630–CNRS
F-75014 Paris Cedex
France

AMS Subject Classification (2000): 01A60, 83-03, 83-06

Library of Congress Cataloging-in-Publication Data
The universe of general relativity / A.J. Kox, editors, Jean Eisenstaedt.
   p. cm. – (Einstein studies ; v. 11)
   Includes bibliographical references.
   ISBN-13 978-0-8176-4380-5 (alk. paper)
   ISBN-10: 0-8176-4380-X (alk. paper)
     1. Relativity (Physics)–History–Congresses. 2. General relativity
(Physics)–History–Congresses. 3. Gravitation–History–Congresses. 4.
 Cosmology–History–Congresses. 5. Unified field theories–History–Congresses. 6. Einstein, Albert,
1879-1955. I. Kox, Anne J., 1948- II. Eisenstaedt, Jean III. Series.

QC173.5.U55 2005
530.11–dc22   2005047817

ISBN-10 0-8176-4380-X   e-IBSN 0-8176-4454-7   Printed on acid-free paper.
ISBN-13 978-0-8176-4380-5

Printed in the United States of America.   (TXQ/EB)

9 8 7 6 5 4 3 2 1

www.birkhauser.com

This volume is dedicated to the memory of Peter Havas

Contents

Preface

A century ago, in 1905, Albert Einstein published, "On the Electrodynamics of Moving Bodies," in which the foundations were laid for the Special Theory of Relativity. Ten years later his relativistic theory of gravitation and the General Theory of Relativity appeared. Fifty years ago, Einstein passed away in Princeton.

In the 1980s, John Stachel, then Editor of the *Collected Papers of Albert Einstein*, brought together a group of historians, philosophers, physicists, and mathematicians who had one thing in common: a lively interest in the history and foundations of the theories of relativity. At a meeting in 1986 at Osgood Hill, this group met for the first time to discuss the prehistory, development, reception, and other aspects of relativity. It was the beginning of a valuable tradition. Since then every three or four years a meeting has been organized during which historical and foundational issues in general (and special) relativity have been discussed. Osgood Hill was followed by Luminy in 1988. Then came Johnstown (1991), Berlin (1995), Notre Dame (1999), and finally Amsterdam (2002), the proceedings of which are presented in this volume (supplemented with some papers from the preceding meeting).

Once again these articles clearly show that an historical approach can lead to new insights into the development and elaboration of relativity. The prehistory of special relativity and an early attempt at a relativistic theory of gravitation are covered in papers by John Stachel and Shaul Katzir, respectively. The birth and early history of general relativity are the topics of the papers by Jürgen Renn, John Norton, and Christoph Lehner. They are followed by Daniel Kennefick's contribution, which focuses on one of the major problems in general relativity: the problem of motion. That general relativity has close ties to mathematics becomes clear from Katherine Brading's paper on the conservation laws and Robert Rynasiewicz's article on the axiomatization of the theory.

Five papers then follow on cosmology, a topic of research that owes much of its existence to General Relativity. George Gale, Helge Kragh, and José Sánchez-Ron take a closer look at the work of Herbert Dingle and Willem de Sitter, George Gamov, and George McVittie, respectively, while Chris Smeenk discusses an aspect of more contemporary cosmology.

As is well known, during the second part of his active life Einstein pursued the goal of creating a unified field theory to unify gravitation and electromagnetism. That he was not the only one searching for a unified theory becomes clear from the contributions of Ulrich Majer and Tilman Sauer on the work of David Hilbert and of Hubert Goenner on the early history of unified field theory. Next, Daniela Wünsch studies the work of Einstein and Theodor Kaluza, which is not just of historical value but in recent years has provoked much interest. Finally, James Mattingly presents some considerations on the necessity of quantum gravity.

The volume concludes with three contributions on a more "personal" level. First, Milena Wazeck uses the recently-discovered papers of the physicist Ernst Gehrcke to provide a view of the reception of relativity by the general public. Next, the two renowned relativists Josh Goldberg and Ted Newman share with us their recollections of the time when they were both actively contributing to the further development of the General Theory of Relativity.

A.J. Kox, Amsterdam
J. Eisenstaedt, Paris
August 2005

The Universe of General Relativity

1

Fresnel's (Dragging) Coefficient as a Challenge to 19th Century Optics of Moving Bodies

John Stachel

Center for Einstein Studies, Boston University, Boston, MA 02215, U.S.A.;
stachel@buphy.bu.edu

1.1 Introduction

It has been suggested that, during the latter half of the 19th century up to about 1890, the optics of moving bodies was considered to be a more-or-less unproblematic branch of physics. In view of the continuing success of Fresnel's formula for the dragging coefficient (hereafter called Fresnel's coefficient) in explaining all new experimental optical data to order v/c, "There were simply no major problems to solve here, or so it was generally thought" (Buchwald 1988, 57).

These words are the summation of the following quotation: "In 1851 Armand Fizeau was able to measure the Fresnel "drag" coefficient, and in 1873 Wilhelm Veltmann demonstrated that no optical experiment with a terrestrial source of light can, to first order, detect motion through the ether if the drag coefficient obtains. Consequently, to this degree of accuracy, Fresnel's original theory which requires a very slight transport of the ether by transparent bodies was quite satisfactory (ibid.)."

(Schaffner 1972) includes a similar comment: "...Fresnel was able to formulate a simple and elegant explanation of Arago's results on the basis of the wave theory of light; an explanation which not only accounted for aberration effects then known but which was subsequently confirmed in a number of ways throughout the nineteenth century (ibid., 24)."

As we shall see, both Buchwald and Schaffner conflate the continued empirical success of Fresnel's formula with the ultimately unsuccessful attempts by Fresnel and others to find a satisfactory theoretical explanation of the formula. I maintain that:

1) On the basis of contemporary documentation, one can demonstrate that, by the first decades of the second half of the 19th century, that is before the Michelson and Michelson–Morley (M-M) experiments, the empirical success of Fresnel's formula in explaining all first-order experiments actually created a critical situation within the optics of moving bodies.[1]

2) The challenge presented by Fresnel's formula was the first indication of the breakdown of classical (Galilei–Newtonian) kinematics, and could have led directly to the search for a new kinematics.

3) The way in which the challenge of this first crisis was met by Lorentz, on the basis of Maxwell's electrodynamics, the stationary ether hypothesis, and the old kinematics, exerted a tranquilizing influence that served to postpone the search for a new kinematics until a new critical situation in the electrodynamics of moving bodies arose, largely due to the results of the second-order M-M experiments, and was resolved in 1905 by Einstein.

1.2 Arago and the Emission Theory

Although the story is mainly concerned with the wave theory of light, I shall start it with Dominique-François Arago's work on the emission theory. In 1810, Arago, still an adherent of this theory, decided to test a hypothesis that seemed to him "both natural and probable on its basis," namely that "stars of differing magnitude can emit [light] rays with different speeds."

Arago's test of this hypothesis was based on the refraction of these rays by a prism, and he found that, to the accuracy of his experiment, rays from all stars were refracted by the same angle. He concluded that "light moves with the same speed no matter what the body from which it emanates." He regarded this conclusion as so important that he later cited it (Arago 1830) as one of the main reasons why "the emission theory now has very few partisans." Indeed, "Arago became a vocal critic of the Newtonian emission theory and, by 1816, an ardent supporter of the undulatory theory" (Hahn 1970, 201).

Arago noted that his conclusion, viz. that the speed of light is independent of the speed of its source, depends on what he called "Newton's principle:" Light beams entering a prism with different speeds are refracted through different angles; and he decided to test this principle. Again he used a prism, but while his earlier experiment had compared light from different stars at the same time, he now used the prism to compare light from the same star at different times of the year. Since the earth moves around the sun during the course of a year, he expected the velocity of light relative to the prism to change and hence, on the basis of "Newton's principle," the angle of refraction to change. But he found, as Fresnel 1818 summarized Arago's results, "that the motion of the terrestrial globe has no noticeable influence on the refraction of rays that emanate from the stars." The situation is summarized in the following table:

Arago's Experiments—Is the Null Result a Problem for:

Starlight refracted by:	Emission Theory	Wave Theory-Stationary Ether
1. Different Stars-Same Time	Yes	No
2. Same Star-Different Times	Yes	Yes

Table 1.1. Emission vs. Wave Theory

On the basis of the emission theory, Arago could only explain this new result by invoking what Fresnel called a "quite strange hypothesis that is quite difficult to accept" (viz., while light travels at many different velocities, the eye is sensitive only to rays traveling at one speed). Having become well acquainted with Fresnel's work on the wave theory by the mid 1810s, Arago asked the latter if the wave theory could provide an explanation. Fresnel now attempted to explain both aberration and Arago's second prism experiment (the wave-theoretical explanation of the first is obvious, as we shall see below).

1.3 Fresnel and the Wave Theory

Fresnel's work on the wave theory was based on the hypothesis of a stationary or immobile ether. On this basis, the explanation of Arago's first experiment is obvious: the speed of propagation of a wave in a medium is independent of the velocity of the source of the wave. The explanation of aberration in a vacuum is also fairly simple (for details, see, e.g., Janssen and Stachel 1999). But two results:

1) that the angle of aberration remains the same in a telescope using lenses or even filled with water;

2) the null result of Arago's second experiment;

cannot be explained without an additional hypothesis attributing some mobility to the ether within a moving medium. Fresnel showed that both of these results can be explained on the assumption that a medium moving through the stationary ether only drags light propagating through it with a fraction of the medium's speed. If the index of refraction of the medium at rest is n, then Fresnel defined the dragging coefficient

$$f = (1 - 1/n^2)$$

and assumed that light propagating in the medium is dragged along with a velocity

$$v_{drag} = f v_{med}.$$

What the dragging coefficient accomplishes is summarized in the following Table:

Wave Theory: Stationary Ether	
Without Dragging Coeff.	With Dragging Coeff.
ABERRATION WITH OPTICAL MEDIUM	
Problem	No Problem
ARAGO'S EXPERIMENT	
Problem	No Problem

Table 1.2. Why Fresnel needs f

What physical explanation does Fresnel offer for the value of the dragging coefficient?

[I]t is only a part of this medium [the ether] which is carried along by our earth, namely that portion which constitutes the excess of its density over that of the surrounding ether. By analogy it would seem that, when only a part of the medium is displaced, the velocity of propagation of the waves can only be increased by the velocity of the center of gravity of the system (Fresnel 1818a, 631; translation from Schaffner 1972, 129, translation modified).

Since the speed of a wave in an elastic medium and (inversely) on the elasticity of the medium, Fresnel's explanation amounts to assuming that the elasticity remains the same in the prism and the ether and that only the density varies between them. In a note later added to the letter, he admits that other hypotheses regarding the elasticity are equally possible, but adds:

But whatever the hypothesis one makes concerning the causes of the slowing of light when it passes through transparent bodies, one may always ... mentally substitute for the real medium of the prism, an elastic fluid with the same tension as the surrounding ether, and having a density such that the velocity of light is precisely the same in this fluid and in the prism, when they are supposed at rest; this equality must still continue to hold in these two media when carried along by the earth's motion; these, then, are the bases upon which my calculation rests (Fresnel 1818b, 836; translation from Schaffner 1972, 134–135, translation modified).

This is the first, but hardly the last time that we shall come upon a disturbing problem: the lack of uniqueness in explanations of Fresnel's coefficient. It has been suggested, notably by Veltmann (see below), that Fresnel first found the value of the coefficient that explained the anomalous experimental results, and then cooked up a theoretical explanation for this value.

During the course of the nineteenth century, various hypotheses about the motion of the ether were introduced to derive the value of f. Even if one assumes that only the density of the ether varies from medium to medium, various possibilities about its state of motion inside a moving body were proposed, all cooked up to lead to the same value of f. To cite only three:

1) A part of the ether moves with the total velocity of the moving body (Fresnel 1818a).

2) All of the ether is dragged along with a part of the velocity of the moving body (Stokes 1846).

3) Various portions of the ether move with all velocities between zero and the total velocity of the moving body (Beer 1855).

The very fact that such widely differing hypotheses could be invoked to explain equally well the value of f raises a good deal of doubt about all such "mechanical ether" explanations.

Wave Theory: Ether is:

Stationary (Fresnel)   Dragged-Along (Stokes)

ABERRATION-Empty Space

No Problem   Problem but Stokes Solves It

ARAGO'S 2nd EXPERIMENT

Problem—f needed No Problem—f not needed

Table 1.3. Stokes vs. Fresnel

1.4 Stokes Saves the Dragged-Along Ether

Of course, life would be much simpler if one could just assume that the ether is dragged along entirely by a moving body. Fresnel realized this, but remarks that he cannot think of a mechanism that would then explain aberration In Stokes 1845, such an explanation is offered. Without going into detail about Stokes's explanation, suffice it to say that there is a striking difference between his explanation and earlier ones. For Stokes, aberration involves a real bending of light beams as they pass from empty-space ether into a moving medium, which even an observer at rest would see; while both for the corpuscular and immobile-ether wave theories, aberration is a sort of optical illusion, apparent only to a moving observer. Stokes theory had a curious history over the next half-century (see Janssen and Stachel 1999 for a bit of the story), but its attractiveness was immediately apparent, and the question was soon raised: Who needs Fresnel's dragging coefficient? As Table 1.3 shows, it seems that nobody did. Two equally good hypotheses about the relation between the ether and ponderable matter immobility with Fresnel's coefficient and total dragging without it both seemed available to explain all the known experimental facts in the optics of moving bodies.

Stokes commented:

> This affords a curious instance of two totally different theories running parallel to each other in the explanation of phenomena. I do not suppose many would be disposed to maintain Fresnel's theory, when it is shewn that it may be dispensed with, inasmuch as we would not be disposed to believe, without good evidence, that the ether moved quite freely through the solid mass of the earth. Still, it would have been satisfactory, if it had been possible, to have put the two theories to the test of some decisive experiment (Stokes 1846, 147).

1.5 Fizeau Forces Fresnel's Formula on Physicists

So things stood until 1850, when an apparently decisive experiment was performed. As Ketteler 1873 (a historical review) reports:

[S]uddenly (1850) Fizeau's famous experiment, by means of which the "entrainment" of the ether by a moving transparent medium was actually proved, brought light into this chaos, and now Fresnel's viewpoint gained a firmer foundation and with it new adherents.

Fizeau 1851 reports the results of this experiment, which was taken to demonstrate conclusively the need for f. He measured the speed of light in a moving optical medium, water in his case (see Janssen and Stachel 1999) for a discussion, by splitting a beam of light into two beams, one of which traveled through a tube of running water in the sense of the water's motion, the other in the opposite sense. His results may be summarized as follows. Let

$$c_{\text{med}} = c/n = \text{the speed of light w.r.t the medium,}$$
$$v_{\text{lab}} = \text{the speed of the medium w.r.t the laboratory,}$$
$$c_{\text{lab}} = \text{the speed of light w.r.t. the laboratory.}$$

From interference effects between the two beams, Fizeau drew the conclusion that:

$$c_{\text{lab}} = c_{\text{med}} \pm f v_{\text{lab}},$$

the sign depending on whether the water is flowing in the same or opposite direction to that of the light propagation.

Now, even adherents of Stokes' theory needed to invoke f to explain Fizeau's results. The dragging coefficient seemed unavoidable! Stationary ether theories once again became the favored ones—"stationary" being interpreted to include dragging effects in moving media, of course.

1.6 Formula Yes! Explanation No!

In spite of its empirical validation, many leading experts in the field, starting with Fizeau himself, and including Ketteler, Veltmann, Mascart, Poincaré, Potier and Lorentz (all before 1890), carefully distinguished between the empirical success of the formula and the dubious nature of Fresnel's—and all other—explanations based on the motion of the ether, in whole or in part. Here are some representative samples:

Fizeau (1851): The success of this experiment seems to me to entail the adoption of Fresnel's hypothesis, or at least of the law that he found to express the change in the speed of light resulting from the motion of bodies; for although this law has been verified ... Fresnel's conception would appear so extraordinary, and in several respects so difficult to accept, that one would require still more proofs and a deepened examination by mathematical physicists [*géomètres*], before accepting it as the expression of the way things really are.

Ketteler (1873): That indeed the speed of propagation of light undergoes a modification corresponding to Fresnel's theory as a result of translation [of the medium] has been experimentally confirmed by Fizeau's experiments with moving fluids. It is one thing to simply acknowledge this modification, another to accept Fresnel's conception of the way in which it comes about.

Veltmann (1873): Fresnel sought to bring this result [i.e., Arago's] into harmony with the wave theory, saw himself thereby compelled to adopt a particular hypothesis, that indeed, as concerns its physical basis, itself again offered insurmountable difficulties, yet for the rest accomplished its aim.

Mascart (1872): In any case, to be rigorous, it must be stated that Fizeau's experiment only verified that the dragging of the [light] waves by moving media is in agreement with [Fresnel's] formula (1) and that one can replace Fresnel's hypothesis by any other hypothesis that will finally lead to the same formula, or a slightly different one.

Mascart (1893): The considerations that guided Fresnel are insufficient; the formula to which he was led by a happy intuition only has an empirical character, which should be interpreted by theory.

Poincaré (1889): We do not know any satisfactory theory to justify that hypothesis [i.e., a hypothesis that would lead to Fresnel's formula].

Lorentz (1886): It will be the task of the theory of light to explain [*rendre compte*] the value that observations give for the dragging coefficient.

1.7 Further Empirical Success Brings Increasing Theoretical Doubt

Indeed, the very empirical successes of Fresnel's formula made ever more evident the inadequacy of all explanations of it based on partial or total dragging of the ether inside a moving optically transparent body. Two further results made this crystal clear:

1) Veltmann (1870) demonstrates experimentally that Fresnel's formula must be applied using the appropriate (different) index of refraction for each color of light. This means that, however the ether moves, it must move differently for each frequency of light. But what happens when white light (or indeed any mixture of frequencies) passes through a transparent medium?

2) Mascart (1872, 1874) demonstrate that, in a birefringent medium, the differing indices of refraction for the normal and extraordinary rays must be used in applying Fresnel's formula. Again, if an explanation of Fresnel's coefficient in terms of a moving ether is given, then in a birefringent medium the ether must be capable of sustaining two different motions at the same time.

But if there was no further progress in explaining Fresnel's formula between 1850 and 1880, there was great progress in understanding its theoretical implications.

1.8 From Compensation to Relative Motion

As we have seen, Fresnel originally introduced f to explain the absence of expected effects of the earth's motion through the ether. This mutual cancellation of effects that, by themselves, would produce evidence of this motion came to be referred to as "compensation" (the first use of this term that I have found is in Fizeau 1851) of the expected effects of this motion, which combined to produce a total null effect in each special case.

Veltmann (1870) introduced a new viewpoint that transcends the use "of this hypothesis [i.e., Fresnel's formula] ... for the explanation of one or another special observation and indeed always by means of a so-called compensation." [In the 1873 version, he was more explicit: "a compensation of various ... changes in the direction of the wave normals from those that had been demonstrated at rest."]

"This viewpoint is simply that of relative motion... Fresnel's hypothesis is thus nothing more than the necessary and sufficient condition for the applicability of the laws that follow from the wave theory for the refraction of the rays in media at rest to the relative rays in moving media."

Veltmann argues that Fresnel actually arrived at his formula by realizing it was needed to explain Arago's results. "The considerations by mean of which Fresnel attempted to give [his formula] a physical foundation are worthless and therefore remain unconsidered here" (1873).

In order to explain interference phenomena (such as the results of Fizeau's experiment), Veltmann showed that Fresnel's formula can be used to prove the following theorem:

> "In order to traverse a closed polygon, light always requires the same time, whether the medium be at rest or has any parallel motion that is very small in comparison to the speed of light" (1873).

Around this time, Mascart (1874) formulated what we may call the optical principle of relativity: "The translational motion of the earth has no appreciable influence on optical phenomena produced by a terrestrial source, or by light from the sun; these phenomena do not provide us with a means of determining the absolute motion of a body, and relative motions are the only ones we are able to determine."

1.9 Potier and Time

Potier (1874) gives a reinterpretation of Fresnel's formula that rids it of its dependence on the index of refraction by emphasizing the time intervals involved in the transmission of light. He showed that:

> If a body is in motion, the time that light takes to travel the distance l between two points A and B belonging to the body is increased by lu/V^2 by virtue of the motion, u being the component of the velocity of the body in the direction of the line AB, V being the speed of propagation of light in vacuum (1874).

While he had only shown his result to follow from Fresnel's formula by neglecting terms of order $(u/V)^2$, he pointed out that his result "alone would rigorously provide the explanation of the observed phenomena," and suggests that "for speeds comparable to the speed of light ... Fresnel's law, exact for small speeds, could thus be supplemented by this purely empirical statement" (1874).

1.10 The Local Time

(Poincaré 1905) is an obituary of Potier, who had been one of Poincaré's teachers. In it, he commented on Potier's work on optics of moving bodies:

> Aberration and Fizeau's experiments show us that the ether is not carried along by matter; how does it happen then that this relative motion of the ether and the earth cannot be demonstrated by any optical experiment? Potier made a considerable step forward in answering this question; and it was necessary to wait for Lorentz before a new step was taken that has brought us so close to the solution that we are almost touching it.

What did Poincaré consider to be Potier's "considerable step forward?" He does not say explicitly, but I believe that it was a step towards the concept of what Lorentz later named "the local time," a concept that Poincaré was the first to give a physical interpretation.

Let me try to justify this claim. Let:

Δt = the time interval for light to travel between the points A and B
in some optical medium at rest,

$\Delta x = l$ be the distance between points A and B,

$\Delta t'$ = the time interval for light to travel between A and B
when the medium is moving with velocity u.

In this notation, Potier's formula becomes:

$$\Delta t' = \Delta t + (u.\Delta x)/V^2,$$

or

$$\Delta t = \Delta t' - (u.\Delta x)/V^2,$$

where V is the velocity of light in vacuum. We see at once that, formally, this is the same as Lorentz' (1895) expression for the local time.

1.11 Comments On This Result and Some Speculations

1) Since the medium is arbitrary, by varying its index of refraction n the velocity of light in the medium can (in principle) be made to vary between 0 and V:

$$V_{med} = V/n, \quad n \geq 1.$$

So the time interval $\Delta t = l/V_{med}$ can assume any value that makes the events at A and B causally connectible (i.e., in special-relativistic language, that keeps the space-time interval between them timelike or null).

2) Suppose we make the assumption that Δt, the time interval between two events at A and B when the medium is at rest, is always the time interval between these events as measured in a frame of reference in which the medium is at rest, even when the medium is in motion [Note that, if we do not make this assumption, it follows that a simple time measurement could detect the earth's motion through the ether.]

We thereby give a physical interpretation of the local time that implies that the time interval between a pair of events depends on the frame of reference in which it is measured, and hence the need for a non-Galilei–Newtonian kinematics.

3) Newton–Galilean kinematics would yield:

$$\Delta x' = \Delta x + u.\Delta t'.$$

If one were to introduce a modified Newton–Galilean formula:

$$\Delta x' = \Delta x + u.\Delta t,$$

then even without any calculation, it is easy to see that the exact relativistic law of addition of velocities would follow from the equations for $\Delta t'$ and $\Delta x'$: both lack the factor, so their quotient will be exactly the same as if the γ factor were there!

4) I have no evidence that, in introducing the local time as a formal mathematical device, Lorentz was influenced by Potier's work; but Poincaré's comment cited above makes it seem likely that Poincaré was influenced by Potier in first giving a physical interpretation to Lorentz's local time:

"If one starts to consider seriously the idea that the time interval between two events at different places might be different when measured in different frames of reference, one is led to reflect on the need to synchronize clocks at rest in each frame in order to measure such time intervals" (Poincaré 1898).

Poincaré was the first to interpret the local time as the time that clocks would read in a moving frame if light rays were used to synchronize them on the assumption that the (vacuum) velocity of light is V, even relative to the moving frame (Poincaré 1900).

5) Lorentz's (1892) success in deriving Fresnel's coefficient on the basis of Maxwell's equations and the hypothesis of a stationary ether served to divert attention from the kinematic aspects of the problem. He soon (1895) gave a simplified derivation, based on the concept of the local time, that did not explicitly involve electromagnetic theory; but by that time the close association between Fresnel's coefficient and Maxwell's theory seems to have been taken for granted.

6) Even Einstein was still so much under the spell of Lorentz's interpretation that he failed to notice the kinematic nature of Fresnel's formula, resulting from direct application of the relativistic law of combination of relative velocities; it was left for Laue to make this observation in 1907.

Is it fantastic to imagine that someone might have been led to develop some or all of these kinematical responses to the challenge presented by the situation in the optics of moving bodies around 1880, given that an optical principle of relative motion had been formulated by Mascart? Perhaps no more fantastic than what actually happened: Einstein's development around 1905 of a kinematical response to the challenge presented by the situation in the electrodynamics of moving bodies, given that an electrodynamic principle of relative motion had already been formulated by Poincaré.

Acknowledgment. Research was done and I gave a talk on the subject of this paper while a guest of the Max-Planck-Institut für Wissenschaftsgeschichte in Berlin. I thank the Director of Department I, Dr. Jürgen Renn, for his ever-ready hospitality and encouragement during my many stays there. A fuller account of this work, containing many more quotations from the contemporary literature, will be issued as a Preprint of the Institute.

References

Arago, François (1810). Mémoire sur la vitesse de la lumière, lu à la prémière classe de l'Institut, le 10 décembre 1810. Académie des sciences (Paris). *Comptes rendus* **36** (1853), 38–49.

— (1830). Eloge historique d'Augustin Fresnel. In *Augustin Fresnel Oeuvres complètes*, 3 vols. H. de Senarmont, E. Verdet, and L. Fresnel, eds., Imprimerie impériale, Paris, 1866–1870, vol. 3, 475–526.

Beer, August (1855). Über die Vorstellungen vom Verhalten des Aethers in bewegten Mitteln. *Annalen der Physik* **4**, 428–434.

Buchwald, Jed Z. (1988). "The Michelson Experiment in the Light of Electromagnetic Theory Before 1900." In *The Michelson Era in American Science 1870–1930, Cleveland, OH 1987*. Stanley Goldberg and Roger H. Stuewer, eds. New York, American Institute of Physics, 55–70.

Fizeau, Hyppolite (1851). Sur les hypothèses relatives à l'éther lumineux, et sur une expérience qui parâit démontrer que le mouvement des corps change la vitesse à laquelle la lumière se propage dans leur intérieur. Académie des sciences (Paris). *Comptes rendus* **33**, 349–355.

Fresnel, Augustin (1818a). Lettre d'Augustin Fresnel à François Arago sur l'influence du mouvement terrestre dans quelques phénomènes d'optique. *Annales de chimie et de physique* **9**: 57–66, 286. Reprinted in *Oeuvres complètes*, 3 vols. H. de Senarmont, E. Verdet, and L. Fresnel, eds., Imprimerie impériale, Paris, 1866–1870, vol. 2, 627–636.

— (1818b): Note additionelle à cette lettre. *Annales de chimie et de physique* **9**, 286. Reprinted in *Oeuvres complètes*, 3 vols. H. de Senarmont, E. Verdet, and L. Fresnel, eds., Imprimerie impériale, Paris, 1866–1870, vol. 2, 636.

Hahn, Roger (1970). Arago, Dominique Francois Jean. In *Dictionary of Scientific Biography*, vol. 1. Charles Coulston Gillispie, ed., Scribners, New York, 200–203.

Janssen, Michel and Stachel, John (1999). The Optics and Electrodynamics of Moving Bodies. To appear in *Going Critical*, vol. 1, *The Challenge of Practice*. Kluwer, Dordrecht.

Ketteler, Eduard (1873). *Astronomische Undulationstheorie oder die Lehre von der Aberration des Lichtes*. P. Neusser, Bonn.

Lorentz, Hendrik Antoon (1886). Over den invloed, dien de beweging der aarde op de lichtverschijnselen uitoefent, *Koninklijke Akademie van Wetenschappen* (Amsterdam). Afdeeling Natuurkunde. Verslagen en Mededeelingen 2, 297–372. French translation, "De l'influence du mouvement de la terre sur les phenomènes lumineux." *Archives néerlandaises des sciences exactes et naturelles* **21** (1887), 103–176.

— (1892). La theorie electromagnétique de Maxwell et son application aux corps mouvants. *Archives néerlandaises des sciences exactes et naturelles* **25**, 363–552.

— (1895). *Versuch einer Theorie der elektrischen und optischen Erscheinungen in bewegten Körpern.* Brill, Leiden.

Mascart, Éleuthère Élie Nicolas (1872). Modifications qu'éprouve la lumière par suite du mouvement de la source lumineuse et mouvement de l'observateur. *Annales Scientifiques de l'École Normale Supérieure*, 2ème série, 1, 157–214.

— (1873). *Mémoire manuscrit déposé à l'Académie des Sciences pour le Grand Prix des Sciences Mathématiques de 1872 sur l'épigraphe Nihil*, 20. Cited from (Pietracola 1992, 86).

— (1874). Modifications qu'éprouve la lumière par suite du mouvement de la source lumineuse et mouvement de l'observateur (deuxième partie). *Annales Scientifiques de l'École Normale Supérieure*, 2ème série, **3**, 363–420.

— (1893). *Traité d'Optique*, vol. 3, chap. 15. Gauthier-Villars, Paris.

Pietracola Pinto de Oliveira, Mauricio (1992). *Élie Mascart et l'optique des corps en mouvement*. Thèse présentée à l'Université Denis Diderot (Paris 7).

Poincaré, Henri (1889). *Leçons sur la théorie mathématique de la lumière, professées pendant le premier semestre 1887–1888*. Jules Blondin, ed., Carr et Naud, Paris.

— (1898). La mesure du temps. *Revue de métaphysique et de morale* **6**, 1–13.

— (1900). La théorie de Lorentz et le principe de la réaction. In *Receuil de travaux offerts par les auteurs à H. A, Lorentz, professeur de physique à l'université de Leiden, à l'ocassion du 25me anniversaire de son doctorat le 11 décembre 1900*. Martinus Nijhoff, The Hague, 252–278.

—- (1905). M. A. Potier, *L'Éclairage Électrique 43*. Reprinted as the Preface to (Potier 1912).

Potier, Alfred (1874). "Conséquences de la formule de Fresnel relative à l'entraînement de l'éther par les milieux transparents. *Journal de physique* **3**, 201–204. Reprinted in (Potier 1912).

— (1912). *Mémoires sur l'électricité et l'optique*. A. Blondel, ed. Gauthier-Villars, Paris.

Schaffner, Kenneth (1972). *Nineteenth-Century Aether Theories*. Pergamon Press, Oxford.

Stokes, George Gabriel (1880). *Mathematical and Physical Papers*, vol. 1, Cambridge Univiveristy Press, Cambridge.

— Stokes, George Gabriel (1845). On the Aberration of Light. *Philosophical Magazine* **27**, 9. Cited from (Stokes 1880, 134–140).

— (1846). On Fresnel's Theory of the Aberration of Light. *Philosophical Magazine* **28**, 76. Cited from (Stokes 1880, 141–147).

Veltmann, Wilhelm (1870). Fresnels Hypothese zur Erklärung der Aberrations-Erscuheinngen, *Astronomische Nachrichten* **75**, 145–160.

— (1873). Über die Fortpflanzung des Lichtes in bewegten Medien. *Annalen der Physik* **150**, 497–535.

Notes

[1]By "critical situation," I mean a feeling expressed by an important segment of the physics community that something is amiss in their field of expertise: a mismatch between either experimental results and theoretical explanations, as in the two critical situations mentioned here; or between the accounts offered by different theories in some area, to which both should be applicable, as in the current critical situation in the field of quantum gravity.

2

Poincaré's Relativistic Theory of Gravitation

Shaul Katzir

Bar Ilan University, Jerusalem, Israel; `shaulk@h2.hum.huji.ac.il`

As early as 1904, the French mathematician, physicist and philosopher Henri Poincaré recognized that the acceptance of the principle of relativity for electrodynamic systems requires a change in the Newtonian theory of gravitation. Accordingly, in his relativistic theory formulated a year later he suggested a modified law of gravitation, which was compatible with the principle of relativity, i.e., with the inability to differentiate between systems in uniform motion and at rest.[1] Poincaré maintained the structure of the classical gravitation theory but made minimal changes to make it "Lorentz covariant", thus ensuring its compatibility with the relativity principle. This approach differed from those adopted by most physicists who tackled the problem of subsuming gravitation in the new relativistic physics. In particular, it was in contrast to the approach that led Einstein to the general theory of relativity. Thus, Poincaré's approach represents an alternative course, which was not adopted. Historical examination of his theory and its consequences sheds light on the reasons for preferring other approaches.

Poincaré was the first to point out the need to reconsider the classical theory of gravitation because of the relativity principle, and to formulate a relativistic force of attraction. He included a discussion of gravitation in all his treatments and surveys of the theory of relativity after the publication of his worked-out theory in 1906. Nevertheless, the historical literature has given little attention to his treatment of gravitation,[2] with the one notable exception of an article by Scott Walter (forthcoming).[3] That Poincaré's theory had only a small influence on the later developments in general relativity explains why this theory is not treated in histories of general relativity. Yet, it does not justify the neglect of this work by historians of Poincaré's considerations of the relativity principle.

This article fills this gap in the historical literature with an examination of Poincaré's relativistic theory of gravitation: its origins, its main consequences, and the later treatment of its approach by others. It is complemented by Walter's article, which focuses on Poincaré's novel mathematical techniques and their subsequent development.

2.1 Implications of the Relativity Principle on Newton's Law

Poincaré first considered the need to change the law of universal gravitation in a lecture of September 1904 on the current state of theoretical physics. The principle of relativity and Lorentz's recently published theory of the electrodynamics of moving systems (Lorentz 1904) were central topics of the lecture. Although Lorentz had not make his theory fully compatible with the relativity principle, Poincaré interpreted it in relativistic terms (Poincaré 1905, 123–147). He explained how the combination of three assumptions introduced by Lorentz, namely the introduction of local time, the existence of a length contraction and the assumption that all forces transform in the same way make it impossible to differentiate between uniform motion and rest. Lorentz had introduced the latter assumption to maintain equilibrium of forces in all inertial systems of reference. Thus, elastic systems in static equilibrium will be in equilibrium in any frame of reference.

Whereas Lorentz was evidently concerned with elastic forces (Lorentz 1904, p. 22), Poincaré, in considering the force of gravitation, formulated another requirement for the forces, which stemmed from his concept of local time. Unlike Lorentz, Poincaré regarded local time as a physical rather than an purely auxiliary variable. According to him, local time is the time measured by watches synchronized by light signals; it is the only time that can be measured (Poincaré 1900, p. 483). The procedure of synchronization with signals propagating at the speed of light was crucial for the theory, i.e., for ensuring the theoretical impossibility to detect absolute motion.

> What would happen, [he asked] if one could communicate by non-luminous signals whose velocity of propagation differed from that of light? If, after having adjusted the watches by the optical procedure, one wishes to verify the adjustment with the aid of these new signals, then deviations would appear that would render evident the common translation of the two stations. And are such signals [which move faster than light] inconceivable, if one admits with Laplace that universal gravitation is transmitted a million times more rapidly than light? (Poincaré 1905, p. 134)

In order to eliminate the possibility of measurements by signals faster than light, the gravitational attraction should propagate no faster than light. This required a change in the Newtonian theory, which assumes instantaneous attraction. Hence a modification of the theory of gravitation needed to be included in any relativity theory.

2.2 Poincaré's Relativistic Theory

Indeed, in the relativity theory that Poincaré suggested a year later, he included a modified theory of gravitation. In July 1905 he completed "*Sur la dynamique de l'électron*" (published in January 1906), his famous presentation of relativistic physics, in which no difference exists between stationary and uniformly moving systems (Poincaré 1906).[4] This theory is based on two premises: the relativity principle, which Poincaré raised to the status of a postulate and the validity of Maxwell's equations. Of the two premises the former is the central one. Every statement of the theory is shown to be compatible with the postulate. The constancy of the speed of light is only a result of the two premises. Due to the invariance of Maxwell's equations under Lorentz transformations, these transformations are adopted as the transformations between one spatial system of coordinates and another (coordinate transformation). Their compatibility with the principle of relativity is the justification for their use. From the transformation of the spatial coordinates Poincaré derived the relativistic transformations for velocity, charge density, the electromagnetic fields, the variation of mass and volume of a moving electron, etc. He did not derive the mass energy equivalence relation, although it is a logical consequence of his theory.

Poincaré's theory of relativity is not restricted to any specific ontological view of nature. In particular, the theory is not based on the hypothesis that all of nature can be explained by electromagnetism (Katzir forthcoming). In his paper, Poincaré showed the need for an additional force, non-electromagnetic in origin, which holds the electron together, both in motion and at rest (sometimes called Poincaré's stress) and showed its compatibility with the relativity principle.[5] Another non-electromagnetic force is that of gravitation, to which the paper's last section is devoted.[6] He explained the reason for its treatment:

> Lorentz's theory [as elaborated by Poincaré] fully explains the impossibility of finding evidence of absolute movement, if all forces were electromagnetic in their origin. However, there are forces, which one cannot attribute to an electromagnetic origin, for example gravitation. There can be two systems of bodies that produce equivalent electromagnetic fields, that is, generate the same action on electric bodies and currents, while these two systems do not generate the same gravitational action on Newtonian masses. The gravitational field is therefore distinct from the electromagnetic one. Lorentz was therefore obliged to complement his hypothesis by supposing that *forces of all origins, and particularly gravitation, are affected by a transformation* (or if one prefers, by the Lorentz transformation) *in the same way as electromagnetic forces* (Poincaré 1906, pp. 538–539 italics in the original).

The modification of the two-centuries-old theory of gravitation stemmed from the need to reconcile it with the theory of relativity, which originated in the study of electrodynamics.

One might assume that in modifying the law of attraction, Poincaré was motivated by problems with the application of Newton's law, with which he was familiar. Indeed, he had a long-standing interest in the law of gravitation: he had been working on celestial mechanics at least since 1882, publishing more on that subject than in any other branch of the natural sciences, including the three celebrated volumes on *The New Methods of Celestial Mechanics*. At the outset of that work he stated that "the final aim of celestial mechanics is to resolve this great question of knowing whether Newton's law by itself explains all the astronomical phenomena" (Poincaré 1892, p. 1). His answer was affirmative. For example, in 1907 he concluded that "we do not have [in astronomical phenomena] any serious reason to modify Newton's law" (Poincaré 1953

p. 265). This was his conclusion in a course of lectures on the validity of Newton's law with respect to current astronomical data.

Nevertheless, slight disagreements between Newton's law and astronomical observations led a few astronomers to suggest modifications of the law. As Poincaré pointed out, the more serious divergence from Newton's law was the unexplained part of the advance of Mercury's perihelion. The unexplained part was merely 10 percent of the total advance of the planet's perihelion, while the influence of the other planets explained the remaining part,[7] but the accuracy of the astronomical observations and of celestial mechanics was good enough to indicate the need to explain the discrepancy. Two basic strategies were employed to solve the puzzle: the first suggested a modification of Newton's law (making it dependent on velocities, adding an extra term of distance etc.). This is, for example, the position of Zenneck in 1901 in his general discussion of the theory of gravitation in the *Enzyklopädie der mathematischen Wissenschaften* (Zenneck 1902). The second strategy was to maintain Newton's law and to look for an unaccounted influence of known or unknown material in space. Poincaré adopted the latter approach. He resolved the anomaly by assuming a gravitational influence of a material ring between Mercury and the Sun.[8]

2.3 The Relativistic Treatment of Gravitation

Poincaré's elaboration of the relativistic force of attraction in his 1905 paper is based on five conditions:

1. The existence of a Lorentz invariant function of the four coordinates and the velocities that define the law of propagation of attraction. This requirement implies a finite velocity of attraction, as entailed by the principle of relativity.

2. The above-mentioned assumption that the Lorentz transformation affects the gravitational and the electromagnetic forces in the same manner.

3. For bodies at rest the force law should coincide with Newton's law.

4. The chosen solution will be the one that least alters Newton's law for small velocities. This derived from the need to account for astronomical data in the same manner as Newton's law. It shows both that Poincaré was concerned with the empirical consequences of the theory, and that he did not find any empirical reason to modify Newton's law.[9]

5. The time variable in the mathematical expressions will always be compatible with the known **physical** fact that it takes time for the attraction to travel from one body to another. The combination of this requirement with the invariance condition ensures that the velocity of propagation of the gravitational force will not exceed the speed of light. (Poincaré 1905, pp. 539–540)

Tacitly, Poincaré assumed that the gravitational mass of a body is independent of its velocity. The gravitational mass is always equal to the rest mass, the "experimental mass," as Poincaré called it. Thus, the variation of the inertial mass of a body with velocity does not affect the force of gravitation. He avoided explicit discussion of the gravitational mass by deriving the mathematical expressions of the force for two masses. On other occasions, the first already in 1904, he pointed out the two alternatives:

> The mass [he wrote] has two aspects: it is at the same time a coefficient of inertia and a gravitating mass that appears as a factor in the Newtonian law of attraction. If the coefficient of inertia is not constant, could the attracting mass be? This is the question. (Poincaré 1905, p. 139)

In 1904 he left the question open. A year later he implicitly chose a constant gravitational mass. Later, in 1908 in a popular article, he claimed that no experiment can determine which of the two alternatives is the right one (Poincaré 1908, p. 577). Yet he did not discuss the consequences of a non-constant gravitational mass and probably did not investigate in detail whether such an assumption would not have empirical consequences. The mathematical simplicity of the assumption of a constant gravitational mass was probably the primary reason for its adoption by Poincaré. In addition, it creates an analogy a between gravitation and electrodynamics, in which the charge—the analogue of the gravitational mass—does not vary with velocity. This allowed the use of results from electrodynamics in the discussion of gravitation (like the transformations of force and charge/gravitational mass density). Moreover, the constant mass hypothesis implied a new relativistic force similar to Newton's law, which Poincaré regarded as empirically satisfactory and which he wished to modify as little as possible.[10]

Poincaré's mathematical elaboration is based on four-dimensional Lorentz invariance and the Newtonian approximation.[11] The concept of invariance is explicitly based on the group properties of the Lorentz transformations, demonstrated in a previous section of the paper. The invariance is the mathematical expression of the fact that there is no difference between bodies in uniform motion and bodies at rest.[12] Poincaré derived the force law in three steps. First, he found invariant functions of the coordinates and velocities. Then he wrote the force as a function of these invariants and some unspecified coefficients, thereby ensuring the compatibility of the force with the relativity principle. Lastly, he found specific expressions for the force through a comparison with the Newtonian approximation for small velocities. Yet the comparison did not lead to a unique expression for the force law and Poincaré offered an alternative.

So, first Poincaré wrote down four homogeneous invariant functions of the coordinates and velocities of two gravitating masses.

$$\sum x^2 - t^2, \quad \frac{t - \sum x\xi}{\sqrt{1 - \sum \xi^2}}, \quad \frac{t - \sum x\xi_1}{\sqrt{1 - \sum \xi_1^2}}, \quad \frac{1 - \sum \xi\xi_1}{\sqrt{(1 - \sum \xi^2)(1 - \sum \xi_1^2)}} \qquad (2.1)$$

where x is the coordinate of the attracted body, ξ and ξ_1 are the x-components of the velocities of the attracted and the attracting bodies respectively, and the summation is over all components. The units are chosen so that the speed of light is equal to one. Consequently, the time t has the dimension of length (Poincaré 1905, p. 542). Further considerations involving the propagation of the attraction and the invariants led him to conclude that the force must propagate at the speed of light. Any other assumption would lead to a violation of causality through the occurrence of an influence of the future (Poincaré 1905, p. 544). The first invariant is therefore identically zero, and $-r = t$. The four invariants are thus written as (Poincaré 1905, p. 547):

$$0, \; A = -k_0(r + \sum x\xi), \quad B = -k_1(r + \sum x\xi_1), \quad C = k_0 k_1(1 - \sum \xi\xi_1) \quad (2.2)$$

where

$$k_0 = \frac{1}{\sqrt{1 - \sum \xi^2}}, \quad k_1 = \frac{1}{\sqrt{1 - \sum \xi_1^2}}.$$

Next he expressed the force law as a multiplication by coefficients of four systems that behave like the spatial coordinates under the Lorentz transformation. These are the coordinate system itself, the components of the force multiplied by k_0, the velocity components of the attracted mass multiplied by k_0 and those of the attracting mass multiplied by k_1. Poincaré expressed the relation in four equations, one for each component of the force and one for the power. The first of these is

$$X_1 = x\frac{\alpha}{k_0} + \xi\beta + \xi_1\frac{k_1}{k_0}\gamma \quad (2.3)$$

where X_1 is the x-component of the force for two masses. Because the coefficients α, β and γ are invariant, the whole expression is. Employing a relativistic relation between any force, velocity and power, Poincaré derived a relation between the coefficients and the invariants of equation (2.2):

$$-A\alpha - \beta - C\gamma = 0. \quad (2.4)$$

Poincaré chose a simple solution:

$$\beta = 0, \; \gamma = \frac{A\alpha}{C}.$$

This is not a unique solution; an alternative would have led to a different force law. Next Poincaré employed the approximation of low velocities, neglecting terms containing the squares of velocities. Then $\alpha = -\frac{1}{r^3}$; in the same order of approximation this is also the value of $1/B^3$ (which he took equal to α). This was not the only possibility: other combinations of the variables give the same value in this approximation. Inserting the invariants A, B, C (equation (2.2)) instead of β and γ in equation (2.3), he obtained four equations for the four components of the generalized force, the first of which is

$$X_1 = \frac{x}{k_0 B^3} - \xi_1 \frac{k_1}{k_0} \frac{A}{B^3 C}. \quad (2.5)$$

Poincaré did not write the force as a direct function of positions and velocities. Expressed in vectorial form in terms of these variables the force that a unit mass of the attracting body exerts on a unit mass of the attracted body is written as[13]

$$\vec{F} = -\frac{1}{k_0 k_1^3 (r + \vec{r}\vec{v}_1)^3} \left[\vec{r} + \frac{\vec{v}_1 (r + \vec{r}\vec{v})}{1 - \vec{v}\vec{v}_1} \right] \tag{2.6}$$

where $\vec{v}$ and $\vec{v}_1$ are the velocities of the attracted and the attracting body, respectively, $\vec{r}$ the distance between the bodies and $\vec{F}$ the gravitational force for two masses.

This gravitational force resembles Lorentz's electromagnetic force. It consists of two terms, one parallel to the line that connects the two bodies, and the other parallel to the velocity of the attracting body. The former depends on the position of the bodies and is analogous to the electrostatic force, while the latter depends on the velocity of the attracting force and resembles the magnetic force. However, unlike the magnetic term of the electromagnetic force, this force is parallel rather than orthogonal to the velocity. "To complete the analogy" (Poincaré 1905, p. 549) Poincaré elaborated an alternative force law, based on a different expression for $1/r^3$. He wrote down a general expression for functions that are equal to $1/r^3$ in the chosen approximation:

$$\frac{1}{B^3} + (C - 1) f_1(A, B, C) + (A - B)^2 f_2(A, B, C) \tag{2.7}$$

where f_1 and f_2 are arbitrary functions and A, B and C are the invariants (2.2). A simple replacement of $\frac{1}{B^3}$ by $\frac{C}{B^3}$ led to an alternative force law. Poincaré then wrote the three components of this force again as functions of the invariants. In vectorial notation in terms of distances and velocities the force for two masses is

$$\vec{F} = -\frac{1}{k_1^2 (r + \vec{r}\vec{v}_1)^3} [(\vec{r} + r\vec{v}_1) + \vec{v} \times (\vec{v}_1 \times \vec{r})]. \tag{2.8}$$

The analogy between this force law and the force between two moving electrons is clear; the first expression in the brackets parallels the effect of the electric field, the second that of the magnetic field (Poincaré 1905, p. 549). Electrodynamics in general and the Lorentz force in particular provided Poincaré with a model in the elaboration of a relativistic force law. Yet the derivation of the theory did not rely on the electrodynamic model. The method of invariance, which Poincaré used, neither derived from nor had a parallel in electrodynamic theory.

While the second term in the Lorentz force is a function of the magnetic field, in Poincaré's force it is a direct vectorial product of the velocity and the distance. Like classical celestial mechanics, the new gravitational theory is also a theory of mass points and not of a field. In this respect the only difference is that the attraction takes a finite time to propagate rather than being instantaneous. Poincaré made no attempt to explain the manner in which the attraction propagates. He did not suggest any mechanism or field for that purpose. He surely knew of earlier unsuccessful attempts at a field theory of gravitation, inspired by Maxwell's electromagnetism. That was not his aim here: he did not try to base gravitation on new foundations. His was a modification of Newton's theory of gravitation, based on the classical theory, rather than a new independent theory.

Instead of a field, Poincaré referred to the propagation of the attraction for which he introduced the term "gravitational wave."

Recall [he clarified] that when we talk about the position or the velocity of the attracting body, we mean its velocity or position at the moment in which the gravitational wave leaves it; for the attracted body, on the contrary, we mean its position or velocity at the moment the gravitational wave, assumed to propagate at the speed of light, reaches it. (Poincaré 1905, p. 548)

The term "gravitational wave" merely emphasizes that the force propagates in a finite time, similarly to light waves.[14] Others also employed the term to express that the attraction propagates at a finite speed.[15] The result that gravitation propagated at the speed of light answered the question that Poincaré had raised in 1904: a theory of gravitation in which the force propagates at the speed of light is possible, in contrast to the old claim of Laplace. As Poincaré explained:

Laplace showed that the propagation [of gravitation] is either indeed instantaneous or much more rapid than the speed of light. But Laplace had examined the hypothesis of a finite velocity of propagation *ceteris non mutatis*; here, on the contrary, this hypothesis is complicated with many others, and there may be more or less perfect compensation between them. (Poincaré 1905, p. 544)

The compensation is not perfect; the new force law predicts results slightly different from Newton's law. Poincaré knew that a "more or less perfect" agreement is not enough. He therefore concluded the paper with an open question:

[T]he prime question we are faced with is whether these [force laws] are compatible with astronomical observations; the deviation from Newton's law is in the order of ξ^2 [the velocity squared], which means 10,000 times smaller than if it were of the order of ξ, that is if the propagation took place at the speed of light, *ceteris non mutatis*; it therefore permits us to believe that it [the deviation] will not be too great. However, only a detailed discussion can teach us that. (Poincaré 1905 p. 550)

Poincaré did not carry out such a discussion; he did not write another research paper on relativity theory. In a university course of 1906–1907 and in later popular presentations he supplied some numerical results on the effects of the relativistic force of attraction. The numerical results he cited show that he employed his second expression of the force law (equation (2.8)),[16] which emphasizes the analogy to electrodynamics. Eventually a detailed discussion of the astronomical consequence of the new force law was published in 1911 by Willem de Sitter. De Sitter found the theory compatible with the observational data (see below sec. 2.5.c).

2.4 Poincaré's Later Treatment of the Law of Attraction

In his later popular presentations of the new relativistic physics, Poincaré always included gravitation as an integral part. He also discussed the relativistic laws of gravitation in a course on Newton's force (Poincaré 1953). There he discussed for the first time its consequences for Mercury's perihelion. For his students he clarified the main differences between the relativistic formula and Newton's law: the propagation of the force at the speed of light and its dependence on the velocity of the masses. On that occasion Poincaré for the first time referred to gravitational radiation due to accelerated neutral matter in analogy to electrodynamic radiation. "However [he clarified], this term is absolutely negligible: actually, the acceleration of the celestial bodies is practically zero" (Poincaré 1953, p. 245).[17] In 1908, in what is probably the first published reference to gravitational radiation, he wrote explicitly that the theory predicts "acceleration waves" due to motions of masses (Poincaré 1908, p. 579). A year later at Göttingen he talked about "dissipation of energy" due to the acceleration of the celestial bodies;[18] Half a year later at Lille he elaborated:

> [T]he acceleration of celestial bodies has consequences like electromagnetic radiation: a dissipation of energy that will make itself felt in a decrease of their velocities.[19] I actually said, that every time an electron undergoes a sudden change of velocity, radiation appears. However, this word 'sudden' lacks precision. If the change is slow, if the acceleration is small, there will still be radiation, but this radiation will be very weak . . . the radiation will be imperceptible, [yet] it does exist and little by little dissipates the living force [the kinetic energy] of the planets. (Poincaré 1909, p. 176)

Poincaré always concluded his discussions of the relativistic force of attraction with a reference to its empirical consequences. For example, in 1908 he wrote:

> In conclusion, *the only appreciable effect on astronomical observations will be a motion of Mercury's perihelion, in the same manner as the one that was observed without being explained, but notably smaller.*
> This cannot be regarded as an argument for the new dynamics, because one will always have to find another explanation for the larger part of Mercury's anomaly; yet this can be regarded even less as an argument against it (Poincaré 1908, p. 581, emphasis in the original).

The advance of Mercury's perihelion that Poincaré cited was $7''$ per century,[20] much less than the $39''$ needed to account for the observed discrepancy. Even if the force law had yielded a value closer to the unexplained one, it would have been a weak argument for the theory. Since Poincaré's approach did not result in a unique force law, it could not have unequivocally determined the magnitude of the advance. One could choose an alternative force law in which the advance is either larger or smaller. This made the exact quantitative results of a particular force law less significant. That the force of attraction remained undefined made Poincaré's treatment of gravitation less attractive to many physicists who preferred alternative approaches that led to more restricted laws. That the new treatment of gravitation had only insignificant observable consequences can explain why it was not further elaborated by Poincaré. Still, the new force law had profound theoretical implications, both in abandoning the Newtonian view of attraction and in raising the question of including gravitation in the emerging relativistic physics.

2.5 Later Discussions and Elaborations of Poincaré's Approach

Poincaré's relativity theory was generally ignored and only few people mentioned his contributions in their writings.[21] However, the few who referred to it did not fail to mention the relativistic force of gravitation. This force was later studied by the German mathematician Hermann Minkowski, who suggested an alternative derivation for it, and the Dutch astronomer Willem de Sitter, who examined its observable consequences on motion in the solar system.

a. Minkowski's treatment of gravitation Minkowski is well-known today for his four-dimensional space-time formulation of relativity theory.[22] As mentioned above, a four-dimensional formalism was first employed in this context by Poincaré in 1905. In addition Minkowski used the concept of invariance and its connection to the relativity principle, which played a central role in Poincaré's derivations.[23] Poincaré's work was also the starting point for Minkowski's theory of gravitation. In his first address on the relativity principle in November 1907, Minkowski reported on Poincaré's demonstration of the possibility of a gravitational attraction propagating at a finite speed and compatible with the relativity principle, against Laplace's earlier claim (Minkowski 1916, pp. 381–382).

In an Appendix to a further study of relativity of 1908 Minkowski treated gravitation "in a totally different manner" from Poincaré (Minkowski 1908, p. 109). He neither stated conditions that the law had to satisfy nor supplied a qualitative description of the law's new properties. He developed the force law in a highly abstract geometrical manner based on his geometrical representation of the relativity theory, without any appeal to experience. His formulation seems to derive from purely mathematical considerations. The force is a "space-time" vector, a property that ensures its covariance. Yet it is expressed by geometrical magnitudes, which have to be translated to be physically meaningful (Minkowski 1908, pp. 109–111).[24] This force depends on the "rest masses" of the two bodies, the spatial distance between them and the velocity of the attracting mass. It has two components equal in magnitude: one in the direction of the velocity four-vector of the attracting mass (at the moment of transmitting the attraction), the other in a direction perpendicular to the velocity four-vector of the attracted mass. The velocity of the attracted mass appears explicitly in the expression only when it is written in classical three-dimensional space. Minkowski briefly showed that the predictions of this law differ from those of Newton only in negligible amounts. For the Earth the extra term is multiplied by a factor of 10^{-8}. From this he concluded that astronomical observations can pose no objection to this law.

b. Sommerfeld's and Lorentz's opinions Poincaré's and Minkowski's similar approaches to the relativistic force of attraction was discussed by Sommerfeld circa 1910. In a long paper on relativity he discussed the force of gravitation, among other things. He showed that Minkowski's law is equivalent to the first of the two alternatives (Sommerfeld 1910, pp. 684–689) suggested by Poincaré (equation (2.6)). He pointed out what he considered a theoretical problem of both formulations: they apply an action that propagates in finite time and thus contradict the direct principle of action and reaction. In electromagnetism, Sommerfeld claimed, one has the field that accounts for the temporary deficit of reaction, but neither Minkowski nor Poincaré had a theory of a gravitational field. Poincaré probably did not consider this a weighty objection, since he had also questioned the validity of the reaction principle in electrodynamics. He was reluctant to attribute reaction to a nonmaterial entity like a field or even the ether (Poincaré 1900, Darrigol 1995, pp. 17–23). However, the objection did convince other physicists. Most physicists did not adopt Poincaré's attitude towards the principle of action and reaction. Clearly, Sommerfeld had pointed out an important source of discontent with Poincaré's approach: its failure to suggest a field theory. The desire for a field theory of gravitation was not new, but the need for a new gravitation theory in accordance with relativity increased the ambition to make it a field theory. Moreover, the finite velocity of the attraction suggested a field. Still, Sommerfeld concluded that although the relativity principle makes Newton's old law inadmissible (*unzulässig*), he found no reason to abandon it, due to the negligible difference between its predictions and those of the relativistic law.

Lorentz expressed a different position in a public discussion of relativity at Göttingen in the same year (Lorentz 1910, especially pp. 216–220). Lorentz, like Poincaré, adopted the second expression for the law suggested in Poincaré's paper (equation (2.8)). That expression predicted small, but observable, differences with Newton's law. Lorentz reported on the primary result for the advance of Mercury's perihelion of de Sitter, his colleague in Leiden.[25] According to de Sitter, Lorentz chose the second expression of the law (equation (2.8)), "because the corresponding Newtonian force does not contain the velocity" of the attracted mass (de Sitter 1911, p. 397). This property made it similar to the electrodynamic force. That was probably a major reason for its adoption by Lorentz.[26] He elaborated on the similarity between the electrodynamic force and the new gravitational force law, and showed that the latter denies any evidence of absolute motion.

c. Empirical examination of the gravitational force law De Sitter in 1911 carried out the only in-depth examination of the consequences of the Lorentz covariant theories of gravitation that assume a constant gravitational mass (de Sitter 1911). De Sitter's aim was to discover the consequences of relativity theory in astronomy. He wrote:

> What is the law of force that must replace Newton's law, and what is the motion of the planet under this law? So far as this differs from ordinary Keplerian motion, we shall have to consider the question whether the differences are large enough to be verified by observation. (de Sitter 1911, 390)

For the force law that should replace Newton's law, de Sitter followed Poincaré. He quoted the first expression suggested by Poincaré in terms of invariants (equation (2.5)). Employing a four-dimensional formalism, he wrote the invariants in terms of the spatial coordinates and the velocities. Following Poincaré, de Sitter multiplied the force law by an invariant to get an alternative force law (equation (2.10)). Minkowski's law is equivalent to the first expression, de Sitter pointed out, while Lorentz and (though de Sitter did not mention it) Poincaré chose the second one (de Sitter 1911, pp. 393–397). De Sitter elaborated on the implications of both force laws, focussing on the deviation from the Keplerian solution for the orbit. He gave the solution for some approximations, like taking the attracted mass negligible, but did not restrict the discussion to these issues. His conclusions are definitive: the deviation of Minkowski's law from Newton's is negligible, while Poincaré's alternative law predicts a non-negligible secular advance in the orbits' perihelia. De Sitter reported on the expected advance of the perihelia of the planets and the comets Encke and Halley. Of all of them the most important is that of Mercury, for which he found $7''15'$ per century (de Sitter 1911, pp. 398–405). He also examined the consequences of the two force laws for the theory of the moon, but found that the deviation from the predictions of Newton's law "is well within the limits of uncertainty of the observed value." He concluded:

> We are thus left with the motion of the perihelion of Mercury as the only effect which reaches an appreciable amount. Unfortunately this same motion presents the well-known excess of observation on theory, which has been explained by Seeliger by the attraction of the masses forming the zodiacal light (ibid., p. 408).

Seeliger suggested that the advance is caused by a distribution of mass around the sun, whose gravitational influence is equal to that of two ellipsoids, one inside Mercury's orbit and the other outside that of Venus. With these two ellipsoids he accounted fully for the discrepancy in the classical theory of the planets (Seeliger 1906). De Sitter pointed out the way in which the hypothesis should be modified following the new results of the relativistic force law. This account, he concluded, "would be on the whole very satisfactory." Later, he carried out the detailed calculations and in 1913 he published values for the mass densities of the two ellipsoids that followed from the relativistic force law. From these values he concluded that "from the secular variations of the elements—and consequently from the planetary motions generally—we can therefore derive no argument either for or against the principle of relativity" (de Sitter 1913, p. 302).

The failure to confirm or refute the relativistic force law by observational data did not originate only in Seeliger's flexible hypothesis, but also in the freedom one has in choosing a relativistic gravitational force law. In 1911 de Sitter explained:

> The two laws are the only ones that have been actually proposed, but we can, without violating the principle of relativity, multiply the force by any power of C [the invariant in equation (2.2)], and consequently any (positive or negative or even fractional) multiple of the quantities [of the planets' perihelia] will be in agreement with that principle (de Sitter 1911, p. 406).

Thus, a force compatible with Poincaré's relativity theory can explain the total of the unexplained advance of Mercury's perihelion. Such an explanation would create a problem with Venus's perihelion, however, since it would increase all perihelium motions. Still, it seems that one could have tried to explain the whole advance of Mercury's perihelion by a relativistic force law, slightly modifying the masses of the planets in accordance with the new law. Unsurprisingly, no one made such an attempt. Why should one make an effort to solve a problem that is already solved? De Sitter showed that the relativistic force law is compatible with the astronomical observations and that was enough. In contrast with the explanation of Mercury's advance by·

general relativity, which is a necessary result of the theory and the known astronomical data, an explanation of the whole perihelion advance by some "Poincaré's law" would have been based on the result sought rather than on intrinsic properties of the theory. In this case it is better to rely on Seeliger's satisfactory solution. General relativity could have been refuted by its prediction of Mercury's motion;[27] Due to its free parameter, Poincaré's theory could not be refuted since one could always change its results by multiplication with an invariant. Therefore, while its prediction of the perihelion's advance confirmed general relativity, such a prediction could not have confirmed Poincaré's theory.[28]

2.6 Possible Objections to Poincaré's Attitude

De Sitter showed that Poincaré's formulations of the force of attraction satisfied both the astronomical observations and the relativity principle. Both could not be used as arguments against the theory; yet other considerations did lead physicists to prefer different theories. As pointed out by Sommerfeld, the lack of a field was an argument against the theory. Alternative approaches had additional advantages for some physicists. Before commenting shortly on these approaches, I wish to examine the relation between inertial and gravitational mass in Poincaré's theory, a relation that might have seemed problematic to other physicists. This part is rather speculative, since I have no direct evidence that contemporary physicists examined Poincaré's suggestion in such a manner. However, at least a few physicists expressed similar considerations concerning theories of gravitation in general, so it is plausible that they also examined Poincaré's treatment of the subject.

Poincaré did not adopt the view that energy possesses inertia. At least until 1904 he viewed an electromagnetic wave as a projectile with no mass. Energy, he wrote, is not matter (Poincaré 1900, 1905, p. 135, Darrigol 1995, pp. 17–31). Still, the transformation of mass into energy and vice versa was a logical consequence of his relativistic theory.[29] In the treatment of gravitation he assumed a constant gravitational mass, i.e., unlike the inertial, the gravitational mass of a body is the same in any (inertial) frame of reference. The Lorentz transformation does not alter the latter's value; in a sense it is invariant. Energy, however, depends on the frame of reference. So, in this approach it cannot possess gravitation. This suggests that material bodies always conserve their gravitational mass. In particular the gravitational mass of a radioactive body that radiates energy and loses inertial mass should be conserved in its material parts. Thus, the ratios of inertial and gravitational mass in a body before and after radioactive decay would not be the same.

In 1905, when Poincaré elaborated his examination of gravitation, the relation between inertial and gravitational mass in radioactive matter had not been examined in the laboratory yet. However, two such experiments were carried out in the following five years. The empirical validity of the equivalence of inertial and gravitational mass of radioactive matter was the subject of the Göttingen prize for 1909. The idea for the experiment was based on a non-relativistic electrodynamic theory, on the assumption that the supposed electrodynamic mass might have no gravitational properties. Only one paper was submitted. The paper (which of course won the prize) was submitted by the Hungarian physicist Eötvös and his colleagues Pekr and Fekee. Eötvös had already succeeded in 1890 in showing that the two masses have the same ratio to the accuracy of $1:2 \times 10^{12}$, yet this was demonstrated only for static (in the laboratory frame) ordinary matter. In 1909 the prize committee announced that the same ratio between inertial and gravitational mass had also been found for radioactive radium-bromide, with an accuracy of $1:2 \times 10^7$. Yet the results were not conclusive, and the experimentalists promised to publish a paper with more exact results (Runge 1909). Eventually they published such a paper in 1922, and the accuracy was reduced to $1:4 \times 10^6$ (Kox 1993). In 1910 Leonard Southerns published his results of "the ratio of mass to weight" for uranium oxide and found them equal with an accuracy of 1 to 2×10^5 (Southerns 1910).

Both experiments were performed on the basis of non-relativistic electrodynamic theories of matter, but their implications for the relativistic theories were clear. In the process of transmutation from uranium to radium the radiated energy per mole was known by the second decade of the century, using the mass-energy transformation, to be equal to 0.02 mass units (Siegel 1978, pp. 341–342). This is about $1:10^4$ of the total mass of the uranium/radium. Both experiments were more exact than this ratio. Though one should take into account the impurity of the elements in the experiment,[30] as well as the inaccuracy of the data (especially in Eötvös's experiment), the experiments gave at least a good indication that the equivalence of inertial and gravitational mass is valid for all static bodies, including all radioactive ones. If the gravitational mass would be conserved and constant, one would expect that in at least one of these experiments (of either uranium or radium) the relation between the gravitational and the inertial mass would be different.

In a survey article on the "New Theories of Gravitation" written in 1914, Max Abraham concluded from Southerns' experiment (he did not refer to Eötvös et al.) that the inertial and gravitational masses are equivalent and thus that "the gravitational mass is proportional to the energy" (Abraham 1915, 481). According to Abraham, the new theory of gravitation should account for the gravitational attraction of energy. Abraham did not discuss Poincaré's work in his survey of the various theories. Perhaps he thought that Southerns' experiment contradicted the assumptions of Poincaré and Minkowski. Yet, the experiments did not necessarily lead to Abraham's conclusion. If one had wanted to keep Poincaré's force law, one could have supposed that the gravitational mass is a non-conserved quantity, which is created and annihilated in every transformation between mass and energy. Obviously, this is an awkward assumption, which runs counter to the inclination towards conservation laws in physics, but it is still a possible way out. Poincaré himself might have been willing to employ such an hypothesis: he was sceptical about the validity of the established conservation laws (Poincaré 1905). However, at least in public, he did not refer to this question. Other physicists, like Abraham, would not have considered such a solution. No one published a criticism of the behaviour of the gravitational mass in Poincaré's approach. Still, the silence does not imply that no one criticized the approach on that ground. Most likely, a few rejected it for that reason, but did not bother to elaborate their critique. Abraham's disregard of Poincaré and Minkowski's approaches clearly shows his opinion about them.

De Sitter's paper marks the apogee of Poincaré's relativistic theory of gravitation. In the same year Einstein published his elaborated attempt at a relativistic theory of gravitation based on the principle of equivalence of gravitational and inertial mass, i.e., all bodies, irrespective of their velocities, fall at the same acceleration in a gravitational field. Most physicists considered Southerns' experiment as a confirmation of this equivalence. The publication of Einstein's paper can be seen as the beginning of the attempts at theories of gravitation that take the principle of equivalence as their basis. Although other scientists interpreted this principle in different ways at the time (Norton 1992), most of them shared the opinion that it had to be satisfied by every relativistic theory of gravitation. In 1912 Nordström, Abraham and Ishiwara all suggested relativistic theories of gravitation that either assumed or obtained the equivalence. The former theory was, like Poincaré's, a Lorentz covariant. However, complying with the equivalence principle and being a field theory, it was more similar to the theories of Einstein and Abraham. The latter attempts not only tried to satisfy the equivalence principle but also to construct a field theory of gravitation rather than merely a law of propagation of attraction in finite time as in Poincaré's and Minkowski's treatments. The importance of the principle of equivalence is evident also in the representation of Mie's theory that violated it (at high temperatures, not at high linear velocities). Mie emphasized that in his theory "the gravitational and the inertial masses are practically indistinguishable" (Mie 1913, p. 50).

After 1911 the interest of the scientific community in gravitation tended more to field theories of gravitation that admit the equivalence principle, as opposed to non-field theories like Poincaré's. These efforts and especially those of Einstein attracted the most attention; the simple theories that assumed constant gravitational mass were neglected. Lorentz's 1914 explanatory article "*La Gravitation*" is a good example of this development (Lorentz 1914). That an expert in electrodynamics, like Lorentz, dedicated an article to a exposition of theories of gravitation to laymen indicates an increase in the importance of the theories of gravitation. This increase followed the more ambitious attempts to construct a field theory of gravitation. Circa 1910 Lorentz was engaged in covariant gravitational theories through his working connections with de Sitter; in 1914 he was interested in Einstein's theory. While in his 1910 survey (Lorentz 1910) Lorentz discussed only Poincaré's and Minkowski's relativistic theories of gravitation, in 1914 he dedicated most of his article to Einstein's (then incomplete) theory. He dedicated twenty pages to the attempts at a theory that posits the equivalence principle and only one to those that do not. Yet he did not reject the latter theories. In his discussion of Einstein's theory Lorentz emphasized the principle of equivalence and its implementation. Einstein's success, the following year, in constructing a field theory of gravitation that admits the equivalence principle and the general principle of relativity made the previous attempts at a simple Lorentz covariant theory of gravitation irrelevant.

2.7 Conclusions

Poincaré suggested a modification of Newton's law of attraction because of the contradiction between that law and the new relativistic physics. A force that transforms from one system of reference to another differently from the electric force, or that propagates faster than light, would reveal absolute velocities in contradiction with the relativity principle. Poincaré's theoretical elaboration of the law of attraction showed that the force of gravitation can be inccorporated in the new relativistic physics. Since his main concern was to resolve the contradiction between the force and the relativity principle (in its electrodynamic interpretation), Poincaré was satisfied with an undefined force law (or laws). This can explain why he neither compared it with the details of astronomical data, nor developed an independent theory. Instead he was satisfied in stating its general agreement with the observations, and left it based on the Newtonian theory. Einstein's general relativity can be seen as an answer to a similar problem: incorporating gravitation in a relativistic theory. Still, the absence of any treatment of gravitation in Einstein's 1905 theory makes its discussion of simultaneity, and thus the whole theory, open to doubts and objections, like those that Poincaré raised in 1904. The treatment of gravitation made Poincaré's relativistic theory of 1905 more complete than Einstein's theory of the same year.

Yet, that one theory is more complete than another is not necessarily an advantage. Poincaré suggested a simple relativistic theory of uniform motion that includes gravitation and agrees with the observational data. Einstein's aim was much more ambitious, and in the long run more successful. The consolidation of gravitation in Poincaré's relativity theory could have discouraged physicists from searching an alternative theory. Poincaré's requirements of the force were the minimum necessary to make it compatible with the relativity principle and the known empirical data. Physicists who saw in the inclusion of gravitation in the new relativistic physics an open question had further requirements that eventually led to a much more successful theory. Being more complete, less revolutionary and in better accordance with classical theory than Einstein's view of gravitation, Poincaré's path would not have led to the general theory of relativity. One could have been satisfied with its undefined force law of attraction that propagates through space.

Poincaré's law of gravitation was compatible with the empirical data in 1905 when he formulated it, in 1911 when De Sitter examined it, and still in 1915 when Einstein published his famous theory. Perhaps ironically for Poincaré, one crucial experiment, or more precisely an observation, refuted his theory. This was Eddington's celebrated observation of the bending of light in 1919, which was made only following the prediction of general relativity. According to Poincaré's theory, energy has no gravity, so an experiment that attributes any property of gravity to light refutes Poincaré's theory. The refuting observation, we should remember, was made fourteen years after the formulation of the theory.

However, when Eddington reported his observational results, no one referred to them as a refutation of the constant gravitational mass theories. The scientific community had abandoned these simple theories after the introduction of the more ambitious attempts at constructing a gravitational field theory compatible with the equivalence principle. It needed no empirical refutation for the rejection of Poincaré's approach. Poincaré's theory did not satisfy several theoretical requirements that many physicists sought. The hypothesis of a constant gravitational mass implies that energy has no gravitational mass, a strange assumption for physicists who adopted the (inertial) mass-energy equivalence. Moreover, the combination of these assumptions led to the conclusion that gravitational mass is not conserved, in contrast to the methodological tendency to conservation rules. Yet probably many physicists did not study these implications. From the point of view of contemporaries two shortcomings of the theory were more conspicuous: the lack of a field and the disregard of the principle of equivalence between gravitational and inertial mass, which was probably regarded as the theory's major disadvantage. In contrast to the attempts during the second decade of the century, Poincaré attempted neither to explain gravitation nor to give a comprehensive theory of its action. The arbitrariness left in his theory suggested that one could demand more from a relativistic theory of gravitation. The requirement that the new theory would be a field theory that satisfies the equivalence principle (which was based on plausible physical assumptions) left less freedom in formulating new relations. Physicists appreciated laws that necessarily followed from the field's equations more than a force law defined by empirical results that it has to explain. Necessity always seems more powerful and revealing than a contingency. Physicists had good reasons to prefer approaches other than Poincaré's. Still, until 1919 neither the empirical data nor any internal contradiction forced them to make that choice.

Acknowledgments. Parts of this paper are based on my M.A. thesis (Katzir 1996). I would like to thank my two supervisors, Ido Yavetz and Jürgen Renn for their help. In doing the research for this paper I enjoyed the hospitality and the helpful intellectual environment of the Max Planck Institute for the History of Science in Berlin in Summer 1996. I am grateful to Tilman Sauer, Jürgen Renn, Jim Ritter and especially to Scott Walter for their advice and criticism of earlier drafts, and to Galia Sartiel for her linguistic help.

References

Abraham, Max. 1915. Neuere Gravitationstheorien. *Jahrbuch der Radioaktivität und Elektronik*, 470–520.

Barrow-Green, June. 1997. *Poincaré and the Three Body Problem*. American Mathematical Society, Providence, RI.

Corry, Leo. 1997. Hermann Minkowski and the Postulate of Relativity. *Archive for History of Exact Sciences*, 51:273–314.

Cuvaj, Camillo. 1968. Henri Poincaré's Mathematical Contribution to Relativity and the Poincaré Stress, *American Journal of Physics*, 36:1102–13.

— 1970. *A History of Relativity—The Role of Henri Poincaré and Paul Langevin*. Ph.D.: Yeshiva University.

Darrigol, Olivier. 1995. Henri Poincaré's Criticism of Fin de Sicle Electrodynamics. *Studies in History and Philosophy of Modern Physics* **26** (1):1–44.

Galison, Peter L. 1978. Minkowski's Space-Time: From Visual Thinking to the Absolute World, in: *Historical Studies in the Physical Sciences* 10:85–121.

Giedymin, Jerzy. 1982. *Science and Convention*. Pergamon Press, Oxford, U.K.

Goldberg, Stanely. 1967. Henri Poincaré and Einstein's Theory of Relativity. *American Journal of Physics* 35: 934–944.

— 1970. Poincaré's Silence and Einstein's Relativity: The role of theory and experiment in Poincaré's Physics. *British Journal for the History of Science* 17:73–84.

Holton, Gerald. 1964. On the thematic Analysis of Science: the Case of Poincaré and Relativity. *Mélanges Alexandre Koyré* 2:257–268.

Katzir, Shaul. 1996. *Poincaré's Relativity theory—Its Evolution, Meaning and its (Non)acceptance*. M.A.: Tel Aviv University.

— (2005). Poincaré's Relativistic Physics and Its Origins, *Physics in Perspective* 7:268-292

Kilmister, C.W. 1970. *Special Theory of Relativity*. Pergamon Press, Oxford, U.K.

Kox, A.J. 1993. Pieter Zeeman's Experiments on the Equality of Inertial and Gravitational Mass. *Einstein Studies* 5:173–181.

Lorentz, Hendrik A. 1904. Electromagnetic phenomena in a system moving with any velocity less than that of light. In *Collected Papers*. The Hague: Martinus Nijhoff, 1934, 5:172–197. Pagination follows the partial reprint in Einstein et al. *The Principle of Relativity*, Dover, New York, 1952.

— 1910. Alte und neue Fragen der Physik. *Physikalische Zeitschrift* 11:1234–1257. Reprinted in *Collected papers*. The Hague: Martinus Nijhoff, 1934, 7:205–245. Page references are to the reprint.

— 1914. La Gravitation. *Scientia* **16** (36):28–59.

Maxwell, Clark G. 1954. *Treatise on Electricity and Magnetism*. 2 vols. 3rd ed. Dover, New York.

Mie, Gustav. 1913. Grundlagen einer Theorie der Materie (Dritte Mitteilung). *Annalen Der Physik* 40:1–66.

Miller, Arthur I. 1973. A Study of Henri Poincaré's 'Sur la dynamique de l'électron.' *Archive for History of Exact Sciences* 10:207–328.

1996. Why did Poincaré not formulate special relativity in 1905, in: *Henri Poincaré Science et Philosophie— Congrès International Nancy France 1994*, Jean-Louis Greffe, Gerhard Heinzmann and Kuno Lorenz, eds., Blanchard, Paris, 69–100.

Minkowski, Herman. 1908. Die Grundgleichungen fr die elektromagnetischen Vorgänge in bewegten Körpern. *Nachrichten — Königlichen Gesellschaft der Wissenschaften zu Göttingen*, 53–111.

— 1916. Das Relativitätsprinzip. *Jahresbericht der deutschen Mathematiker-Vereinigung* 24:372–382.

Norton, John. 1992. Einstein, Nordström and the early demise of scalar, Lorentz covariant theories of gravitation. *Archive for History of Exact Sciences* 45:17–94.

Poincaré, Henri. 1892. Les Methodes Nouvelles de la Mécaniques Céleste, vol. 1. Gauthier-Villar, Paris.

— 1898 Sur la stabilité du système solaire. *Revue Scientifique* 9:609–613.

— 1900. La théorie de Lorentz et le principe de réaction. in: *Poincaré 1954*, 464–488.

— 1902. *La Science et l'Hypothèse*. Flammarion, Paris.

— 1905. *La Valeur de la Science*. Flammarion, Paris. The pagination follows the 1970 edition.

— 1906. Sur la dynamique de l'électron. in: *Poincaré 1954*, 494–550.

— 1908. La dynamique de l'électron. in: *Poincaré 1954*, 551-585.

— 1909. La mécanique nouvelle. *Revue Scientifique* 84:170–177.

— 1910. La mécanique nouvelle. In: *Sechs Vorträge über ausgewählte Gegenstände Aus der reinen Mathematik und mathematischen Physik*, B.G. Teubner, Leipzig and Berlin, 49–58.

— 1910a. Die neue Mechanik. *Himmel und Erde* 23:97–116.

— 1953. Les Limites de la Loi de Newton. *Bulletin Astronomique* 17:121–269.

— 1954. *Oeuvres*, Tome 9. Gauthier-Villars, Paris.

Pyenson, Lewis. 1977. Hermann Minkowski and Einstein's Special Theory of Relativity. *Archive for History of Exact Sciences*: 71–95.

Richman, Sam. 1996. Resolving discordant Results: Modern Solar Oblateness Experiments. *Studies in History and Philosophy of Modern Physics* 27:1–22.

Roseveare, N.T. 1982. *Mercury's Perihelion from Le Verrier to Einstein*. Clarendon Press, Oxford, U.K.

Runge, C. 1909. Benekesche Preisstiftung. *Nachrichten—Königlichen Gesellschaft der Wissenschaften zu Göttingen*, 37–41.

Seeliger, Hugo von. 1906. Das Zodiakallicht und die empirischen Glieder in Bewegung der innern Planeten. *Königlich Bayerische Academie der Wissenschaften (München). Sitzungsberichte* 36:595–622.

Siegel, Daniel M. 1978. Classical-Electromagnetic and Relativistic Approaches to the Problem of Non integral Atomic Masses. *Historical Studies in the Physical Sciences* 9:323–360.

de Sitter, Willem. 1911. On the Bearing of the Principle of Relativity on Gravitational Astronomy, *Monthly Notes of the Royal Astronomical Society* 51:388–415.

— 1913. The Secular Variations of the Elements of the Four Inner Planets, *Observatory*, 36:296–303.

Smith, Crosbie and Wise, M. Norton. 1989. *Energy and Empire—A Biographical Study of Lord Kelvin*. Cambridge University Press, Cambridge, U.K.

Sommerfeld, Arnold. 1910. Zur Relativitätstheorie II: Vierdimensionales Vektoranalysis *Annalen der Physik* 33:649–689.

Southerns, Leonard. 1910. A Determination of the Ratio of Mass to Weight for a Radioactive Substance. *Royal Society of London. Proceeding A* 84:325–344.

Stachel, John. 1995. History of Relativity. in: *Twentieth Century Physics*, vol. 1, L.M. Brown, A. Pais and B. Pippord, eds., Institute for Physics Publishing, Bristol and Philadephia, 249–356.

Walter, Scott. A. 1999. Minkowski, Mathematicians, and the Mathematical Theory of Relativity. In *The Expanding Worlds of General Relativity*, H. Goenner et al. eds., (Einstein Studies 7), Birkhäuser, Basel, 45–86.

— Forthcoming. Breaking in the 4-vectors: the four-dimensional movement in gravitation, 1905–1910. *The Genesis of General Relativity Vol. 3: Theories of Gravitation in the Twilight of Classical Physics; Part I*, J. Renn and M. Schemmel, eds., Kluwer, Dordrecht.

Zahar, Eli G. 1983. Poincaré's Independent Discovery of the Relativity Principle. *Fundamenta Scientiae* 4:147–176.

Zenneck, J. 1902. Gravitation. In *Enzyklopaädie der mathematischen Wissenschaften*, Arnold Sommerfeld, ed., 25–67.

Notes

[1]Poincaré 1906 (the paper was submitted in July 1905). I support the claim that Poincaré suggested in this paper a theory based on the relativity principle in (Katzir forthcoming). For similar interpretations, see (Darrigol 1995, Giedymin 1982 and Zahar 1983). However, for different view see (Cuvaj 1970, Goldberg 1967 and 1970, Holton 1964, Miller 1973 and 1996).

[2]A. I. Miller's detailed study of Poincaré's memoir (Miller 1973), which does not discuss the section of gravitation, is the best example of this neglect, which is common to most treatments of Poincaré and relativity theory. Important exceptions to this neglect are (Cuvaj 1970), which dedicates a brief discussion of less than two pages to the subject (pp. 92–93), and C.W. Kilmister's translation of Poincaré 1906, which includes part of the section on Gravitation (Kilmister 1970, pp. 145–187).

[3]The present article had already been written when I read Walter's paper.

[4]Page numbers in parentheses in the following two sections refer to this publication.

[5]This force is proportional to the electron's volume (Poincaré 1906, pp. 528–529 and 536–538).

[6]Poincaré stated clearly that electromagnetism cannot account for gravitation. In a university course delivered in 1906–1907, he demonstrated that neither other forces of nature nor any mechanism can explain the force of attraction. Lorentz's recent attempt to explain gravitation by electrodynamics and Lessage's older attempt to reduce it to mechanics and all similar attempts are refuted by their prediction of heat accumulation (Poincaré 1953, pp. 215–216, 257).

[7]According to Le Verrier's table from 1859, in which the divergence was first pointed out, the unexplained advance was of 39″ per century, out of a total of more than 565″, most of it explained by the perturbations due to the other planets' influence. For Le Verrier's data and the history of the theories of Mercury's perihelion, see (Roseveare 1982).

[8]Poincaré elaborated on the hypothesis of a material ring in the 1906–1907 university course. Yet, most likely, this was his position also in the summer of 1905. Although a rigorous and detailed discussion of the same suggestion was published only in 1906 by Seeliger (Seeliger 1906, p. 596), it is unlikely that the later publication influenced Poincaré's thought. First, the publication appeared only in December 1906, probably too late to be included in a course of the 1906–1907 autumn semester.

Second, Poincaré assumed one circular ring rather than the two ellipsoids suggested by Seeliger. Third, he neither mentioned Seeliger nor referred to his quantitative results. Instead, he supplied only qualitative estimates and mentioned the suggestion of Newcomb from 1895.

[9]A comparison with Minkowski brings out Poincaré's empirical concern. The former also formulated a relativistic gravitational force law, but based it on geometrical considerations, without mentioning empirical considerations in the elaboration of the force, as if the law was derived only from mathematics (Corry 1997, pp. 286–87).

[10]The constant mass assumption might also be connected to Poincaré's point of view that part of the inertia (but not all) can be seen as a property of the ether. Then the connection between inertial and gravitational mass seems accidental; if gravity is a property of matter alone while inertia depends on the ether as well, why should they vary in the same way? The view that inertia is in some cases a property of the ether does not appear in the 1905 paper. Yet a year earlier, Poincaré had argued that the inertia of the negative electron is a property of the ether, while the positive electron (the proton) has a material mass (Poincaré 1905, 138). He probably maintained this notion a year later. Indeed, in 1908 he had the different opinion that: "What we call mass would seem to be nothing but an appearance, and all inertia would be of electromagnetic origin" (Poincaré 1908, p. 556), but such statements did not appear earlier. Two years later he referred to neutral elementary particles, which cannot get their inertia from the electromagnetic ether (Poincaré 1910a, p. 113).

[11]For a thorough discussion of Poincaré's mathematical elaboration, see (Walter forthcoming).

[12]Poincaré had used the concept of invariance extensively in his mathematical work. In 1886 he formulated a theory of invariant integrals, which he used and elaborated in his celebrated memoir on the three-body problem in 1890 (Barrow–Green 1997, p. 83).

[13]Poincaré discarded the power component in the following equation (p. 549) and discussed only the force. Writing the force equation in vectorial notation therefore makes the equations clearer to the modern reader while retaining their physical meaning.

[14]This wave is not connected to gravitational radiation and general relativity's gravitational waves. Some authors have regarded them erroneously as radiation waves (see, e.g., Cuvaj 1968). In later writings (discussed below), Poincaré referred to gravitational radiation, which is analogous to general relativity's wave.

[15]See for example (Zenneck 1902, p. 48).

[16]In the popular addresses he informed his audience neither about which of his alternative expressions for the law he had used, nor of the methods by which he obtained quantitative results. A lecture in which he probably derived the relativistic law of gravitation is missing from the notes, so the historian should defer to his choice on the basis of the numerical results.

[17]Poincaré was led to this conclusion by his use of "Langevin's waves of velocity and acceleration," suggested earlier only for electrodynamics (Poincaré 1953, pp. 241–245).

[18]Dissipation of kinetic energy was not a new idea in celestial mechanics. The phenomenon of the tides was a paradigmatic example for such dissipation in the solar system, and was discussed extensively (in connection with heat dissipation) in the second half of the nineteenth century, especially in Britain (Smith and Wise 1989). Though in his study of the three-body problem Poincaré concluded that the solar system is dynamically stable, in 1898 he claimed that it is unstable due to energy dissipation. This is a result of friction in the planets caused by gravitational tidal forces and by non-gravitational forces like magnetic attraction (Poincaré 1898).

[19]This sentence appears in the texts of the lectures at Göttingen (Poincaré 1910, p. 57) and at Lille; the rest of the quote appears only in the text of the Lille lecture.

[20]This is also the value reported in his course a year earlier (Poincaré 1953, p. 245). On later occasions he reported lower values of $6''$ (Poincaré 1909, p. 177) and $5''$ (Poincaré 1910a, p. 115). Whether the changes originated from a better approximation or from carelessness about the issue is unknown. De Sitter's later calculation that gave a value of $7''15'$ makes the latter possibility likely.

[21]For a preliminary discussion of the reception of Poincaré's theory and its reasons, see chapter V of (Katzir 1996). I do not consider references to "Poincaré's stress" as a treatment of Poincaré's theory. Though Poincaré introduced the concept in his relativistic memoir, it is independent of his relativistic theory. Most of the physicists who referred to it did not refer to Poincaré's theory, or to other sections of his memoir.

[22]On the history of Minkowski's relativity theory, see (Galison 1978, Pyenson 1977, Corry 1997 and Walter 1999).

[23]On the influence of Poincaré's paper on Minkowski's work and the differences in their approaches, see (Katzir 1996, pp. 93–96) and (Walter forthcoming).

[24]For details on Minkowski's force law, its geometrical representation and its derivation, see Corry 1997, pp. 286–291.

[25]The value of $6''69'$ he gave was based on approximations and was corrected later by de Sitter.

[26]Perhaps Lorentz still toyed with the idea to explain gravitation through electromagnetism. He had made a (failed) attempt in that direction a decade earlier. Among others, Poincaré had refuted this explanation of gravitation (Poincaré 1953, pp. 251–257; 1908, pp. 584–585).

[27]Robert Dicke claimed in 1964 that his observation of the sun's oblateness can be seen as such refutation. Since this oblateness adds a $4''$ in a century to the advance of Mercury's perihelion, it therefore shows that general relativity predicts an advance larger than the observed one (Richman 1996).

[28]On this point John Stachel wrote: "Many special-relativistic theories of gravitation—scalar, vector and tensor—have been and continue to be proposed as competitors

of general relativity. In a sense, the problem is too easy: given a special-relativistic theory with several free parameters, it is rather easy to choose their values so that the theory gives the same predictions as general relativity for the three 'classical tests' of the latter" (Stachel 1995, p. 284).

[29]For a derivation of the transformation within the context of Poincaré's theory, see (Katzir 1996, pp. 114–115).

[30]However, due to their high specific weight, the radioactive elements form the dominant contribution to the mass.

3

Standing on the Shoulders of a Dwarf: General Relativity—A Triumph of Einstein and Grossmann's Erroneous Entwurf Theory[1]

Jürgen Renn

Max-Planck-Institut für Wissenschaftsgeschichte, Wilhelmstraße 44, 10117 Berlin, Germany;
renn@mpiwg-berlin.mpg.de

3.1 General Relativity as an Heroic Achievement

When a single figure plays such a distinctive historical role as Einstein did in the emergence of general relativity it becomes almost unavoidable to tell this story, be it drama or comedy, in theatrical terms. The narratives of the history of general relativity thus refer to blunders and breakthroughs, to fatal errors and the dawning of truth. They characterize this history as the drama of a lonely hero, as a comedy of errors, or even as the irresistible rise of a slick opportunist. Dramatic narratives tend to emphasize the achievements of great heroes and to neglect the minor figures; they favor the mysticism of great ideas (or great failures) and usually ignore their tedious elaboration. In this form, apparent mistakes, while presenting little interest in themselves, provide the contrast that makes the victory of truth appear all the more triumphant.

As an example of such dramatic narratives, let me quote from Kip Thorne's fascinating account of recent developments in general relativity which, however, presents David Hilbert as the true hero of the story (Thorne 1994):[2]

> In autumn 1915, even as Einstein was struggling toward the right law, making mathematical mistake after mistake, Hilbert was mulling over the things he had learned from Einstein's summer visit to Göttingen. While he was on an autumn vacation on the island of Rugen in the Baltic the key idea came to him, and within a few weeks he had the right law—derived not by the arduous trial-and-error path of Einstein, but by an elegant, succinct mathematical route.

The work undertaken on the history of general relativity, pursued by participants of a collaborative research project that began in Berlin in 1991, has involuntarily contributed to the thrill of this story:[3] the insight that Einstein already formulated the correct field equations in linearized form in 1912 and then discarded them, his similar treatment of the gravitational lensing effect in the same year (Renn, Sauer, and Stachel 1997), and our finding that Hilbert did not actually discover the field equations, but rather first formulated a non-covariant version of his theory, which he modified only *after* the publication of Einstein's theory of general relativity (Corry,

Renn, and Stachel 1997) are all results of this joint research effort and add further dramatic turns to an already exciting plot.

The focus here is on what is generally seen as perhaps the most boring period of Einstein's search for a generalized theory of relativity, the time between spring 1913 and fall 1915, in which he firmly stuck with the erroneous "Entwurf" theory he published together with Marcel Grossmann before the end of June 1913 (Einstein and Grossmann 1913).

According to the dramatic narratives of the emergence of general relativity, this period was one of stagnation. It was the calm interval between two major storms, Einstein's tragic struggle with and eventual rejection of generally covariant field equations in the winter of 1912/1913, and the shocking revelation of fatal errors in the "Entwurf" theory that led immediately to its demise and then to a triumphant, if gradual, return to generally covariant field equations in the fall of 1915.

Based on the results of a joint research effort and an alternative approach to the history of science, this period will be presented here from a new perspective. From the point of view of an historical epistemology, the apparent stagnation between 1913 and 1915 can be considered a period in which new knowledge was assimilated to a conceptual structure still rooted in classical physics. As a result of this assimilation of knowledge, this conceptual structure became richer, both in terms of an increasingly extended network of conclusions that it made possible, and in terms of new opportunities for ambiguities and internal conflicts within this network. It was this gradual process of enrichment that eventually created the preconditions for a reflection on the accumulated knowledge which, in turn, induced a reorganization of the original knowledge structure. The enrichment of a given conceptual structure by the assimilation of new knowledge and the subsequent reflective reorganization of this enriched structure are the two fundamental cognitive processes which explain the apparent paradox that the preconditions for the formulation of general relativity matured under the guidance of a theory that is actually incompatible with it.

The results achieved on the basis of the "Entwurf" theory should therefore perhaps not be understood as so many steps in the wrong direction, whereupon it appears that their only function was to make the deviation from the truth evident, but rather as instruments for accumulating and giving new order to this knowledge. It obviously makes little sense to consider one of these processes as being more central than the other since both are essential to the development of scientific knowledge.

This perspective on the genesis of general relativity, as flowing "out of the spirit of the 'Entwurf' theory," also leads to a new evaluation of what are usually cast as the stepchildren in the heroic narratives of the history of science: namely the "erroneous" approaches and theories, the boring periods of tedious elaboration of such theories, and the faceless minions in their service. It will become clear from my account that it was precisely the insistence with which Einstein labored to plug the holes of the erroneous "Entwurf" theory that made his approach so much more successful than Hilbert's. It will hopefully also become clear that the inconspicuous contributions to this theory by Grossmann, Bernays, and Besso were crucial in overturning it.

This counter-story proceeds in five acts. The next act attempts to destroy four legends on the history of general relativity: a breakthrough in late 1915, a pitfall in early

1913, a period of stagnation between 1913 and 1915, and the almost simultaneous discovery of general relativity by Einstein and Hilbert.

3.2 The Legend of a Breakthrough in Late 1915

After more than two years of intensive work on his "Entwurf" theory, Einstein suddenly abandoned this theory on the 4th of November 1915 with the publication of a short paper in the *Sitzungsberichte* of the Prussian Academy (Einstein 1915). In this and subsequent papers, as well as in his correspondence, Einstein himself gave the reasons for his abandoning of the "Entwurf" theory (CPAE 8, Doc. 153). The "Entwurf" theory could not explain the perihelion shift of Mercury, the earliest astronomical touchstone of general relativity; it did not allow treatment of a rotating system as being equivalent to the state of rest, and hence did not satisfy Einstein's Machian heuristics, and finally, a flaw was discovered in the derivation of the theory.

From the point of view of later general relativity, each of these three arguments seems to represent a major blunder that in itself would have sufficed to reject the "Entwurf" theory. Accordingly, historians of science are disputing which of these arguments was first or decisive in leading to the demise of this theory. On the other hand, they tend to leave unquestioned the assumption that it must have been one of these three or perhaps a fourth stumbling block that led to its downfall. This assumption fits well with the philosophical idea of progress being due to falsification and also to the historical topos of a dramatic turn in early November 1915, initiating the true birth of general relativity.

A closer look at the historical evidence, however, makes this assumption doubtful. Indeed, it can be shown that the "Entwurf" theory survived *all* the blunders listed above. As has become clear from research notes of Einstein and Besso, they knew at least from mid-1913 that the "Entwurf" theory failed to explain the perihelion shift of Mercury.[4] Recently discovered additional notes documenting the Einstein–Besso collaboration in 1913–1914 show that Besso warned Einstein in August 1913 that the Minkowski metric in rotating coordinates is not a solution to the "Entwurf" field equations. Einstein seems to have accepted this conclusion for a while but thought in early 1914 that he had found an argument showing that this metric had to be a solution (CPAE 8, Doc. 47). Besso questioned this result (CPAE 8, Doc. 516) but Einstein did not listen.[5] Furthermore, when Einstein found out, around mid-October 1915, that his mathematical derivation of the "Entwurf" theory did not work, he nevertheless continued initially to stick to this theory as is made evident by a new demonstration he sent to H.A. Lorentz (CPAE 8, Doc. 129). The stubbornness with which Einstein held on to the "Entwurf" theory is the same characteristic that guided his entire search for a relativistic theory of gravitation. But, in the face of so many counter arguments, for which reasons did he cling so stubbornly to the "Entwurf" theory and what caused him finally to change his mind?

In order to answer these questions, a short review of how Einstein reacted to each of the three "Entwurf" theory problems listed above may be appropriate. This is most easily done for the problem of the perihelion shift of Mercury. The same research notes containing Einstein and Besso's calculation of the perihelion shift of Mercury on the basis of the "Entwurf" theory also document them checking whether its main competitor, Nordström's theory of gravitation, would yield the correct result—which turned out not to be the case. Mercury's perihelion shift was thus not a criterion for choosing between the alternative theories available (CPAE 5, Doc. 14).

The situation is more complicated for the question of rotation. The Machian idea that the inertial effects in a rotating system may actually be due to the interaction with distant masses had been an important element of Einstein's heuristics. It motivated his search for a theory with a generalized relativity principle in which a rotating frame of reference can be considered as being equivalent to an inertial frame with gravito-inertial forces. By mid-1913 he knew, however, that the Minkowski metric in rotating coordinates is *not* a solution to the "Entwurf" field equations, and that therefore the state of rotation cannot be considered as being equivalent to the state of rest. Nevertheless, on the basis of general considerations in the course of his further elaboration of the "Entwurf" theory in 1914, Einstein convinced himself that this theory did, after all, comply with his Machian heuristics (CPAE 5, Doc. 514). He believed that he had actually reached the goal of a generally relativistic theory of gravitation in spite of the fact that the "Entwurf" field equations are not generally covariant. In fact, he interpreted the conditions on the covariance properties of the theory that he had meanwhile identified *not* as restrictions on possible solutions for the metric tensor but only as restrictions of the *coordinate systems* for representing a given solution (CPAE 8, Doc. 47, 80). Influenced by these general considerations, Einstein tended to forget his earlier finding that the Minkowski metric in rotating coordinates is not a solution of the "Entwurf" field equations and rediscovered this fact only in September 1915 (Earman and Janssen 1993). The fact that this "oversight" did not constitute a sufficient reason for abandoning the "Entwurf" theory is made evident by his development of a new derivation of the theory in October of the same year (CPAE 8, Doc. 129).

The third problem with which the "Entwurf" theory was confronted was the flaw in its derivation from general principles. In the winter of 1912/1913, Einstein had developed the theory, jointly with Marcel Grossmann, by starting from a cautious generalization of Newtonian gravitation theory and of a special relativistic expression for energy-momentum conservation. We have called this strategy, along which Einstein hoped to eventually reach an implementation of his Machian heuristics—without ever losing touch with the secure knowledge of classical and special-relativistic physics—his "physical strategy." He turned to this strategy after having first followed what we have called his complementary "mathematical strategy," which started by immediately implementing his Machian heuristics in terms of the absolute differential calculus, and then aimed at recovering, within this framework, a representation of the familiar knowledge on gravitation and energy-momentum conservation.

After the physical strategy had led Einstein to the formulation of the field equations of the "Entwurf" theory, it was natural for him to turn around and attempt to derive these equations following the mathematical strategy, which is precisely what he undertook in 1914. In a 1914 review paper, he published a lengthy and complicated demonstration in which the "Entwurf" field equations were derived from a general variational principle without explicitly introducing the requirement that the resulting theory should incorporate the classical knowledge on gravitation (CPAE 6, Doc. 9).

In October 1915, however, Einstein discovered that this demonstration actually does not uniquely determine the "Entwurf" field equations, but instead only provides a general mathematical framework for formulating a theory of gravitation. While he was naturally disappointed by this discovery, it also clearly did not represent a reason for abandoning the "Entwurf" theory. After all, the flaw in the 1914 demonstration did not affect the earlier justification of the theory in Einstein and Grossmann's original paper. Einstein even found a way of repairing his 1914 demonstration by supplementing it with an assumption representing the classical knowledge on gravitation.

In summary, all three objections to the "Entwurf" theory, which in hindsight appeared to mark its decline, if not its demise, emerge, on closer inspection, as failures only of the more ambitious and more problematic parts of Einstein's heuristics. In particular, these parts included his goal to find an astronomical confirmation of his new theory of gravitation in observations already available, his hope to realize the Machian idea of conceiving rotation as rest, and his expectation that the "Entwurf" theory could also be derived through mathematical strategy. On the other hand, these objections did not touch upon what had been the firm foundation of the "Entwurf" theory from the beginning: its roots in the knowledge of classical and special relativistic physics. It is therefore clear that the abandonment of the "Entwurf" theory in early November 1915 is not properly characterized as a "breakthrough" in the commonly accepted sense: that is, the demise of a faulty theory followed by the gradual dawning of the correct one. Before we return to the question of what it was that eventually changed Einstein's mind on the "Entwurf" theory, we first have to tackle another legend on the history of general relativity: the legend of a pitfall in early 1913.

3.3 The Legend of a Pitfall in Early 1913

The legend of a pitfall in early 1913 is structurally related to that of a breakthrough in late 1915. While the previous legend conveys the elimination of errors, this one imparts their introduction. Even before Einstein's calculations in the Zurich notebook (CPAE 4, Doc. 10) had been reconstructed by the members of the project mentioned previously, it had long been known that as early as 1912/13 Einstein had come close to formulating the final field equations of general relativity, at least for the source-free case. The analysis of the Zurich notebook has made the situation even more dramatic because it revealed the existence of an entry representing the linearized form of the definitive full gravitational field equations. It thus seems that Einstein must have been detracted from the correct path by some error, for otherwise he would have preserved these field equations instead of rejecting them in favor of those of the "Entwurf" theory. Several hypotheses have been advanced concerning the nature of this "error."

From remarks in the 1913 "Entwurf" paper as well as from later recollections by Einstein it was clear that he must have encountered a problem with recovering Newtonian gravitation theory from a relativistic gravitation theory based on the Riemann tensor.[6] In fact, when one forms the Ricci tensor, which represents a natural candidate for the left-hand side of gravitational field equations, one finds that, for weak static gravitational fields, it does not reduce to a form that immediately lends itself to a recovery of the Newtonian limit. From a page of the Zurich notebook we know that Einstein was indeed disappointed to find what he called "disturbing terms," in addition to the one from which the Newtonian limit can be derived (CPAE 4, Doc. 10, 233). In the modern understanding of general relativity, one can eliminate these additional terms by choosing an appropriate coordinate condition. What is therefore more plausible than to assume that Einstein was, in 1912/13, not yet aware of the possibility of picking an appropriate coordinate condition allowing the transition to the Newtonian limit, at least if one admits that Einstein could have been guilty of such a trivial error? A glance at the Zurich notebook shows, however, that this explanation cannot work because precisely the coordinate condition that we would introduce today, the harmonic condition, appears only a few pages later (CPAE 4, Doc. 10, 244–245).[7]

Which other "error" is then responsible for Einstein's publication of the "Entwurf" theory, other than a theory based on the Riemann tensor? There can be no doubt that, from a modern point of view, Einstein's thoughts on gravitation during this period were plagued by "errors." He assumed, for instance, that for weak gravitational fields the metric tensor becomes spatially flat and that it can be represented by a diagonal matrix with only one variable component, corresponding to the classical gravitational potential. He was convinced that the Newtonian limit could only be attained for weak static fields of this type.[8] This was a plausible assumption for Einstein for a number of reasons and hence offered historians the opportunity to accuse him of a non-trivial error since, in fact, the Newtonian limit also works fine with off-diagonal terms in the metric tensor because these have no effect on the equation of motion that describes the relevant weak-field effects. Were Einstein's faulty expectations concerning the Newtonian limit the reason why he discarded all candidates for the left-hand side of the field equations that were based on the Riemann tensor and decided in favor of the "Entwurf" theory instead?

From what we have learned from the reconstruction of the Zurich notebook, problems with the Newtonian limit were indeed the reason why in 1912 he rejected field equations based on the Einstein tensor. He realized that the trace term of this tensor would give rise to weak static fields incompatible with his expectations. We also know, however, that these expectations cannot have been the reason why he discarded other candidate gravitation tensors based on the Riemann tensor.

In particular, in the Zurich notebook, Einstein examined a tensor, covariant under unimodular coordinate transformations, for which he did not encounter this problem (CPAE 4, Doc, 253–254). And indeed, it was this tensor that constitutes the basis for the new gravitation theory with which Einstein replaced the "Entwurf" theory in early November 1915— it has therefore been dubbed the "November tensor." At this point he had not yet abandoned his expectations concerning the Newtonian limit and obviously found the November tensor in agreement with them. But why then did he discard this candidate in the winter of 1912/13? Did this rejection involve another fatal error that as yet has not been recognized?

Again, various hypotheses have been proposed to explain this apparent "pitfall." The coordinate condition with the help of which the Newtonian limit of the November tensor is attained turns out to be incompatible with the Minkowski metric in rotating coordinates (Norton 1989). Was this the reason why Einstein rejected this candidate? But why on earth should he have expected that the same coordinate condition could be employed for deriving both the Newtonian limit and the Minkowski metric in rotating coordinates, *if* he really understood coordinate conditions in a modern sense as the freedom to choose coordinates appropriate to a particular physical situation? And what if, in the end, he did *not* understand coordinate conditions in this way? After years of pondering the reasons for Einstein's rejection of the November tensor, this question seems to have brought us back to square one, namely the hypothesis about Einstein's ignorance of coordinate conditions in the modern sense.

It therefore comes almost as a shock to realize from indications in the Zurich notebook that Einstein's understanding of coordinate conditions was indeed different from the modern one. Why else should he have applied coordinate transformations to a coordinate condition, as he did with the condition reducing the November tensor to a form appropriate for the Newtonian limit (CPAE 4, Doc. 10, 252–253)? From a modern point of view this makes no sense.

But which error was it that induced Einstein to perform this strange operation? Did he see coordinate conditions as a set of equations on the same level as the field equations? They would thus guarantee in all admissible coordinate systems that the field equations keep the form that Einstein had recognized as being appropriate for obtaining the Newtonian limit—without, in the words he uses in his notebook, the "disturbing terms." But is this not just another way of simply incriminating Einstein of being ignorant of the freedom to choose a coordinate system in a generally covariant theory?

Perhaps there is some deeper error involved here, one that cannot simply be identified in the calculations of the notebook because it is more of a conceptual, if not metaphysical nature. What if Einstein had been guilty of believing in the famous hole argument at the time of the Zurich notebook, and, if not guilty of that, then at least of the commitment to the physical reality of coordinate systems underlying this argument?[9]

The evidence available makes it, in my view, implausible that this was indeed Einstein's pitfall in early 1913. If he committed an error conceptually close to the hole argument, then it becomes incomprehensible why, as the historical documents indicate, Einstein only formulated this argument as late as summer 1913, and from then on regarded it as the life belt of the "Entwurf" theory, while, before that, he considered its lack of being generally covariant as a shameful dark spot. The recently found document mentioned earlier offers additional documentary evidence suggesting that it was unlikely that he committed such an error. This document, written by Michele Besso and dated 28 August 1913, contains what probably represents the record of an exchange with Einstein and shows the hole argument in *statu nascendi*—without the hole (Janssen 2005). The basis of the argument against general covariance is the construction of distinct solutions of the field equations in the same coordinate system—in contradiction to the requirement of uniqueness. If such a construction was available at the end of August 1913 and could be used as an argument against general covariance, why should Einstein not have used it before as a defense of the "Entwurf" theory? In fact, however, he celebrated the discovery of the hole argument in his correspondence later in the year as a new and important achievement.[10]

In summary, the "hole-argument error" was not the original sin that marked the death of the November tensor and the birth of the "Entwurf" theory. More generally speaking, it seems that searching for the fatal error supposedly responsible for the pitfall of early 1913 may be something like the hunt for the white elephant in Mark Twain's famous story: despite successful reports of detectives claiming to have seen the missing elephant all over the country, and in spite of announcements that they were ready to capture it, the poor beast had been lying dead in the cellar of the New York Police headquarters since the beginning of the chase.[11]

3.4 The Legend of a Period of Stagnation between 1913 and 1915

We may now take a leaf from Twain's book and turn away from where the action is apparently taking place and instead take a closer look at an area where apparently nothing is happening: the supposed period of stagnation between early 1913 and late 1915. The traditional picture of this period can, with little exaggeration, be summarized in a single sentence: Einstein wasted his time elaborating the erroneous "Entwurf" theory and invented misleading arguments to support it. How does this picture change if we assume that neither the November tensor nor the "Entwurf" theory were ever definitively rejected in this period? The supposed period of stagnation would then become, objectively, a period of contest between two rival theories, even though, during this period, the contest did not surface openly. While it is clear which contender was, for Einstein, the stronger candidate in the beginning and which was stronger in the end, the question of what changed this balance of power becomes the decisive one and can only be answered by reconsidering what happened during this period.

From our reconstruction of the calculations in the Zurich notebook, Einstein's criteria for comparing candidate gravitational tensors have become clear. He checked whether the Newtonian limit could be obtained from them, whether they allow for energy-momentum conservation, and whether and to which extent they imply a generalization of the relativity principle (Renn and Sauer 1999). These heuristic guidelines did not, however, function like knockout criteria since their precise expression depended on the formalism used and on the degree of its elaboration. It was hence necessary for Einstein to check, on the concrete level of his calculations, whether his various criteria were compatible with each other, or whether, for instance, one had to be restricted or modified in order to allow for the implementation of the other. What was the situation in this respect for the November tensor in the Zurich notebook? It apparently allowed for an extension of the relativity principle since the November tensor is covariant under unimodular coordinate transformations. The requirement of energy-momentum conservation turned out, on the level of the weak field approximation, to be compatible with and even equivalent to the coordinate condition necessary for obtaining the Newtonian limit: a test which previous candidates had failed.

In spite of this positive record, however, some problems remained. For instance, does the requirement of energy- momentum conservation imply a restriction on the admissible coordinate transformations, as Einstein expected on the basis of earlier experiences? In the weak field approximation, this could be easily checked by exploring the transformational behavior of the expression which represents energy-momentum conservation and also the coordinate condition for the Newtonian limit (CPAE 4, Doc. 10, 252–253). Here then is a plausible explanation for the curious fact that Einstein seems to have explored the behavior of the coordinate condition for the November tensor under coordinate transformations. What he actually explored was the transformational behavior of an equation implied by energy-momentum conservation, a procedure that is also familiar from his work on the "Entwurf" theory (CPAE 6, Doc. 2 and Doc. 9). The outcome of this exploration was not very promising since it indicated that not even simple standard cases were included in the class of admissible coordinate transformations.

Naturally, however, a calculation in the weak field limit must remain inconclusive so that what Einstein really needed to do at some point was to formulate energy-momentum conservation for the full November tensor field equations. But, as his calculations in the Zurich notebook indicate, in view of its technical challenges he did not actually pursue this task. Eventually, instead of deriving an expression for energy-momentum conservation from the field equations, he turned around and took such an expression as the starting point for his search for appropriate field equations. Following this strategy, he finally arrived at the "Entwurf" field equations. This reconstruction is confirmed by Einstein's later recollections referring to difficulties with establishing energy-momentum conservation for candidates based on the Riemann tensor, recollections which until now found no place in the reconstruction of Einstein's discovery of the field equations.[12]

But let us return to the supposed period of stagnation. On closer inspection, it turns out that practically all the technical problems Einstein had encountered in the Zurich notebook with candidates derived from the Riemann tensor were actually resolved in this period, in the course of his examination of problems associated with the "Entwurf" theory. In order to deal with the issue of its unclear transformational properties, for instance, Einstein and Grossmann developed, at the suggestion of the mathematician Paul Bernays, a variational formalism for this theory (CPAE 6, Doc. 2 and Doc. 9). As a by-product, this variational formalism made it possible to derive an expression for energy-momentum conservation for any given Lagrangian and hence also for a theory based on the November tensor, provided that it can be reformulated in terms of a Lagrangian formalism.

But there was more. When Einstein and Besso calculated the perihelion shift for the "Entwurf" theory, Besso found that only the 4-4 component of the metric tensor for a weak static field is relevant in the equations of motion and made a note to that effect on the back of a letter from Guye to Einstein, dated 31 May 1913.[13] Besso thus effectively removed *the* major stumbling block that had prevented Einstein in 1912/13 from accepting the correct field equations of general relativity. If our so-called "period of stagnation" had any heroes, then their names were Grossmann, Bernays, and Besso.

With so many new tools at his disposal, why did Einstein not immediately return to the candidates based on the Riemann tensor? This was simply because the open contest between his candidates was temporarily suspended. It was a matter of perspective. With the publication of the "Entwurf" theory in early 1913, he had concluded an exploratory phase and entered a defensive phase of his work. Only when problems began to accumulate for the "Entwurf" theory did his perspective change causing him to switch back to the explorative stance. The effect of these problems was hence not to refute the "Entwurf" theory, but to trigger a process of reflection in which the new technical possibilities could now be brought to bear on a reevaluation of the candidates that had earlier been excluded from the contest. The suddenness of the apparent breakthrough of late 1915 was hence not induced by new factual insights that somehow popped up like a Jack-in-a-box, but was generated by a process of reflection on results that had accumulated over the past two years.

3.5 The Legend of Hilbert's Discovery of the Field Equations

In the introduction it was claimed that the genesis of general relativity can be described as the result of a double process, an assimilation of knowledge to a structure largely shaped by ideas from classical physics and the subsequent reflective reorganization of this structure. But before this hypothesis may become acceptable as an explanatory account of a scientific revolution, we must, in conclusion, tackle yet another legend, that of Hilbert's almost simultaneous discovery of the field equations as described in the beginning. If it were indeed true that there was a royal road to general relativity, paved by superior mathematical competence, then whatever the period of stagnation might have meant for Einstein's achievements, *sub specie eternitatis* it would be nothing but an unnecessary detour.

Recent findings (based on an analysis of the proofs of Hilbert's first paper) have shown that Hilbert did not actually anticipate Einstein in finding the field equations of general relativity.[14] What is much more important, the theory which Hilbert expounded in his proofs is remarkably similar in structure to Einstein's "Entwurf" theory. In particular, both in the "Entwurf" theory and in the proof version of Hilbert's theory, covariance properties are determined by the requirement of energy-momentum conservation. But while Hilbert's theory may be mathematically more sophisticated, Einstein's "Entwurf" theory turns out to be a much more realistic theory of gravitation. Its notion of energy-momentum, for instance, is based on special-relativistic continuum dynamics, while Hilbert's notion of energy merely represents an attempt to establish a connection with Mie's speculative theory of matter. The "Entwurf" theory was supported by the classical knowledge on gravitation by way of its Newtonian limit, while this question was not even tackled by Hilbert. In short, as far as its support by the available physical knowledge is concerned, Hilbert's theory is at best comparable to some of the early candidates in the Zurich notebook that Einstein decided not to publish.

Einstein's elaboration of his "Entwurf" theory, on the other hand, not only extended the formalism and hence the network of possible conclusions but also augmented the occasions for confronting these conclusions with the physical knowledge incorporated into the theory. The tensions thus created during the so-called stagnation phase gave Einstein the opportunity to reflect upon a reorganization of his theory, while for Hilbert, as the subsequent revisions of his theory testify, a similar tension was created not by internal conflicts but by the challenge with which Einstein's results confronted his framework. Rather than representing a detour, the stagnation period was hence precisely what justifies the claim that it was Einstein and his collaborators and not Hilbert who founded general relativity.

Acknowledgements. As mentioned in the beginning, this paper is based on the results of a joint research project with Michel Janssen, John Norton, John Stachel, and Tilman Sauer. In addition it has substantially benefited from insights, comments, and advice from Michel Janssen, Matthias Schemmel, and Peter Damerow.

References

Corry, Leo, Renn, Jürgen and Stachel, John (1997). Belated Decision in the Hilbert-Einstein Priority Dispute. *Science* **278**, 1270–1273.

Earman, John, and Janssen, Michel (1993). Einstein's Explanation of the Motion of Mercury's Perihelion. In *The Attraction of Gravitation: New Studies in the History of General Relativity. Einstein Studies* Vol. 5. John Earman, et al., eds. Birkhäuser Boston, 129–172.

Einstein, Albert (1915). *Zur allgemeinen Relativitätstheorie. Königlichen Preussischen Akademie der Wissenschaften*, (Berlin) *Sitzungsberichte*, 778–786. [CPAE 6, Doc. 24.]

— (CPAE 4). *The Collected Papers of Albert Einstein*. Vol. 4. *The Swiss Years: Writings, 1912–1914*. Martin J. Klein et al., eds. Princeton University Press, Princeton, 1995.

— (CPAE 5). *The Collected Papers of Albert Einstein*. Vol. 5. *The Swiss Years: Correspondence, 1902–1914*. Martin J. Klein et al., eds. Princeton University Press, Princeton, 1993.

— (CPAE 6). *The Collected Papers of Albert Einstein*. Vol. 6. *The Berlin Years: Writings, 1914–1917*. A. J. Kox, et al., eds. Princeton University Press, Princeton, 1998.

— (CPAE 8). *The Collected Papers of Albert Einstein*. Vol. 8. *The Berlin Years: Correspondence, 1914–1918*. Robert Schulmann et al., eds. Princeton University Press, Princeton, 1998.

Einstein, Albert and Grossmann, Marcel (1913). *Entwurf einer verallgemeinerten Relativitätstheorie und einer Theorie der Gravitation*. Teubner, Leipzig. [Reprinted with added comments in *Zeitschrift für Mathematik und Physik* **62** (1914): 225–261]. [CPAE 4, Doc. 13 and Doc. 26.]

Fölsing, Albrecht (1997). *Albert Einstein: a biography*. Translated from the German by Ewald Osers. Viking, New York, London.

Janssen, Michel (2005). What Did Einstein Know and When Did He Know It? A Besso Memo Dated August 1913. In *The Genesis of General Relativity*. Vol. 2, *General*

Relativity in the Making: Einstein's Zurich Notebook, Commentaries and Essays. Jürgen Renn, ed. Springer, Dordrecht.

Norton, John (1989). How Einstein Found His Field Equations, 1912–1915. In *Einstein and the History of General Relativity*. Don Howard and John Stachel, eds. Birkhäuser Boston, 101–159.

— (2005). What was Einstein's Fateful Prejudice? In *The Genesis of General Relativity*. Vol. 2, *General Relativity in the Making: Einstein's Zurich Notebook*. Part II. Jürgen Renn et al., Springer, Dordrecht.

Renn, Jürgen, ed. (2005). *The Genesis of General Relativity* (4 vols.). Springer, Dordrecht.

Renn, Jürgen, Tilman Sauer, and John Stachel (1997). The Origin of Gravitational Lensing. A Postscript to Einstein's 1936 Science Paper. *Science* **275**, 184–186.

Renn, Jürgen and Tilman Sauer (1999). Heuristics and Mathematical Representation in Einstein's Search for a Gravitational Field Equation. In *The Expanding Worlds of General Relativity*, Einstein Studies Vol. 7, Hubert Goenner et al., eds. Birkhäuser Boston, 87–125.

Renn, Jürgen and Stachel, John (1999). *Hilbert's Foundation of Physics: From a Theory of Everything to a Constituent of General Relativity*. Max Planck Institute for the History of Science, Berlin, Preprint 118. Available online at: http://www.mpiwg-berlin.mpg.de/Preprints/P118.pdf.

Renn, Jürgen, Peter Damerow, and Simone Rieger (2001). Hunting the White Elephant: When and How did Galileo Discover the Law of Fall? (with an Appendix by Domenico Giulini). In *Galileo in Context*, Jürgen Renn, ed., Cambridge University Press, Cambridge, 29–149.

Stachel, John (1989a). Einstein's Search for General Covariance, 1912–1915. In *Einstein and the History of General Relativity, Einstein Studies*, Vol. 1. Don Howard and John Stachel, eds., Birkhäuser, Boston, 36–100.

— (1989b). The Rigidly Rotating Disk as the 'Missing Link' in the History of General Relativity. In *Einstein and the History of General Relativity*, *Einstein Studies*, Vol. 1. Don Howard and John Stachel, eds. Birkhäuser Boston, 63–100.

Thorne, Kip. S. (1994). *Black Holes and Time Warps: Einstein's Outrageous Legacy*. W. W. Norton, New York.

Twain, Mark (1882). *The Stolen White Elephant*. James R. Osgood and Co., Boston.

Notes

[1] This paper was presented at the Fifth International Conference on the History and Foundations of General Relativity held at the University of Notre Dame, July 8–11, 1999.

[2] See also (Fölsing 1997).

[3] For publications representing the output of this joint research project see, e.g., (Renn and Sauer 1999). For a comprehensive publication see (Renn 2005).

[4]See (CPAE 5, Doc. 14) and for a historical discussion, see (Earman and Janssen 1993).

[5]See (Janssen 2005, section 3) for a detailed discussion of this episode.

[6]See the pioneering studies of John Norton (1989) and John Stachel (1989b).

[7]This was first noted in (Norton 1989).

[8]For a historical discussion, see (Norton 1989; Stachel 1989b).

[9]This interpretation has been developed by John Norton in the context of our collaboration and will be expounded in detail in (Norton 2005). For discussions of Einstein's Hole Argument see, (Stachel 1989a).

[10]See Einstein to Ludwig Hopf, 2 November 1913 (CPAE 5, Doc. 480), Einstein to Paul Ehrenfest, before 7 November 1913 (CPAE 5, Doc. 481); Einstein to Paul Ehrenfest, second half of November 1913 (CPAE 5, Doc. 484); Einstein to Paul Ehrenfest, second half of November 1913 (CPAE 5, Doc. 484, 568).

[11]See (Renn, Damerow, and Rieger 2001; Twain 1882).

[12]See, e.g., Einstein to Michele Besso, 10 December 1915 (CPAE 4, Doc. 14, 392) and Einstein to H.A. Lorentz, 1 January 1916.

[13]This is documented by a page in the Einstein–Besso manuscript, written by Michele Besso on the back of a letter to Einstein; the page can be dated to June 1913 when both worked together in Zurich (CPAE 4, Doc. 14, 392).

[14]See (Corry, Renn, and Stachel 1997) and for a detailed account (Renn and Stachel 1999).

Before the Riemann Tensor:
The Emergence of Einstein's Double Strategy

Jürgen Renn

Max-Planck-Institut für Wissenschaftsgeschichte, Wilhelmstraße 44, 10117 Berlin, Germany;
renn@mpiwg-berlin.mpg.de

4.1 The Paradox of General Relativity

This paper represents the third in a series based on joint work with Michel Janssen, John Norton, Tilman Sauer, and John Stachel on the genesis of general relativity as documented by the Zurich Notebook and other sources. The first paper (Renn and Sauer 1999) identifies the heuristics guiding Einstein's search for the gravitational field equation between 1912 and 1915, in particular what we call his "double strategy." The second paper, also published in this volume, analyzes the sense in which Einstein's work between 1913 and 1915 on the erroneous Entwurf theory created, paradoxically, the preconditions for formulating his final theory. This paper returns to the beginning of this development and will show how the crucial heuristic strategy that guided Einstein's work throughout those years emerged in the first place. I claim that this strategy actually took on its specific form before Einstein encountered the most important mathematical tool in his search for the field equation: the Riemann tensor. In order to substantiate this claim, it is necessary to examine Einstein's heuristics once more, and interpret it from a point of view that owes much to ongoing research at the Max Planck Institute for the History of Science on an historical epistemology of scientific knowledge, as well as to a theoretical framework originally developed with Tilman Sauer, and to an intensive collaboration with Michel Janssen on the early part of the Zurich notebook (Renn 2005).

The epistemological framework

From the perspective of an historical epistemology, the genesis of general relativity confronts us with a paradox: How was it possible for Einstein to formulate a theory that turned out to be amazingly suited to interpreting empirical knowledge which was unknown at the time of its creation (such as the expansion of the universe) and that involved substantial conceptual novelties (such as understanding gravitation as the curvature of space-time) and all this on the basis of knowledge still anchored in the older conceptual foundation of classical physics? Such a development can hardly be described in terms of formal logic. In fact, if the knowledge on gravitation relevant to the emergence of general relativity were structured as a deductive system in the sense of formal logic, it would suffice for one of the premises to be wrong for the entire building to collapse. But as Einstein's investigative pathway strikingly illustrates, in contrast to the inferences of formal logic, scientific conclusions can be corrected. Even when knowledge is subjected to major restructuring, science never starts from scratch as would be the case for a system structured according to formal logic and whose premises are no longer acceptable. In fact, scientific knowledge and also, of course, what can be termed the "shared knowledge" of large domains of human experience, transmitted over generations, is not simply lost when scientific theories are restructured. In the case at hand, it was mainly the shared knowledge of classical physics that needed to be preserved and exploited in a conceptual revolution that gave rise to a relativistic theory of gravitation whose far-reaching physical consequences, which eventually changed our understanding of the universe, were largely unknown at the time of the theory's creation.

What is therefore required in order to adequately describe the cognitive dynamics of the genesis of general relativity is an account of the underlying shared knowledge that illuminates, first, how past experiences can shape inferences about a matter on which only insufficient information is available, and, second, how conclusions can be corrected without always having to start from scratch. In order to satisfactorily account for these features in the case of Einstein's search for the gravitational field equation, it has turned out to be useful to introduce concepts from cognitive science, in particular the concepts of "mental model" and "frame."[1] A mental model is conceived here as a knowledge structure possessing slots that can be filled not only with empirically gained information but also with "default assumptions" resulting from prior experience. These default assumptions can be substituted by updated information so that inferences based on the model can be corrected without abandoning the model as a whole. Information is assimilated to the slots of a mental model in the form of "frames" which are understood here as "chunks" of knowledge with a well-defined meaning anchored in a given body of shared knowledge.

Conceiving the shared knowledge of classical and special relativistic physics in terms of mental models and frames makes clear how this knowledge could serve as a resource in Einstein's search for the gravitational field equation. In fact, essential relations between fundamental concepts such as field and source largely persist, even though the concrete applications of these concepts may differ considerably, as in the case of a classical vs. relativistic field equation. This structural stability turned the concepts and principles of classical and special relativistic physics into heuristic orientations when Einstein entered unknown terrain, for instance, when encountering a new expression generated by the elaboration of a mathematical formalism. None of these expressions in themselves constituted a new theory of gravitation. Only by complementing them with additional information based on the experience accumulated, not only in classical and special relativistic physics, but also in the relevant branches of mathematics, could such expressions become candidates for a gravitational field equation embedded in a full-fledged theory of gravitation. In the language of mental models, such past experience provided the default assumptions necessary to fill the gaps in the emerging and necessarily incomplete framework of a relativistic theory of gravitation. It was precisely the nature of these default assumptions that allowed them to be discarded again in the light of novel information—provided, for instance, by the further elaboration of the mathematical formalism—without, however, having to abandon the underlying mental models which could thus continue to function as heuristic orientations.

4.2 The Lorentz Model

The mental model of a field theory

The mental model that was crucial in Einstein's search for the gravitational field equation was shaped largely by prior experiences with the classical gravitational potential governed by the Poisson equation and by the treatment of electromagnetic fields, which had taken on its most developed form in Lorentz's theory of electromagnetism (Lorentz 1904a, 1904b). For this reason it may be called the "Lorentz model." The Lorentz model describes in terms of a field how the environment is affected by the matter considered to be the "source" of the field, and how this field then determines, in turn, the motion of matter, now conceived as a "probe" exposed to the field. A mathematical representation of physical processes interpreted according to this model therefore necessarily comprises two parts, a field equation describing how a localized source creates the global field, and an equation of motion describing how the global field determines the motion of a localized probe.

In order to apply the Lorentz model of a field theory to the case of a relativistic theory of gravitation, one had to identify an appropriate mathematical representation of both the gravitational field and of its source, and to find a generalization of the classical Poisson equation that was at least Lorentz-invariant. The structure of the as yet unknown field equation, as suggested by the Lorentz model, may be represented by the symbolic equation **GRAV(POT) = MASS**. The familiar quantities from classical physics, the scalar gravitational potential and the gravitational mass, represented the original defaults for the slots **POT** and **MASS**, respectively, while the classical default setting for the operator **GRAV** was the Laplace operator.

When Einstein took up his systematic search for a relativistic gravitational field equation in mid-1912, however, it had become clear from both his own research and that of contemporaries such as Abraham and Laue that these original default settings were no longer acceptable.[2] Indeed, the experience of the years between 1907 and 1912 had suggested new default assumptions. In particular, the metric tensor was now taken to represent the gravitational potential and hence became the canonical concretization of **POT** in the Lorentz model. The default setting for the metric tensor was, in turn, the spatially flat metric suggested by Einstein's experiences with his theory of static gravitational fields.[3] Similarly, the energy-momentum tensor became the standard setting for **MASS**, which in turn took the special case of dust as its default case (CPAE 4, Doc. 10, 10). These two key ingredients of the gravitational field equation had the appealing feature of being generally covariant objects and therefore embodied the expectation that the field equation itself would also take the form of a generally covariant tensorial equation, thus allowing Einstein to realize his ambition of creating a generalized relativity theory.

It is only for the third component of the Lorentz model, the differential operator **GRAV**, that the situation was more complicated. At the beginning of his search, not only was Einstein largely ignorant of the mathematical techniques necessary for constructing suitable candidates, but the many requirements to be imposed on an acceptable candidate effectively prevented the selection of an obvious default assumption for a differential operator **GRAV** compatible with all these requirements.

The heuristic requirements[4]

In fact, finding a field equation turned out to be the most challenging task Einstein ever tackled in his struggle for a relativistic theory of gravitation. First of all, he was confronted with the daunting mathematical problem that the representation of the gravitational potential by the metric tensor requires a field equation not for a single function but for a ten-component object. Second, Einstein could not avoid taking into account that the action of the gravitational field under ordinary circumstances was well known and satisfactorily described by Newton's law of attraction. The relativistic field equation of gravitation therefore had to yield the same results as this law under appropriate circumstances, a requirement we have referred to as the "correspondence principle." Third, the new field equation obviously had to be compatible with the well-established knowledge on energy and momentum conservation as well, a requirement we have labeled the "conservation principle."

Double strategy and default settings

How did Einstein's heuristics actually bring these resources together, and how did they effectuate the process of knowledge integration required to identify an acceptable field equation? How could the Lorentz model ever be harmonized with the physical requirements embodied in the correspondence principle, the conservation principle, and, of course, in his generalized principle of relativity? And how could these structures of physical knowledge ever be matched with the representational tools offered by mathematics? The response that emerged from our earlier work was that Einstein pursued two complementary heuristic strategies, one physical and the other mathematical. Einstein's "physical strategy" took the Newtonian limiting case as its starting point and then turned to the problem of the conservation of energy and momentum in order to finally examine the degree to which the principle of relativity is satisfied. His "mathematical strategy," on the other hand, is characterized by the fact that he took the principle of relativity as a starting point in order to then check the question of the Newtonian limiting case, and to finally make sure that the conservation of energy and momentum was also satisfied.

What is the meaning of this double strategy from the viewpoint of historical epistemology? The answer to this question can be found in the nature of the Lorentz model as a mental model embedded in the shared physical knowledge of the time. Due to

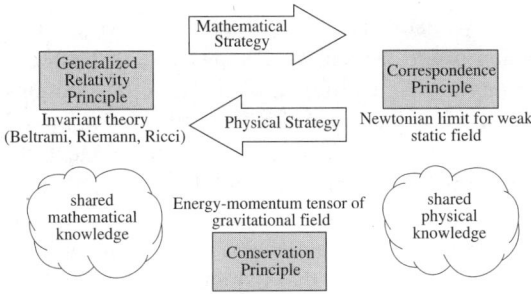

Fig. 4.1. Einstein's heuristic strategies

its epistemic architecture, the Lorentz model is in fact not just an abstract scheme into which knowledge about the properties of the gravitational field could be more or less successfully pressed. Rather it represents the crucial cognitive motor of Einstein's heuristics. The default settings of the Lorentz model made it possible to complement, if only by as yet uncertain and in hindsight often problematic information grounded in classical knowledge, the dim and shaky picture emerging in the course of Einstein's attempt to construct a relativistic field equation of gravitation.

As a mental model anchored in an elaborate body of physical and mathematical knowledge, the Lorentz model also offered the resources for constructing a mathematical representation of a candidate field equation, embodying both the structure of the model and its default assumptions. If such a candidate then turned out to be incompatible with other requirements to be imposed on a gravitational field equation, a fact that could now be checked on the level of an explicit representation, it did not necessarily follow that the theory built upon this candidate was thus falsified such that a new attempt had to begin from scratch. Instead it usually sufficed to merely adjust one or the other default assumptions, replacing them with the information newly gained in the course of exploring the given candidate field equation.

The most obvious starting point for constructing and exploring a candidate gravitational field equation was, in any case, an object with a well-defined physical meaning, constructed on the basis of the Lorentz model and with default settings rooted in classical and special relativistic physics. The approach of starting from such an object and then modifying it according to the heuristic requirements to be imposed on a candidate gravitational field equation is precisely what we have called the "physical strategy." Given the nature of its starting point, the physical strategy complies immediately with the correspondence principle, but it is not obviously clear whether a candidate constructed according to this strategy also satisfies the conservation principle and the generalized principle of relativity. The physical strategy is thus comparable to the synthetic approach of traditional Euclidean geometry, proceeding from what is "known" (in this case: gravitation as understood in classical physics) to the construction of the "unknown" (in this case: gravitation as it must be understood according to a generalized theory of relativity).[5]

The complementary "mathematical strategy" starts with embedding the Lorentz model within a higher-order mathematical knowledge structure, from which it then inherits default assumptions of a different kind. These assumptions are rooted in the shared mathematical knowledge of the time, e.g., the derivation of such expressions as the Ricci tensor from the metric tensor. Also in this context, the model may thus serve to guide the construction of concrete candidate gravitational field equations that now, from the outset, make perfect sense as mathematical objects and comply, in particular, with the generalized principle of relativity. It remains to be seen, however, whether these candidates satisfy the other requirements placed on a field equation, in particular the heuristic expectations rooted in physical knowledge as represented by the principles of correspondence and conservation. The mathematical strategy is hence comparable to analytic geometry, which proceeds from the "unknown" (in this case: a mathematical object whose physical meaning is unclear) to the "known" (in this case: a physically meaningful field equation of gravitation).

In summary, in both the case of the physical and of the mathematical strategy, the available knowledge led, via default assumptions of the Lorentz model, to concrete mathematical representations of candidate field equations which made it possible to check whether they fulfilled and were compatible with the criteria an acceptable field equation has to satisfy. The alternation between physical and mathematical strategy fostered the assimilation of both physical and mathematical resources to the basic model of a field equation. The eventual success of Einstein's search for the gravitational field equation was thus the result of a particularly efficient way of exploiting these shared knowledge resources.

4.3 Learning from a dilemma

The role of reflection

How did this efficient heuristics, Einstein's double strategy, take on the specific form we see at work in the main part of the Zurich Notebook where Einstein deals with the Ricci tensor, the Einstein tensor, the so-called November tensor, and finally with the Entwurf operator? The emergence of this truly pathbreaking strategy can only be understood if yet another cognitive mechanism is taken into account, a mechanism that, in a sense, is complementary to the one involved in the application of this strategy. In fact, the mechanism by which the exploitation of shared knowledge resources took place has been considered as yet only from a single perspective, that of incorporating physical and mathematical resources into the basic model of a field equation. While the assimilation of physical and mathematical knowledge to the Lorentz model is basically a top-down process guided by the relatively stable and high-level cognitive structures at the core of Einstein's heuristic criteria, a reflection on the experiences resulting from such an assimilation, including its failures, could trigger a corresponding bottom-up process. Such a process could induce an accommodation of the high-level structures, including the Lorentz model itself, to the outcome of these experiences, or could result in new higher-level structures operating on a strategic level, that is, guiding the implementation of Einstein's heuristic requirements in terms of what in cognitive science is called "procedures." It is, as will be argued here, precisely such a process that triggered the emergence of Einstein's successful double strategy in an early "tinkering phase" of the research documented in his Zurich Notebook.

The tinkering phase in Einstein's Zurich Notebook

The earliest notes on gravitation in the Zurich Notebook represent a stage of Einstein's search for the field equation in which he had hardly any sophisticated mathematical tools at hand that would allow him to construct candidates fitting the framework provided by the Lorentz model (CPAE 4, Doc. 10, 201ff.). Even his knowledge of the metric tensor and its properties were still rudimentary. Only gradually did he find ways of exploiting his knowledge of vector analysis for his search. Eventually he familiarized himself with the scalar Beltrami invariants as another instrument that allowed him to tinker with the few building blocks at his disposal, that is, the metric as a representation of the gravitational potential, the four-dimensional Minkowski formalism, and his theory of the static gravitational field. In spite of the staggering lack of mathematical sophistication characterizing this early tinkering phase, not to mention his failure to produce a promising candidate for the field equation, it is precisely in this period that Einstein acquired essential insights that shaped his research in subsequent phases of work, in particular his double strategy.

In the following, Einstein's attempt to assimilate knowledge about the static gravitational field to a metric formalism will first be outlined. I will then concentrate on the dilemma resulting from his construction of two incompatible default settings in the Lorentz model. It will be argued that the experience he had when attempting to resolve this dilemma caused him to devise a procedure for constructing and examining candidate gravitation tensors, a procedure that was to become essential for the mathematical prong of his double strategy. This procedure involves, on one hand, the identification of a physically meaningful default setting for the left-hand side of the gravitational field equation, an object we have called the "core operator" (Renn and Sauer 1999, 102). In the weak-field limit, this reduces to the ordinary d'Alembertian and allows the construction of a weak-field equation compatible with the correspondence principle. The procedure involves, on the other hand, a method for turning a mathematically meaningful default setting into a physically acceptable candidate gravitation tensor. This method makes use of coordinate restrictions limiting the validity of the relativity principle (in contrast to "coordinate conditions" in the modern understanding of general relativity). While such a method seems strange from a modern perspective, it actually determined Einstein's understanding of his theory of gravitation until the fall of 1915 (Renn 2005).

How did Einstein's procedure emerge in the course of his research? Here it will be argued that it was his reflection on the experiences of the tinkering phase that led to what one might describe as a "chunking" of his trials, alternately using physically and mathematically plausible default settings, in the procedure at the heart of his double strategy.

Assimilating knowledge about the static gravitational field to a metric formalism[6]

Before explaining the emergence of this procedure, let us look briefly at the beginning of Einstein's research on gravitation as documented by the notebook. His first attack constitutes an attempt at assimilating knowledge about the static case to a metric formalism, concentrating on two slots of the Lorentz model for a field equation, the slot for the gravitational potential and the slot for the differential operator. Einstein's key problem was that the default-settings for these two slots, representing his earlier experiences with implementations of this model, did not match. While the default setting for the gravitational potential was represented by a spatially-flat metric *tensor*, the default setting for the differential operator was the left-hand side of his 1912 field equation involving merely a *scalar* gravitational potential. Was there any way of bridging this gap between a scalar differential operator and a tensorial potential?

A mathematical toy model as a new starting point[7]

The mismatch between the default-settings for two of the slots of the Lorentz model of a field equation, that for the differential operator and that for the gravitational potential, left Einstein with two principal options for proceeding. He could try to somehow build, from whatever knowledge was at his disposal, an appropriate differential operator applicable to the metric tensor. Alternatively, he could tentatively explore variations of the default-setting for the gravitational potential, thus creating "toy-models" in the sense of manifestations of the Lorentz model with purposefully simplified default-settings. Even if that could mean temporarily shelving the insight that the gravitational potential is represented by the metric tensor, it might still be possible to gain knowledge from exploring such "toy-models" that could help to construct a real candidate field equation.

When, at some point during his work on the notebook, Einstein became familiar with the second Beltrami invariant as a generalization of scalar differential operators, it must have immediately appealed to him as a mathematically plausible setting for the differential operator slot, since a field equation formulated with its help would satisfy the heuristic requirement of the generalized principle of relativity from the outset. But choosing this setting also posed a problem: it was incompatible with filling the potential slot by the metric tensor as the Beltrami invariant was applicable only to scalar functions. In a sense, a scalar field equation formulated in terms of the second Beltrami invariant represents the counterpart to the scalar field equation of Einstein's 1912 static theory: while the latter constitutes an initial, physically plausible default-setting for the Lorentz model, the former represents an equally plausible initial default-setting rooted in mathematical knowledge. In both cases, the resulting field equations were merely starting points for further investigations that had to establish contact with knowledge not yet embodied in these initial default-settings.

It therefore comes as no surprise that Einstein attempted to understand the conditions under which a generally covariant scalar field equation, formulated in terms of the second Beltrami invariant, reduces to the ordinary Poisson equation. In fact, such a reduction must be possible if the candidate (or rather "toy") field equation is to comply with the correspondence principle. It turned out that the implementation of this heuristic principle requires an additional hypothesis on the choice of the coordinates supplementing the field equation. Essentially by inspection, Einstein was able to identify the harmonic coordinate restriction as a condition that would ensure that the Beltrami field equation would reduce to the ordinary Poisson equation for weak gravitational fields, eliminating disturbing terms. In other words, the exploration of a toy field equation taught Einstein that a candidate field equation obtained from a mathematical default-setting might require an additional coordinate restriction in order to be viable from a physical point of view as well.

This sequence—to first pick a generally covariant candidate and then reduce it to a familiar physical format, that is, to get rid of disturbing terms by imposing a coordinate restriction—was to become the basic procedure of Einstein's mathematical strategy. What was still lacking for this strategy to emerge fully was a more realistic target than the classical Poisson equation, a target that involved the true setting for the potential-slot: the metric tensor. This missing piece was found after Einstein had made a fresh attempt directed at creating a physically more meaningful candidate.

A physical toy model as a new starting point[8]

Einstein's exploration of the Beltrami invariant left him, in the end, uncertain about how to get from a mathematically plausible scalar differential equation to a tensorial field equation that was both mathematically *and* physically plausible. Reflecting on this problem, he now started out from a physically-plausible "toy" field equation. Instead of taking a simplified default-setting for the potential-slot of the Lorentz model in order to explore a mathematical toy model, he chose a simplified default-setting for the differential operator slot while keeping the realistic setting for the potential slot, i.e., the metric tensor.

Einstein's broad experience with the tools of vector analysis and their use in physics in fact made it easy for him to write down a straightforward translation of the ordinary Laplacian operator into a differential operator capable of acting on the metric tensor, the core operator. But while even his limited familiarity with Beltrami invariants must have made it obvious that the core operator could hardly represent a generally covariant object, the way in which it was constructed made it equally clear that a field equation based on it satisfies the correspondence principle. For this reason, the core operator became the default-setting for all Einstein's subsequent attempts to implement this principle and with it both the starting point for his physical strategy and the target of his mathematical strategy.

The challenge was now to confront the core operator with the other heuristic requirements to be imposed on a field equation and, in particular, to explore its relation to mathematical knowledge about coordinate transformations. Einstein therefore began to check the transformational behavior of the core operator. But he quickly discovered that the analysis of explicit coordinate transformations of the core operator became rather involved and offered hardly any general insight into its covariance properties.

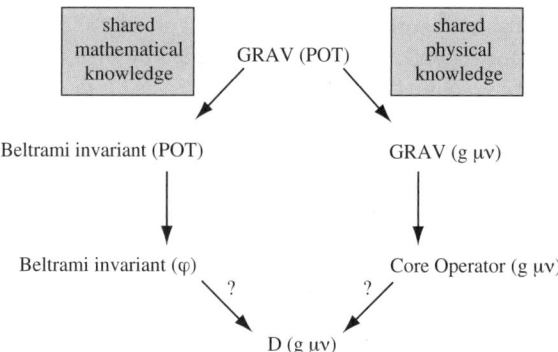

Fig. 4.2. The dilemma: the Beltrami invariant and the metric tensor as incompatible fillings of the slots of the Lorentz model

What he learned from this attempt was merely the possibility of also taking into account coordinate transformations that explicitly depend on the metric tensor—again a heuristic insight with far-reaching implications for his further research.

Identifying the core operator as the target of the mathematical strategy[9]

Einstein's attempt to start from a physically satisfactory candidate gravitation tensor had thus failed to yield any tangible results. He therefore turned again to the Beltrami invariants, this time, however, under new conditions. Earlier, he had unsuccessfully attempted to connect the second Beltrami invariant with physical knowledge on static gravitational fields. At that point, however, he was not yet in possession of the core operator, which now offered a new and more promising target for a transition from a mathematically well-defined object to a physically acceptable candidate gravitation tensor. But first of all, he had to establish a relation between the Beltrami invariants, applicable to scalar functions, and the realistic setting of the potential slot: the metric tensor. For this purpose, he focused on the determinant of the metric tensor. Indeed, if unimodular coordinate transformations are assumed, the determinant of the metric becomes a scalar function and can hence be inserted into the Beltrami invariants.

Einstein tried next to extract the core operator from the scalar expression resulting from inserting the determinant of the metric into the Beltrami invariant. Apart from a term involving the contraction of the core operator, however, he also found an additional term that was difficult to interpret. This disturbing remaining term posed a problem analogous to the one Einstein first encountered when comparing a mathematical toy model based on the second Beltrami invariant with the ordinary Laplace operator. This analogy thus suggested taking up the idea of introducing a coordinate restriction as an additional hypothesis under which a mathematically plausible expression reduces to a physically acceptable one.

In any case, Einstein must have hoped to infer the transformational behavior of the core operator by analyzing the behavior of the remaining term under coordinate transformations. In this way, a bridge would have been built between the transformational behavior of the mathematically well-defined second Beltrami invariant and the physically plausible core operator. Unfortunately, however, although the remaining term essentially representing the difference between the Beltrami invariant and the contracted core operator was a simpler expression than the core operator itself, it turned out to be still too complex for an evaluation of its transformational behavior. While this unsuccessful attempt terminated Einstein's use of the Beltrami invariants in the course of his research documented in the Zurich notebook, it did establish a heuristic procedure that, as puzzling as it may appear from the perspective of modern general relativity, was consistently applied by Einstein even when he eventually learned about the Riemann tensor from Marcel Grossmann.

The procedure that resulted from this experience takes a covariant object as its starting point and then attempts to extract a candidate gravitation tensor compatible with the correspondence principle by imposing an additional coordinate restriction. Typically, such a candidate gravitation tensor would be represented by the core operator plus some harmless correction terms. Examining the transformational behavior of the coordinate restriction then allows, together with knowledge about the covariant starting point of the procedure, the inference of the transformational behavior of the candidate gravitation tensor.

The genesis of this procedure from the challenge of filling the slots of the Lorentz model of a field equation in a mutually compatible way illustrates a typical learning experience encountered by Einstein in his search, making it evident that this search did not simply consist of the elimination of unworkable alternative candidates. In fact, both the identification of the core operator and its conjointment with a covariant object were results that substantially changed the conditions of his further search quite independently from accepting or discarding a specific candidate. The role of the Beltrami invariants illustrates this seemingly paradoxical feature of Einstein's research: While the Beltrami invariants played no role whatsoever in formulating the final field equation, they were crucial in triggering the higher- order heuristic insights that paved Einstein's way to this solution.

References

Abraham, Max (1912). Berichtigung. *Physikalische Zeitschrift* **13**,176.

Büttner, Jochen et al. (2001). Traces of an Invisible Giant: Shared Knowledge in Galileo's Unpublished Treatises. In *Largo Campo Di Filosofare: Eurosymposium Galileo 2001*. J. Montesinos and C. Solís, eds. Fundación Canaria Orotava De Historia De La Ciencia, La Orotava, 183–201.

Büttner, Jochen et al. (2003). Exploring the Limits of Classical Physics - Planck, Einstein and the Structure of a Scientific Revolution. *Studies in History and Philosophy of Modern Physics* **34**, 37–59.

Damerow, Peter (1996). Abstraction and Representation. *Essays on the Cultural Revolution of Thinking*. Kluwer, Dordrecht.

Damerow, Peter et al. (2002). Mechanical Knowledge and Pompeian Balances. In *Homo Faber: Studies on Nature, Technology, and Science at the Time of Pompei*. Giuseppe Castagnetti and Jürgen Renn, eds. L'Erma di Bretschneider, Rome, 93–108.

Davies, Robert B. (1984). *Learning Mathematics: The Cognitive Science Approach to Mathematics Education*. Croom Helm, London/Sydney.

Einstein, Albert (1912). Zur Theorie des statischen Gravitationsfeldes. *Annalen der Physik* **38**, 443–458.

— (CPAE 4). *The Collected Papers of Albert Einstein*. Vol. 4. *The Swiss Years: Writings, 1912–1914*. Martin J. Klein et al. eds., Princeton University Press, Princeton, 1995.

Gentner, Dedre, and Albert L. Stevens, eds. (1983). *Mental Models*. Erlbaum, Hillsdale, N.J.

Laue, Max (1911a). *Das Relativitätsprinzip*. Friedrich Vieweg und Sohn, Braunschweig.

(1911b). Zur Dynamik der Relativitätstheorie. *Annalen der Physik* **35**, 524–542.

Lorentz, Hendrik A. (1904a). Maxwell's elektromagnetische Theorie. In A. Sommerfeld, ed., *Encyklopädie der mathematischen Wissenschaften, mit Einschluss ihrer Anwendunge*, vol. 5/II., Teubner, Leipzig, 63–144.

— (1904b). Weiterbildung der Maxwellschen Theorie. Elektronentheorie. In A. Sommerfeld, ed., *Encyklopädie der mathematischen Wissenschaften, mit Einschluss ihrer Anwendungen*, vol. 5/II. Teubner, Leipzig, 145–280.

Minsky, Marvin L. (1987). *Society of Mind*. Heinemann, London.

Norton, John (1989). How Einstein Found His Field Equations, 1912–1915. In *Einstein and the History of General Relativity*. Don Howard and John Stachel, eds., Birkhäuser Boston, 101–159.

— (1992). Einstein, Nordström, and the Early Demise of Scalar, Lorentz Covariant Theories of Gravitation. *Archive for the History of Exact Sciences* **45**, 17–94.

Renn, Jürgen (2000). Mentale Modelle in der Geschichte des Wissens: Auf dem Wege zu einer Paläontologie des mechanischen Denkens. In *Dahlemer Archivgespräche*, vol. 6. E. Henning, ed. Archiv zur Geschichte der Max-Planck-Gesellschaft, Berlin, 83–100.

— ed. (2005). *The Genesis of General Relativity* (4 vols.). Springer, Dordrecht.

Renn, Jürgen and Tilman Sauer (1999). Heuristics and Mathematical Representation in Einstein's Search for a Gravitational Field Equation. In *The Expanding Worlds of General Relativity*, *Einstein Studies*, Vol. 7, Hubert Goenner et al. eds. Birkhäuser Boston, 87–125.

Notes

[1] The concepts have been adopted from cognitive science (see e.g. (Davies 1984; Gentner and Stevens 1983; Minsky 1987, and also Damerow 1996) to historical research on both individual and shared scientific knowledge in the ongoing work at the Max Planck Institute for the History of Science (see the Institute's Research Report 2000–2001, available at http://www.mpiwg-berlin.mpg.de/resrep00_01/index.html). For historical case studies making use of this approach, see e.g. (Büttner et al. 2001; Büttner et al. 2003; Damerow et al. 2002; Renn 2000).

[2] In 1912 Max Abraham was the first person, in the context of a controversy with Einstein, to suggest generalizing the line element of Minkowski's four-dimensional spacetime to include a variable speed of light (as it occurred both in his and in Einstein's gravitational field theories), see (Abraham 1912). This suggestion soon became the basis for Einstein's introduction of the metric tensor as the representation of the gravitational tensor, see (Einstein 1912) and (CPAE 4). Max Laue's work on a relativistic continuum theory, on the other hand, suggested taking the energy-momentum tensor as the default-filling of the source slot of the Lorentz model, see (Laue 1911a, 1911b, as well as CPAE 4) and, for historical discussion (Norton 1992).

[3] See (Einstein 1912) and (CPAE4, Doc. 10, 201), and for historical discussion (Norton 1989).

[4] See, also for the following, (Renn and Sauer 1999).

[5] This comparison was suggested to me by Peter Damerow.

[6] See (CPAE 4, Doc. 10, 201–203).

[7] See (CPAE 4, Doc. 10, 214–217).

[8] See (CPAE, Doc. 10, 216–220).

[9] See (CPAE 4, Doc. 10, 220–223).

5

A Conjecture on Einstein, the Independent Reality of Spacetime Coordinate Systems and the Disaster of 1913

John D. Norton

Department of History and Philosophy of Science, University of Pittsburgh, Pittsburgh, PA 15260 U.S.A.; jdnorton@pitt.edu

Two fundamental errors led Einstein to reject generally covariant gravitational field equations for over two years as he was developing his general theory of relativity. The first is well known in the literature. It was the presumption that weak, static gravitational fields must be spatially flat and a corresponding assumption about his weak field equations. I conjecture that a second hitherto unrecognized error also defeated Einstein's efforts. The same error, months later, allowed the hole argument to convince Einstein that all generally covariant gravitational field equations would be physically uninteresting.

5.1 Introduction

This paper will present elementary accounts of both errors described above. The first will be reviewed in Sections 5.2 and 5.3. The second, the new conjecture, will be motivated in Section 5.4, the hole argument sketched in relevant detail in Section 5.5, and the conjecture itself developed in Section 5.6. Conclusions are in Section 5.7.

By mid-1913, Einstein had come so close. He had the general theory of relativity in all its essential elements. This theory, he believed, would realize his ambition of generalizing the principle of relativity to acceleration. It would harbor no preferred coordinate systems and its equations would remain unchanged under arbitrary coordinate transformation; that is, they would be generally covariant. Yet, in spite of the able mathematical assistance of his friend Marcel Grossmann, this vision of general covariance was slipping away. The trouble lay in his gravitational field equations. He had considered what later proved to be the equations selected in November 1915 for the final theory, at least in the source free case. But he had judged them wanting and could find no generally covariant substitute. So in his joint "Entwurf" paper with Marcel Grossmann,[1] Einstein published gravitational field equations of unknown and probably very limited covariance. This was the disaster of 1913. Nearly three dark years lay ahead for Einstein as he struggled to satisfy himself that these unnatural equations were well chosen. Towards the end of 1915, a despairing and exhausted Einstein returned to general covariance and ultimately to the gravitational field equations that now bear his name.

What had gone wrong? How did Einstein manage to talk himself out of these final equations for nearly three years? Historical scholarship of the last two decades has given us a quite detailed answer to these questions.[2] Much of this answer comes from Einstein's "Zurich Notebook,"[3] a notebook of private calculations that catalogs Einstein's deliberations from his early acquaintance with the new mathematical methods required by his theory, through the evaluation of candidate gravitational field equations to the derivation of the gravitational field equations of the 1913 "Entwurf" theory. One error has long been understood. Whatever else the theory may do, it must return Newton's theory of gravitation in the domain of weak, static fields in which that older theory has been massively confirmed. Einstein made some natural but erroneous assumptions about weak static fields and the corresponding form his gravitational field equations must take in the weak field limit. They made recovery of this Newtonian limit impossible from the natural gravitational field equations.

This error alone does not suffice to explain fully Einstein's misadventure of 1913. For he proved able to find gravitational field equations that were both of very broad covariance and also satisfied his overly restrictive demands for weak, static fields. These equations too were developed in the Zurich Notebook but rejected in 1913 without clear explanation. Einstein must have later judged that rejection hasty, for these same equations were revived and endorsed in a publication of early November 1915 (Einstein 1915) when he returned to general covariance. What explains his 1913 rejection of these equations? What had he found by November 1915 that now made them admissible? Some additional error must explain it.

The problem has been investigated in detail by a research group to which I belong.[4] Several possible explanations have been found. Some are related to hitherto unnoticed idiosyncrasies in Einstein's treatment of coordinate systems when he developed the "Entwurf" theory. My purpose in this paper is to review one of these explanations that I believe will be of special interest to philosophers of space and time. The suggestion is that Einstein was misled and defeated by a fundamental conceptual error concerning the ontology of spacetime coordinate systems that lay hidden tacitly in his manipulations. What makes the account especially attractive is that we need attribute no *new* error to Einstein. It can be explained by the one other major error from this time that Einstein later freely conceded. That was his "hole argument," the vehicle that he would use repeatedly over the next three years to justify his abandoning of general covariance. By his own later analysis, the error of this argument was that Einstein accorded an existence to spacetime coordinate systems independent of the fields defined on them. While the earliest extant mention of the hole argument comes in November 1913, months after the completion of the "Entwurf" paper, I maintain that the error at its core had already corrupted Einstein's earlier attempts to recover the Newtonian limit from his candidate gravitational field equations. To recover this limit, Einstein needed to restrict his theory to specialized coordinates. If we presume that Einstein treated these limiting coordinate systems in the same way as those of the hole argument months later, it turns out that they appear to have an absolute character that contradicts the extended principle of relativity whose realization was the goal of Einstein's theory.

Moreover, the view I conjecture Einstein took of these limiting spacetime coordinate systems effectively precluded his acceptance of virtually all generally covariant gravitational field equations. So the hole argument was not merely a clever afterthought designed to legitimate Einstein's prior failure to find generally covariant gravitational field equations. Rather, in best Einstein tradition, it encapsulated in the simplest and most vivid form the deeper obstacle that precluded Einstein's acceptance of generally covariant gravitational field equations.

In this paper I will not reconstruct the evidential case for these errors in all detail, with its strengths and weaknesses; that has already been done in (Norton, forthcoming). Rather my purpose is to present a primer for those who want a simple, self contained account of how Einstein went wrong and are willing to cede to the citations a more detailed analysis of the extent to which the account can be supported by our historical source material. I will try to explain in the simplest terms possible what these two errors were, why Einstein found them alluring and how they defeated his efforts to find acceptable, generally covariant gravitational field equations. In the decades following Einstein's work, our formulations of general relativity have become far more sophisticated mathematically and more geometrical in spirit. My account will adhere as closely as practical to Einstein's older methods and terminology, for that will keep us closer to Einstein's thought and render the errors in it more readily intelligible.

5.2 The Spatial Flatness of Weak, Static Gravitational Fields

In 1913, Einstein presumed that in weak static fields, his new gravitation theory must deliver Euclidean spaces. His final theory of 1915 allows spatial geometry to differ from the Euclidean in first order quantities even in this limiting case.

Einstein was induced to give up the natural generally covariant gravitational field equations for his "Entwurf" theory by his attempt to relate his new theory with the theory it supersedes, Newtonian gravitation theory. We can see the problem as it appeared to him in 1913 if we compare the two theories. The Newtonian theory of gravitation is based on representing a gravitational field by a single potential ϕ spread over a Euclidean space. Einstein's "Entwurf" theory of 1913 and his final general theory of relativity were built around a quadratic differential form:[5]

$$ds^2 = \sum_{\mu,\nu} g_{\mu\nu} dx_\mu dx_\nu,$$

where ds is the invariant interval between neighboring events with spacetime coordinates x_μ and $x_\mu + dx_\mu$. The coefficients of the metric tensor $g_{\mu\nu}$:

$$\begin{bmatrix} g_{11} & g_{12} & g_{13} & g_{14} \\ g_{21} & g_{22} & g_{23} & g_{24} \\ g_{31} & g_{32} & g_{33} & g_{34} \\ g_{41} & g_{42} & g_{43} & g_{44} \end{bmatrix},$$

now represent the gravitational field as well as the geometry of spacetime. The one gravitational potential of Newtonian theory has been replaced by 16 coefficients. Since the metric tensor is symmetric, we have $g_{\mu\nu} = g_{\mu\nu}$, so that only ten of these coefficients can be set independently. But that is still nine more than in Newtonian theory.

Newtonian theory has enjoyed spectacular confirmation in its domain of application. So, when Einstein's new theory is restricted to this domain, it must return results indistinguishable from those of Newtonian theory.

5.2.1 Weak, Static, Gravitational Fields

Weak gravitational fields differ in quantities of first order of smallness from a Minkowski metric. A static field may be naturally sliced into three dimensional spaces and admits observers that see its geometric properties as time independent.

Newtonian theory prevails in the domain of weak, static gravitational fields and the restriction to this domain appears simple. In a weak gravitational field, the metric tensor differs only in small quantities from the metric tensor of a Minkowski spacetime, the spacetime of special relativity. That is, there is a coordinate system in which the metric can be written as:

$$g_{\mu\nu} = \eta_{\mu\nu} + h_{\mu\nu}, \tag{5.1}$$

where the background Minkowski metric is:

$$\eta_{\mu\nu} = \begin{bmatrix} -1 & 0 & 0 & 0 \\ 0 & -1 & 0 & 0 \\ 0 & 0 & -1 & 0 \\ 0 & 0 & 0 & 1 \end{bmatrix} \tag{5.2}$$

and the weak field perturbation is:

$$h_{\mu\nu} << \eta_{\mu\nu}.$$

If a gravitational field is static, then we can find a coordinate system in which the coefficients $g_{\mu\nu}$ of the metric tensor are not functions of the time coordinate x_4 and the mixed time-space components of the metric vanish: $g_{14} = g_{41} = g_{24} = g_{42} = g_{34} = g_{43} = 0$. The metric has the form:

$$\begin{bmatrix} g_{11} & g_{12} & g_{13} & 0 \\ g_{21} & g_{22} & g_{23} & 0 \\ g_{31} & g_{32} & g_{33} & 0 \\ 0 & 0 & 0 & g_{44} \end{bmatrix}.$$

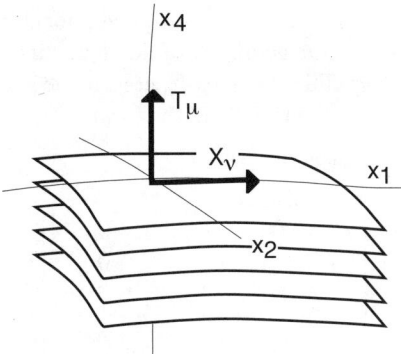

Fig. 5.1. A static spacetime

These algebraic requirements admit a simple geometric interpretation. If an observer's worldline coincides with a curve of constant x_1, x_2, x_3, such as the x_4 coordinate axis, then that observer will see the geometric properties of space as time independent. The vanishing of the time-space components of the metric tensor allow the spacetime to be divided naturally into a family of time indexed three-dimensional spaces as shown in Figure 5.1. Assuming that the coordinate system covers the entire spacetime, each three-dimensional space is chosen by fixing a constant value for the time coordinate x_4; the coordinates x_1, x_2 and x_3 are then the coordinates of the three-dimensional space and the metric of the space is:

$$\begin{bmatrix} g_{11} & g_{12} & g_{13} \\ g_{21} & g_{22} & g_{23} \\ g_{31} & g_{32} & g_{33} \end{bmatrix}. \tag{5.3}$$

These spaces are orthogonal to the observer's world line. That is, a vector tangent to the observer's worldline, such as $T_\mu = (0, 0, 0, 1)$, will be orthogonal to a vector tangent to the three-dimensional space, such as $X_\nu = (1, 0, 0, 0)$, since $\sum_{\mu\nu} g_{\mu,\nu} T_\mu X_\nu = 0$.

5.2.2 Recovering the Newtonian Limit

As the Newtonian domain is approached, Einstein's new gravitation theory must restore Euclidean geometry in three dimensional space. Einstein assumed that exact restoration occurs in weak, static fields since this reduces the ten coefficients of the metric tensor to the single potential of Newtonian theory.

These weak, static fields must now return the two properties characteristic of Newtonian gravitation theory: Euclidean space and a single gravitational potential. In the "Entwurf" paper Einstein presumed that this would happen in the simplest way imaginable. (Einstein and Grossmann 1913, I Sect. 2) In quantities of first order of smallness, there would be just one component of $g_{\mu\nu}$ that was not constant. That would be g_{44} which would represent the Newtonian gravitational potential φ. The components of the metric that return the geometry of the three dimensional spaces would be constants coinciding with Euclidean values. That is, in the relevant coordinate system, Einstein expected weak static fields to be of the form:

$$
\begin{array}{c}
\overbrace{\text{Euclidean metric}}^{} \\
\text{of space}
\end{array}
$$

$$
\left[
\begin{array}{ccc|c}
-1 & 0 & 0 & 0 \\
0 & -1 & 0 & 0 \\
0 & 0 & -1 & 0 \\
\hline
0 & 0 & 0 & g_{44} = 2\varphi
\end{array}
\right]
$$

$$
\underbrace{}_{\substack{\text{Newtonian} \\ \text{gravitational} \\ \text{potential}}}
$$

(5.4)

That the g_{44} corresponds to the Newtonian potential was strongly suggested to Einstein by the equations of motion of a slowly moving point mass in gravitational free fall in the theory.[6] Such a point follows a geodesic in spacetime, a curve of extremal interval s. It is governed by the geodesic equation:

$$
\frac{d^2 x_\mu}{ds^2} + \sum_{\alpha\beta} \begin{Bmatrix} \alpha\beta \\ \mu \end{Bmatrix} \frac{dx_\alpha}{ds} \frac{dx_\beta}{ds} = 0,
$$

where the Christoffel symbols are given as:[7]

$$
\begin{Bmatrix} \alpha\beta \\ \mu \end{Bmatrix} = \frac{1}{2} \sum_\nu \gamma_{\mu\nu} \left(g_{\alpha\nu,\beta} + g_{\beta\nu,\alpha} - g_{\alpha\beta,\nu} \right).
$$

Most terms in the geodesic equation vanish in quantities of the first order of smallness. The derivatives $dx_1/ds, dx_2/ds, dx_3/ds$, in the Newtonian limit in this coordinate system, correspond to the velocity of the mass and are thus each first order small. Thus the only significant component of the second term of the geodesic equation is the term in $dx_4/ds\, dx_4/ds \approx 1$. Because of the vanishing of the time-space components of the metric tensor, the related Christoffel symbols reduce to:

$$
\begin{Bmatrix} 44 \\ i \end{Bmatrix} = \frac{1}{2} \sum_\nu \gamma_{i\nu} \left(g_{4\nu,4} + g_{4\nu,4} - g_{44,\nu} \right) \approx \frac{1}{2} g_{44,i}, \qquad \begin{Bmatrix} 44 \\ 4 \end{Bmatrix} = 0
$$

for $i = 1, 2, 3$. The geodesic equation reduces to:

$$
\frac{d^2 x_i}{ds^2} = -\frac{1}{2} \frac{\partial g_{44}}{\partial x_i}; \qquad \frac{\partial^2 x_4}{ds^2} = 0.
$$

The last of these equations shows that the x_4 coordinate is linearly related to the interval along the mass' trajectory, justifying the interpretation of the x_4 coordinate as time read by a clock—at least at this level of approximation—and the above assumption that $dx_4/ds \, dx_4/ds \approx 1$. The first equation relates the acceleration of the mass to the spatial gradient of g_{44} exactly as in Newtonian theory:

$$\text{Acceleration} = -\text{gradient} \, (g_{44}/2) = \text{gradient} \, (\varphi)$$

affirming the equation of the Newtonian potential ϕ with $g_{44}/2$. Since the remaining coefficients of the metric play no role in this equation of motion, there seemed no obstacle to setting these to the constant Euclidean values.

5.2.3 The Principle of Equivalence

The principle of equivalence delivered Einstein one instance of a static gravitational field, a homogeneous gravitational field. That one instance proved to be spatially flat and Einstein readily generalized the result to all static fields.

Einstein had a stronger motivation for his conclusion that weak static fields are spatially flat. He had begun work on a relativistic theory of gravitation in (Einstein 1907, Part V) with an ingenious idea he later called the "principle of equivalence." That principle supplied a heuristic means of generating a theory of gravitation. It began with one simple case. Einstein considered a Minkowski spacetime, the spacetime of special relativity, and determined how it would look to an observer in uniform acceleration. That observer would see all free objects uniformly accelerated in a direction opposite to that of the observer's acceleration. Since all these objects suffered the same acceleration, their motion conformed to a familiar characteristic of gravitation: all bodies fall alike, irrespective of their masses. It was as if the masses were under the influence of a homogeneous gravitational field. Einstein's principle of equivalence removes the "as if." It asserts the full equivalence of the two cases, a uniform acceleration in Minkowski spacetime and a homogenous gravitational field.

The principle of equivalence supplied Einstein with a relativistic account of one special case of the gravitational field, that of a homogeneous gravitational field. Einstein's development of a theory of static gravitational fields prior to 1913 (Einstein 1907, Part V; 1911; 1912a,b) resided in judiciously generalizing the properties of the homogeneous field to that of arbitrary static fields. For our purposes, the most important property of the homogeneous gravitational field produced by uniform acceleration was that its spatial geometry remained Euclidean. Therefore he assumed that spatial geometry in the presence of an arbitrary static field would also remain Euclidean[8] and this presumption was carried over explicitly to the "Entwurf" theory.

The preservation of Euclidean geometry is seen most clearly if the transformation to uniform acceleration is analyzed within the framework of the "Entwurf" and later theories. We start with a Minkowski spacetime and a coordinate system (X, Y, Z, T) in which the expression for the interval is:

$$ds^2 = -dX^2 - dY^2 = dZ^2 + dT^2.$$

We may represent a transformation from inertial to accelerated motion as a coordinate transformation following Einstein's usual practice.[9] The simplest form of the transformation is given later in (Einstein and Rosen 1935) as

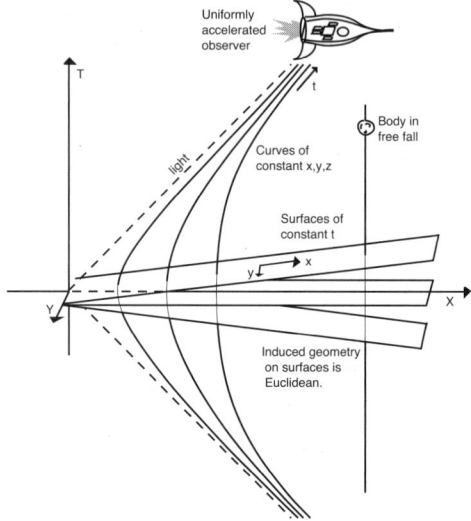

Fig. 5.2. Principle of Equivalence : Uniform acceleration in a Minkowski spacetime . . .

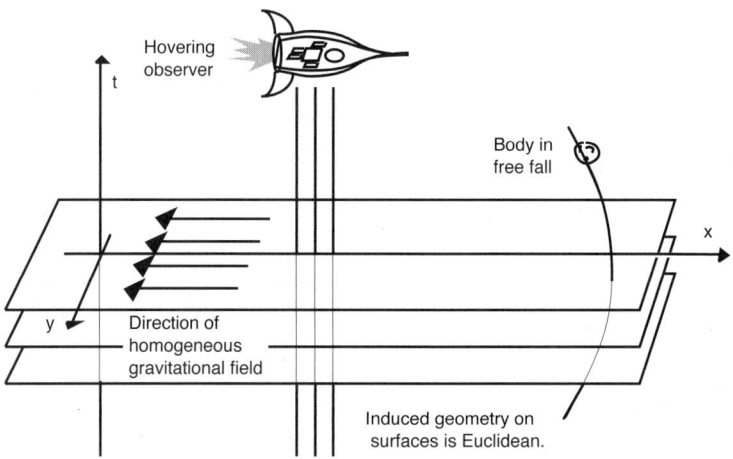

Fig. 5.3. . . . is equivalent to a homogeneous gravitational field

$$X = x \cosh at \qquad Y = y \qquad Z = z \quad T = x \sinh at \qquad (5.5)$$

where a is a constant that measures the magnitude of the acceleration. The expression for the interval transforms to

$$ds^2 = -dx^2 - dx^2 - dz^2 + a^2 x^2 dt^2,$$

from which we recover an expression for the metric:

$$g_{\mu\nu} = \begin{bmatrix} -1 & 0 & 0 & 0 \\ 0 & -1 & 0 & 0 \\ 0 & 0 & -1 & 0 \\ 0 & 0 & 0 & a^2 x^2 \end{bmatrix}.$$

Even though this is not a case of a weak field, it is a static field and it conforms exactly to the expectations encoded in (5.4) that the spaces of such fields be Euclidean.[10] The transformation (5.5) is shown graphically in Figure 5.2 and its reinterpretation as a homogeneous gravitational field in Figure 5.3.

If we apply these expectations to one of the most important weak static fields addressed by the theory, the gravitational field of the sun, we recover Einstein's expectation that its metric tensor is:

$$g_{\mu\nu} = \begin{bmatrix} -1 & 0 & 0 & 0 \\ 0 & -1 & 0 & 0 \\ 0 & 0 & -1 & 0 \\ 0 & 0 & 0 & 1 - \frac{\alpha}{r} \end{bmatrix},$$

were α is determined by the mass of the sun and the coordinate r is fixed as $R^2 = x_1^2 + x_2^2 + x_3^2$.

5.2.4 Contradiction With Einstein's Final Theory

Weak, static gravitational fields are not spatially flat in Einstein's final theory of November 1915.

The modern reader will recognize immediately how seriously Einstein has strayed if this metric is compared with the exact solution for the gravitational field of the sun, the Schwarzschild solution. The three-dimensional space surrounding the sun, even in weak field approximation, does deviate from Euclidean flatness. As Einstein would later ruefully discover, the metric tensor for the field of the sun in first order approximation is given by:

$$g_{\mu\nu} = \begin{bmatrix} -1 - \frac{\alpha x_1^2}{r^3} & -\frac{\alpha x_1 x_2}{r^3} & -\frac{\alpha x_1 x_2}{r^3} & 0 \\ -\frac{\alpha x_1 x_2}{r^3} & -1 - \frac{a x_2^2}{r^3} & -\frac{a x_2 x_3}{r^3} & 0 \\ -\frac{\alpha x_1 x_3}{r^3} & -\frac{\alpha x_2 x_3}{r^3} & -1 - \frac{\alpha x_3^2}{r^3} & 0 \\ 0 & 0 & 0 & 1 - \frac{\alpha}{r} \end{bmatrix}.$$

5.3 The Rejection of the Ricci Tensor

Einstein discarded the Ricci tensor as gravitation tensor since he could find no coordinate condition that would reduce it to a spatially flat Newtonian limit in the case of weak static fields.

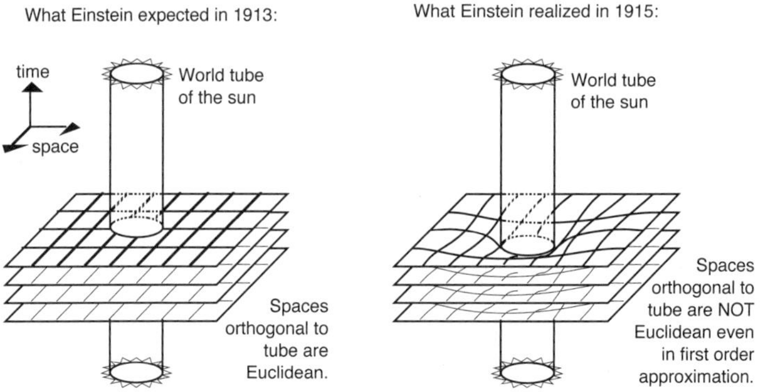

Fig. 5.4. What Einstein expected for the gravitational field of the sun.

Einstein expected weak, static gravitational fields to be spatially flat. Whether this would be so in his theory depends upon the gravitational fields the theory admits. That in turn is decided by the theory's gravitational field equations. In Newtonian theory, the single equation for the single potential φ that selects the admissible gravitational fields is Poisson's equation:

$$\Delta\varphi = \left(\frac{\partial^2}{\partial x^2} + \frac{\partial^2}{\partial y^2} + \frac{\partial^2}{\partial z^2}\right)\varphi = 4\pi G\rho, \tag{5.6}$$

where x, y and z are the Cartesian coordinates of space, G the gravitational constant and ρ the density of matter. Einstein sought a system of ten equations for the ten components of the metric tensor that would be the relativistic analog of this single equation. He expected it to have the form

$$G_{\mu\nu} = kT_{\mu\nu}, \tag{5.7}$$

where k is some constant, $T_{\mu\nu}$ is the stress energy tensor of matter and the gravitation tensor $G_{\mu\nu}$ is composed of terms in the metric tensor and its first and second derivatives.

5.3.1 An Over-Simplified Form of the Gravitational Field Equations for the Weak Field

Einstein assumed a natural form (5.8) for the gravitational field equations in weak field approximation that would return both Poisson's equation of Newtonian theory and spatial flatness in simple cases.

Since Einstein's new theory must revert to Newton's in suitable limiting circumstances, Einstein's choice for gravitational field equations (5.7) must eventually revert to (5.6). To ensure this, Einstein presumed that his gravitational field equations (5.7) must first revert to the equations:[11]

$$\sum_{\alpha,\beta} \frac{\partial}{\partial x_\alpha}\left(\gamma_{\alpha\beta}\frac{\partial \gamma_{\mu\nu}}{\partial x_\beta}\right) + \begin{pmatrix} \text{further terms} \\ \text{that vanish in the} \\ \text{first approximation} \end{pmatrix} = kT_{\mu\nu}, \qquad (5.8)$$

in the case of a weak field (5.1). The motivation for this presumption is clear if one considers the form (5.8) takes in the weak, static field of (5.4) with a source of pressureless, motionless dust:

$$T_{\mu\nu} = \begin{bmatrix} 0 & 0 & 0 & 0 \\ 0 & 0 & 0 & 0 \\ 0 & 0 & 0 & 0 \\ 0 & 0 & 0 & \rho_0 \end{bmatrix}, \qquad (5.9)$$

where ρ_0 is the rest density of the matter. Equation (5.8) then reduces to:

$$\Delta\gamma_{44} = (-k)\rho_0, \qquad (5.8a)$$

and for the remaining terms for which $\mu \neq 4$ or $\nu \neq 4$ (or both):

$$\Delta\gamma_{\mu\nu} = 0. \qquad (5.8b)$$

The first equation (5.8a) is merely the recovery of Poisson's equation (5.6) of Newtonian theory as expected. The second is readily solved in special cases to yield the result that the $\gamma_{\mu\nu}$ are constants whenever $\mu \neq 4$ or $\nu \neq 4$ so that the spatial metric (5.3) is flat. This last conclusion affirmed Einstein's expectation that weak static fields are to be spatially flat; the same result is now recoverable from the natural equation (5.8)—another misleading corroboration of the result.

The special case for solving (5.8b) that Einstein considered in his Vienna lecture (Einstein 1913, Sect. 8) was one in which the components of the metric tensor approached Minkowskian values at spatial infinity. Presumably he imagined the matter distribution ρ_0 to be concentrated into a central island of matter that diluted away completely with distance into an otherwise empty space, else the presumption of Minkowskian values at infinity would not be plausible.[12]

A weakness of Einstein's (1913) recovery of spatial flatness in the weak static field is that it depends on a source matter distribution with stress energy tensor (5.9). This is physically implausible since it is a matter distribution that does not undergo gravitational collapse but has no pressures or other stresses to counteract the collapse. Were the collapse not counteracted by such stresses, then the resulting velocity of the dust would contribute further non-zero terms (other than T_{44}) to the stress energy tensor. If such stresses are present, then they would appear directly as further non-zero terms in the stress energy tensor. In either case, these new non-zero terms would defeat the derivation of (5.8b).

Einstein realized that his inference to spatial flatness was not quite so fragile. As he affirmed in a postcard to Erwin Freundlich of March 19, 1915 (CPAE, vol. 8A, Doc. 63), spatial flatness could still be recovered for the space outside the sun if one assumed that the static mass distribution of the sun was sustained by pressures or stresses. For this case, the individual components of the stress energy tensor $T_{ik}(i, k = 1, 2, 3)$ will in general be non-vanishing, so that (5.8b) must be replaced by:

$$\Delta \gamma_{ik} = (-k)T_{ik}. \tag{5.8c}$$

But, as Einstein noted, the condition of equilibrium entailed the vanishing of the integrals over space $\int T_{ik}dV$, a result due also to Laue, as Einstein also noted. [For more on Laue's analysis, see (Norton 1992, Sect. 9)]. This in turn entailed that the coefficients γ_{ik} adopt Minkowskian values under suitable conditions. While Einstein did not complete the proof, it is easy if we presume spherical symmetry for the matter distribution. From Gauss' theorem, we conclude that each γ_{ik} is constant[13] and thus must everywhere adopt the Minkowskian values presumed at spatial infinity.

5.3.2 The Attempt to Recover Them: The Harmonic Coordinate Condition

Einstein found that the natural choice of gravitation tensor, the Ricci tensor, would yield weak field equations of form (5.8) if he restricted himself to harmonic coordinates. But the Ricci tensor is rejected since the harmonic coordinate condition is incompatible with the spatial flatness Einstein presumed for weak, static fields.

Einstein now needed to find a gravitation tensor for his field equations (5.7) that would revert to (5.8) in the weak field. Grossmann reported the key mathematical result in his part of the "Entwurf" paper (II Sect. 2): one generates "the complete system of differential tensors of the manifold" by covariant algebraic and differential operations on what we now call the Levi-Civita tensor density and the fourth rank Riemann curvature tensor R_{iklm}, where the indices now range over 1, 2, 3 and 4. The natural candidate for the gravitation tensor was the Ricci tensor, the unique first contraction:

$$G_{im} = \sum_{k,l} \gamma_{kl} R_{iklm}$$

$$= \sum_{k,l} \gamma_{kl} \frac{1}{2} \left(\frac{\partial^2 g_{im}}{\partial x_k \partial x_l} + \frac{\partial^2 g_{kl}}{\partial x_i \partial x_m} - \frac{\partial^2 g_{il}}{\partial x_k \partial x_m} - \frac{\partial^2 g_{mk}}{\partial x_l \partial x_i} \right) + \begin{matrix} \text{terms quadratic} \\ \text{in the first derivatives} \\ \text{of the metric.} \end{matrix}$$

This choice is familiar to modern readers since it coincides with the final field equations in the source free case of $T_{ik} = 0$. But, Grossmann reported, this choice must be abandoned since it fails to yield the Newtonian expression $\Delta\varphi$ in the special case of a weak, static field.

Whether the Ricci tensor can reduce to this form depends solely on the four second derivative terms displayed above; the terms quadratic in the first derivatives can be neglected as second order terms in the weak field approximation. Of these four second derivative terms, the first term alone is sufficient to yield a field equation of the form (5.8) in the weak field. To assure recovery of (5.8), the remaining second derivative terms must be eliminated.

What Einstein and Grossmann did not indicate in the "Entwurf" paper was that they knew precisely how this could be achieved. But Einstein's private calculations of the Zurich Notebook do reveal it. If one restricts the spacetime coordinate systems under consideration, these three terms can be made to vanish. In particular, they vanish if one selects the coordinates that satisfy the harmonic condition:[14]

$$\sum_{\alpha,\beta} \gamma_{\alpha\beta} \left\{ \begin{matrix} \alpha\beta \\ \mu \end{matrix} \right\} = 0. \tag{5.10}$$

That the three unwanted second order derivative terms in the Ricci tensor vanish follows from another differentiation of the harmonic condition (5.10) as Einstein shows on p. 37/ 3 6 19L of the Zurich Notebook.[15]

Nonetheless, Einstein and Grossmann report that the Ricci tensor fails to yield the correct Newtonian limit. What had gone wrong? Again the Zurich Notebook supplies the answer as we watch Einstein grapple unsuccessfully with the weak field in the pages following, pp. 38–42/ 3 6 19R-21R. While the harmonic coordinate condition did reduce the gravitational field equations to the appropriate Newtonian limit (5.8), the harmonic condition itself proved objectionable. For Einstein expected the metric to reduce to the spatially flat form (5.4). A short calculation shows that the harmonic condition (5.10) is not satisfied in the coordinate system used in (5.4). Without the harmonic coordinate condition, Einstein could no longer reduce the Ricci tensor to the appropriate Newtonian form. Since he could find no suitable alternative, the Ricci tensor had to be rejected.

It is instructive to see how Einstein's final theory of 1915 avoids inferring the spatial flatness of a weak static field. Following (Einstein 1992, 86–89), we set $g_{\mu\nu}$ equal to the Ricci tensor. Einstein's final field equations of 1915 then do not have the form (5.7) but are:

$$G_{\mu\nu} = k(T_{\mu\nu} - (1/2)g_{\mu\nu}T).$$

First we restrict the equation to harmonic coordinates and then proceed as above for the case of a source of pressureless, motionless dust. In place of (5.8a) and (5.8b) we recover:

$$\Delta g_{44} = (-k)\rho_0 \quad \Delta g_{11} = \Delta g_{22} = \Delta g_{33} = -(-k)\rho_0 \quad \Delta g_{\mu\nu} = 0 \quad \text{all other } \mu, \nu.$$

We see immediately from the second set of equations that the components g_{11}, g_{22} and g_{33} will not in general be constant if ρ_0 is anywhere non-vanishing so Einstein's earlier inference to spatial flatness is blocked.

5.4 The Puzzle of the Second Candidate

In the Zurich Notebook, Einstein found a second gravitation tensor of broad covariance that yielded the appropriate Newtonian limit and the spatial flatness of weak, static fields. It was briefly revived in November 1915. What explains its rejection in 1913 and revival in 1915?

Einstein's presumption of the spatial flatness of weak, static fields was sufficient to preclude his consideration of the Ricci tensor. But it does not explain why he ended up abandoning general covariance in 1913. The field equations he announced in the "Entwurf" paper were of unknown covariance and Einstein could assert at best a near trivial covariance under linear coordinate transformations. In this regard, the Zurich Notebook contains a puzzle. Immediately after the harmonic condition was abandoned, on p. 44/ 3 6 22R Einstein found a reduced form of the Ricci tensor with very broad covariance that could be used as a gravitation tensor and, with a suitable choice of coordinate condition, would yield the equation (5.8) in the weak field. In this instance, the coordinate condition was compatible with the spatially flat metric (5.4), so none of the difficulties we have seen so far preclude acceptance of it as the gravitation tensor. That tensor proved so unobjectionable that Einstein later came to endorse it briefly in publication. When he returned to general covariance in late 1915, but before he realized his error concerning the spatial flatness of weak, static fields, Einstein (1915) published field equations using this very gravitation tensor.

The puzzle is this: why were these equations inadmissible in 1913 but admissible briefly[16] in November 1915. Some additional error must explain it. What was it?

5.4.1 Construction of the New Candidate Gravitation Tensor

Einstein splits off a gravitational tensor of near general covariance from the Ricci tensor and shows how to reduce it to the required Newtonian form by application of a coordinate condition.

While the details of the construction of this gravitation tensor are inessential for the conjecture to follow, it is included here briefly for completeness. Einstein noted that the Ricci tensor could be written as a sum of two parts:[17]

$$T_{il} = \underbrace{\left(\frac{\partial T_i}{\partial x_l} - \sum \begin{Bmatrix} il \\ \lambda \end{Bmatrix} T_\lambda \right)}_{\text{tensor 2nd rank}} - \underbrace{\sum_{\kappa l} \left(\frac{\partial \begin{Bmatrix} il \\ \kappa \end{Bmatrix}}{\partial x_\kappa} - \begin{Bmatrix} i\kappa \\ \lambda \end{Bmatrix} \begin{Bmatrix} l\lambda \\ \kappa \end{Bmatrix} \right)}_{\text{presumed gravitation tensor } T^x il},$$

where the quantity T_i of the first term is defined as $T_i = \partial lg \sqrt{G} / \partial x_i$ with G the determinant of the metric tensor. A unimodular transformation of the spacetime coordinates $x_\alpha \to x'_\beta$ is one for which the determinant $\text{Det} \left(\partial x'_\beta / \partial x_\alpha \right) = 1$. It follows that unimodular transformations preserve G which becomes a scalar. Immediately we have that T_i is a vector under unimodular transformation since it is just the derivative of a scalar. The first quantity in the expression for the Ricci tensor T_{il} proves to be the covariant derivative of this vector and thus also a tensor of second rank under unimodular transformation. Since T_{il} is a generally covariant tensor, it now follows that the second term must also be a tensor under unimodular transformation. Labeling the second term T^x_{il}, Einstein adopted it as the gravitation tensor.

This tensor is not generally covariant, but its covariance is sufficiently broad to support Einstein's ambitions for generalizing the principle of relativity to acceleration. Unimodular transformations include those that set the Cartesian spatial coordinate axes of a Minkowski spacetime into uniform rotation, for example, transformation (5.12) below.

Like the Ricci tensor, the new candidate gravitation tensor contained more second derivative terms in the metric tensor than present in the weak field equation (5.8). As before Einstein eliminated them with a coordinate condition. This time he chose the simple condition:

$$\sum_{\kappa} \frac{\partial \gamma_{\kappa\alpha}}{\partial x_{\kappa}} = 0, \tag{5.11}$$

and was able to show that in coordinate systems in which it holds, a gravitational field equation based on T_{il}^{x} reduces to the desired form (5.8). Finally one can see by inspection that the coordinate condition (5.11) is satisfied in the weak static field (5.4).

5.4.2 Einstein's "fateful prejudice"

Part of Einstein's rejection of this second candidate was due to his "fateful prejudice" concerning the Christoffel symbols. I discount the possibility that the rejection can be explained by the supposition that he was unaware of the standard use of coordinate conditions.

Many factors may have entered into Einstein's decision to abandon this second candidate. In this paper I will discuss just one possibility of special interest. There are others. In his later remarks of November 1915, Einstein blamed the decision on a "fateful prejudice." Its most expansive description came in a letter to Sommerfeld of November 28, 1915 (CPAE, vol. 8A, Doc. 153.). There he reflected ruefully on field equations that used the second candidate tensor:[18]

> I had already considered these equations 3 years ago with Grossmann ... but had then arrived at the result that they did not yield Newton's approximation, which was erroneous. What supplied the key to this solution was the realization that it is not $\sum g^{l\alpha} \partial g_{\alpha i}/\partial x_m$, but the associated Christoffel symbols $\begin{Bmatrix} im \\ l \end{Bmatrix}$ that are to be looked upon as the natural expression for the "components" of the gravitational field. If one sees this, then the above equation becomes simplest conceivable, since one is not tempted to transform it by multiplying out the symbols for the sake of general interpretation.

Our best interpretation of this depends upon insights by (Renn, manuscript) and (Janssen and Renn, forthcoming). They pertain to a difficulty in assuring energy momentum conservation in a theory based on this gravitation tensor. For it to be assured, Einstein required that he be able to define a stress-energy tensor for the gravitational field. The prejudice Einstein outlined induced him to seek an expression for it in terms of the derivatives of the metric tensor. That yielded a calculation so daunting that

Einstein abandoned it. By 1915, after he had developed more powerful variational techniques, Einstein found that this quantity could be expressed more simply in terms of the Christoffel symbols and the difficulty disappeared. For further discussion, see (Norton, forthcoming).

It is quite improbable that this was the only difficulty faced by this second candidate gravitation tensor. Otherwise we must assume that Einstein gave up at his moment of triumph simply because the calculation looked hard. Also we would have no explanation for his remark to Sommerfeld above that the equations did not yield the Newtonian limit. There must have been a further problem of sufficient gravity to thwart Einstein completely.

The pages surrounding the analysis of this second candidate gravitation tensor in the Zurich notebook are concerned with problems of coordinates and covariance. There it becomes clear that Einstein is not using coordinate condition (5.11) and others like it in the now standard way. He was not merely invoking the condition in the special case of the Newtonian limit. (For that usage, we reserve the label "coordinate condition.") Rather he was invoking it universally, so that the resulting reduced form of the gravitational field equations were not just to be used in the weak field limit. They were the theory's gravitational field equations. To distinguish this usage from the standard use, we have come to call equations such as (5.11) used this way "coordinate restrictions." This interpretation of Einstein's use of (5.11) and the label "coordinate restriction" was foreshadowed in (Renn and Sauer 1999, p. 108) and elaborated in (Renn, Sauer et al., forthcoming).

That Einstein sometimes used the requirement (5.11) as a coordinate restriction does not explain why he might think that the second candidate gravitation tensor fails to yield the Newtonian limit. A stronger supposition is needed. We must presume in addition that Einstein was unaware of the other way of using the requirement as a coordinate condition. A case can be made that this awareness defeated recovery of the Newtonian limit. For if Einstein tried to use requirement (5.11) as a coordinate restriction in the attempt to recover the Newtonian limit, the covariance of the final field equations would be reduced to that of requirement (5.11). We shall see below in Section 6.1 that requirement (5.11) has insufficient covariance to support an extension of the principle of relativity. However I do not find this supposed lack of awareness plausible for reasons given in some detail in (Norton, forthcoming).[19] Briefly, it requires Einstein to fail persistently to see that he may impose a restriction on covariance in setting up the special conditions needed for recovery of the Newtonian limit, just as he may impose the assumption of near Minkowskian values for the metric tensor. He must overlook this in spite of his continued insistence that the restriction of covariance Newtonian theory is what distinguishes it fundamentally from his new theory and that covariance principles are his area of greatest insight and expertise. Also Einstein makes no later concession of an error of this type and is very careless in his introduction of coordinate conditions to the point of obscuring their presence, an attitude that is odd if their neglect proved fatal to his earlier efforts.

The alternative conjecture to be developed in the sections following draws on the same base of evidence and does require Einstein to commit an error concerning coordinate systems and coordinate conditions. But the error attributed to Einstein is one that we see him committing unequivocally later and to which he also later admits. The conjecture just requires that he committed the same error earlier and pursued its consequences.

5.5 The Hole Argument

Einstein's other error from this period was his "hole argument," which appeared months later. With it he sought to establish that generally covariant gravitational field equations would be physically uninteresting.

To understand why the gravitation tensor T_{il}^x was inadmissible in 1913 but not in early November 1915, we must locate some new error on Einstein's part—that is, an assumption that Einstein himself would later regard as erroneous. Rather than needing to locate a new, hitherto unknown error, my conjecture is that the error Einstein later conceded in the context of his notorious hole argument can also explain Einstein's earlier rejection of the gravitation tensor T_{il}^x. At the same time, it will reveal just how difficult Einstein had made his search for any admissible, generally covariant gravitation tensor and that the search's failure in 1913 was all but assured until that error was corrected.

Once Einstein had published gravitational field equations of very limited covariance in 1913, he needed to convince his readers and correspondents that this choice was acceptable. After some vacillation,[20] he settled upon the hole argument for this task. While the "Entwurf" paper was published in mid 1913, Einstein does not seem to have had the hole argument in hand until months later. The first unambiguously dated mention of it is in a letter to November 2, 1913, to Ludwig Hopf (CPAE, vol. 5, Doc. 480). In the ensuing year, Einstein published the argument four times, with the final version in (Einstein 1914, 1067) being the clearest.

5.5.1 Outline of the Argument

A generally covariant gravitation theory is inadmissible since a full specification of the metric field outside some small region (the "hole") cannot fix the metric field within it.

The purpose of the argument was to show that a version of Einstein's theory with generally covariant gravitational field equations would violate what he called the "law of causality" (Einstein, 1914, p. 1066). In effect he meant that the theory would be indeterministic. That is, a full specification of the metric field outside some region of spacetime must fail to fix the metric field within that region, no matter how small the region may be.

In slightly simplified form, Einstein's argument proceeded as follows.[21] Let us assume that the metric field in the source free case is governed by generally covariant gravitational field equations $G_{\mu\nu} = 0$ and that we have a solution of these equations $g_{\mu\nu}$ in some coordinate system x_α. Since the field equations $G_{\mu\nu} = 0$ are generally covariant, any arbitrary transform of $g_{\mu\nu}$ will also be a solution of these field

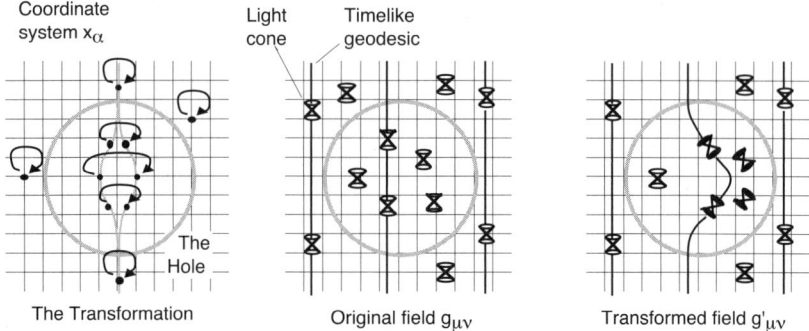

Fig. 5.5. Transforming fields for the hole argument

equations. Consider the following transformation. We select some arbitrary region of spacetime—call it the "hole." The transformation is the identity outside the hole, but comes smoothly to differ from the identity within the hole; it maps a point P with coordinates x_α to a point Q with coordinates $x'_\beta = f_\beta(x_\alpha)$ in the same coordinate system. Outside the hole $P = Q$; inside $P \neq Q$. We apply this active transformation to the metric $g_{\mu\nu}$ and thereby generate another solution of the gravitational field equation $g'_{\mu\nu}$ in the same coordinate system x_α. The transformation is displayed in Figure 5.5 in which the metric field is represented by the light cones and timelike geodesics it induces on the spacetime.

To arrive at the violation of determinism, we begin with the solution $g_{\mu\nu}(x_\alpha)$ in the coordinate system x_α. We imagine this field removed from the coordinate system and then replaced by the transformed field $g'_{\mu\nu}(x_\alpha)$ as shown in Figure 5.6. If we compare the two solutions, we find they agree fully outside the hole since the transformation is the identity there, but they disagree within. That is, if we specify the metric fully outside the hole, we cannot know which field we will find within. This is a failure of determinism so severe that Einstein felt it must be suppressed. That, he urged, was achieved by disavowing the general covariance of the gravitational field equations.

5.5.2 Active Versus Passive Transformations

It proved easy to overlook that Einstein intended the transformation of the hole argument to be read actively so that it left the coordinate system unchanged but spread the metric differently over it.

Einstein's earlier discussion of this construction caused considerable confusion among later commentators and was only made completely explicit in the version in (Einstein 1914, 1067). To generate the active transformation, Einstein first read the transformation passively as a change of coordinate system and it proved easy to overlook the crucial conversion of that passive transformation into an active transformation of the field in just one coordinate system.

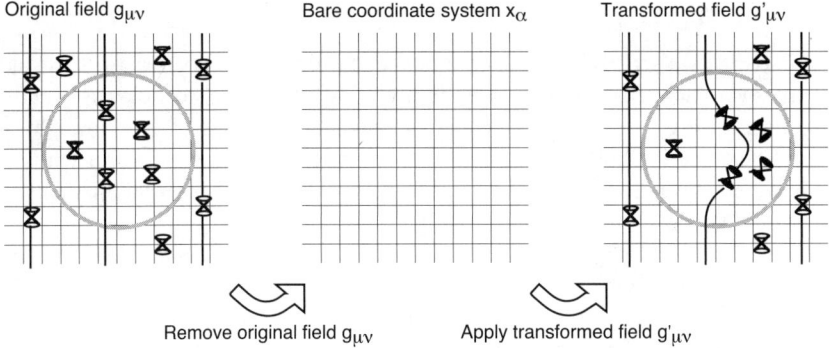

Fig. 5.6. The manipulation of the hole argument

He began by using the transformation represented by the functions f_β as a passive transformation to relabel the point P with new coordinates x'_β. Under this coordinate transformation, the components of the tensor $g_{\mu\nu}(x_\alpha)$ transform to components $g'_{\mu\nu}(x'\beta)$ in the new coordinate system x'_β following the usual law of transformation of tensor components. To proceed to the active transformation, Einstein considered the functional dependence of the transformed $g'_{\mu\nu}(x'_\beta)$ on its arguments, the coordinates x'_β. That functional dependence alone was all that was needed to assure that the field $g'_{\mu\nu}$ satisfies the field equations $G_{\mu\nu} = 0$. Those same functional forms realized in any other coordinate system would then also represent a solution of the field equations. Thus the new field $g'_{\mu\nu}(x_\alpha)$—those same functions but now of the original coordinate system x_α—will also be a solution of the field equations. This new field is the active transform $g'_{\mu\nu}(x_\alpha)$ of the original field $g_{\mu\nu}(x_\alpha)$ with both represented in the original coordinate system x_α.

As a trivial illustration of the conversion to the active transformation, imagine that the functions $g'_{\mu\nu}(x'_\beta)$ just happen to be the constant functions of the arguments x'_β. Then we know that constants $g'_{\mu\nu} = K_{\mu\nu}$ solve the field equations $G_{\mu\nu} = 0$ and that will be true no matter which coordinate system we consider. So, take the original coordinate system x_α and construct a new field in it whose components $g'_{\mu\nu}(x_\alpha) = K_{\mu\nu}$ are those same constants. The new field will be distinct from the original field $g_{\mu\nu}(x_\alpha)$ but will still be a solution of the gravitational field equations.

For another development of the mathematical constructions used in the hole argument, see (Howard and Norton 1993, Sect. 1).

5.5.3 The Erroneous Assumption: Independent Reality of Spacetime Coordinate Systems

Einstein's error, as he later explained, was that he believed that the spacetime coordinate system had an existence independent of the metrical field defined on it so that it made sense for the same coordinate system to host distinct metric field.

Of course Einstein's use of the hole argument is flawed. It does not force us to abandon general covariance. An easier escape simply allows that the two fields $g_{\mu\nu}$ and the active transform $g'_{\mu\nu}$ are distinct mathematical representations of the same physical reality. Therefore the hole argument fails to show that the physically real within the hole is underdetermined; it merely shows failure of determinism for the mathematical structures we choose to describe the one physical reality.

Why do $g_{\mu\nu}$ and $g'_{\mu\nu}$ represent the same physical reality? Since they are transforms of one another they must agree on all invariants. So if elements of physical reality are represented only by invariants, the two fields represent the same physical reality. Einstein's preferred formulation of this escape is to note that two intertransformable systems agree on all point coincidences. For example, if the world consisted just of particles in motion, the intersections of their worldlines, he asserted, would be the only observable and they would be preserved under all transformations. This is Einstein's "point-coincidence argument," best know from his review article, (Einstein 1916, Sect. 3).

For our purposes, however, what is most important is not the correct analysis of the hole argument but the error Einstein committed that prevented him seeing the correct analysis. That error was explained by Einstein to his correspondents late in 1915 and in early 1916. The difficulty pertains to the coordinate system that carries the fields. For example in a letter to his friend Michele Besso a little over a week later on January 3, 1916 (CPAE, vol. 8A, Doc. 178; Einstein's emphasis) he explained:[22]

> There is no physical content in two different solutions $G(x)$ and $G'(x)$ existing with respect to the *same* coordinate system K. To imagine two solutions simultaneously in the same manifold has no meaning and the system K has no physical reality.

The error Einstein identifies concerns what happens at some particular quadruple of values x_α in the coordinate system. The naive reading is that this quadruple picks out a particular physical event in spacetime and that the two solutions $g_{\mu\nu}(x_\alpha)$ and $g'_{\mu\nu}(x_\alpha)$ attribute different metrical properties to that event. This naive reading is mistaken. The quadruple x_α does not pick out any particular physical event until a metrical field is defined on the coordinate system. Only then can it do so. As a result the two solutions $g_{\mu\nu}(x_\alpha)$ and $g'_{\mu\nu}(x_\alpha)$ do not necessarily ascribe different metric properties to the same physical event. Thus a coordinate system is something less than we may naively think. It coordinates with nothing until a metric is defined on it. That is, take the metric off and one is not left with a coordinate system that labels the physical events of reality; that labeling is gone and the coordinate system as a labeling device ceases to function. In Einstein's words the "[coordinate] system ... has no physical reality." We might phrase this more cautiously by saying that the coordinate system has no reality independent of the metric, for the combination of coordinate system and metric certainly do represent aspects of physical reality.

In terms of the construction of the hole argument represented in Figure 5.6, the error is to think that the bare coordinate system x_α remains and continues to label the same physical events once the metric $g_{\mu\nu}$ is removed and that it can then host the new field $g'_{\mu\nu}$. What really would happen if we could somehow remove the metric field $g_{\mu\nu}$ from the coordinate system is shown figuratively in Figure 5.7.

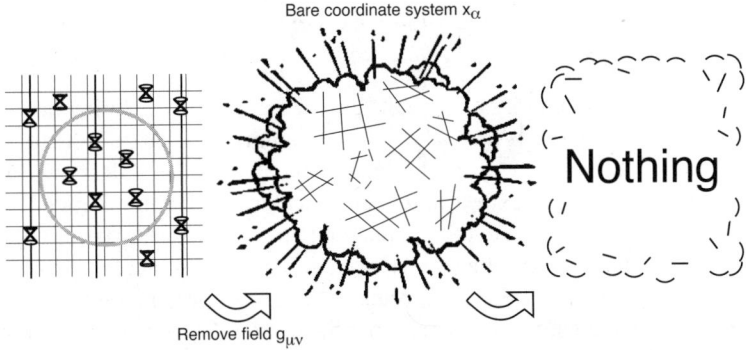

Fig. 5.7. Failure of the hole argument

This rather melodramatic portrayal may well not be so far from the way that Einstein himself visualized his error. Years later, after the sharply positivistic tone in his writing had much blunted, he wrote even more sharply about what happens if we imagine the removal of the metric field. A 1952 appendix "Relativity and the Problem of Space" to his popular text *Relativity* gives his mature view of the issues addressed hastily to his correspondents in late 1915 and early 1916. He wrote (Einstein 1954, p.155; Einstein's emphasis):

> On the basis of the general theory of relativity, on the other hand, space as opposed to "what fills space," which is dependent on the coordinates, has no separate existence. Thus a pure gravitational field might have been described in terms of the g_{ik} (as functions of the coordinates), by solution of the gravitational equations. If we imagine the gravitational field, i.e., the functions g_{ik}, to be removed, there does not remain a space of the type (1),[23] but absolutely *nothing* and also no topological space.

Finally it is important to note the tacit character of Einstein's error. He could not have been consciously aware that his hole argument depended essentially on according an independent reality to the coordinate systems. If he had noticed it, he would surely have rejected the supposition. Einstein's explanations of the transformations in the early expositions of the hole argument are sufficiently hasty to obscure their true character. This reflects his inattention to the presumptions on which they depend.

5.6 The Conjecture

> *If Einstein erroneously accorded the same independent reality to the restricted class of coordinate systems in which the Newtonian limit is realized, then these coordinate systems would adopt the same objectionable absolute properties as the preferred inertial coordinate systems of special relativity, rendering the candidate gravitational field equations inadmissible.*

Why did Einstein so rapidly forsake the gravitation tensor T_{il}^x during his preparation of the "Entwurf" paper? Why did he recall to Sommerfeld that he thought it would not return the Newtonian limit, when the calculation of p. 44/ 3 6 22R shows just how this can be done? We should expect to find clues on the pages of the Zurich notebook surrounding p. 44/3 6 22R where the gravitation tensor T_{il}^x appears. These pages deal with coordinate conditions and their transformation properties. On the pages following p. 44/3 6 22R, one particular transformation is given special attention, the transformation that sets the Cartesian spatial coordinate axes (x, y, z) of the Minkowski spacetime (5.2) into uniform rotation:

$$x' = x \cos \omega t + y \sin \omega t \qquad y' = -x \sin \omega t + y \cos \omega t \qquad z' = z \qquad t' = t.$$
$$(5.12)$$

While that is unremarkable, something more incongruous is on p. 43/3 6 22L, the page facing p. 43/3 6 22R. There Einstein investigates the covariance properties of the requirement (5.11) under non-linear, unimodular transformations. (The rotation transformation (5.12) is a non-linear, unimodular transformation.) That would not be surprising if Einstein was merely using the requirement as a coordinate restriction. But might it also be revealing some defect perceived by Einstein in requirement (5.11) if it is to be used as a coordinate condition?

If the coordinate condition (5.11) is used in the modern way, there would be no point in an investigation of its covariance The coordinate condition is merely used to reduce the gravitational field equations to their Newtonian form in some restricted set of coordinate systems; let us call them x_α^{LIM}. The condition need not have any more covariance than the Galilean covariance of Poisson's equation (5.6). One sees with minimal calculation that the condition (5.11) is not merely covariant under Galilean transformation but under any linear transformation of the coordinates.[24] Might Einstein's concern with the covariance properties of this coordinate condition reveal why he mistakenly thought its use in recovering the Newtonian limit a failure?

5.6.1 The Independent Reality of the Spacetime Coordinate System of the Newtonian Limit

> *Using the same construction as in Figure 5.6, Einstein would find that the limiting coordinate system x_α^{LIM} admits the special relativistic field $\eta_{\mu\nu}$ (5.2) but not the rotation field $g_{\mu\nu}^{\text{ROT}}$ (5.13) because of the insufficient covariance of the coordinate condition (5.11).*

My conjecture is that, in 1913, Einstein may have harbored a different understanding of the coordinate condition (5.11) and the coordinate systems x_α^{LIM} that they pick out. That difference is just the error Einstein later conceded in the context of the hole argument. That is, Einstein treated the coordinate systems x_α^{LIM} as physically real elements within his theory, whose existence is independent of the metric fields defined on them. In particular, this means that it is possible to reproduce with them exactly the construction depicted in Figures 5.5 and 5.6. He could consider one solution of the gravitational field equations in a coordinate system x_α^{LIM}, imagine that field removed and then a new, transformed field applied to the very same coordinate system.

Let us consider this construction in the simple case suggested by Einstein's concern for the rotation transformation (5.12). We begin with the Minkowski metric $\eta_{\mu\nu}$ shown as (5.2), which is the metric Einstein associated with special relativity. [25] We (actively) transform it under the rotation transformation (5.12). The result is the rotation field $g_{\mu\nu}^{ROT}$ whose components are:

$$g_{\mu\nu}^{ROT} = \begin{bmatrix} -1 & 0 & 0 & \omega y \\ 0 & -1 & 0 & -\omega x \\ 0 & 0 & -1 & 0 \\ \omega y & -\omega x & 0 & 1 - \omega^2(x^2 + y^2) \end{bmatrix}. \tag{5.13}$$

The transformation is shown in Figure 5.8, where the metric field is represented as before by the light cones and timelike geodesics it induces on the spacetime. The field $g_{\mu\nu}^{ROT}$ is a rotation field in the sense that free particles follow helical worldlines in the coordinate system x_α that rotate around a central axis.

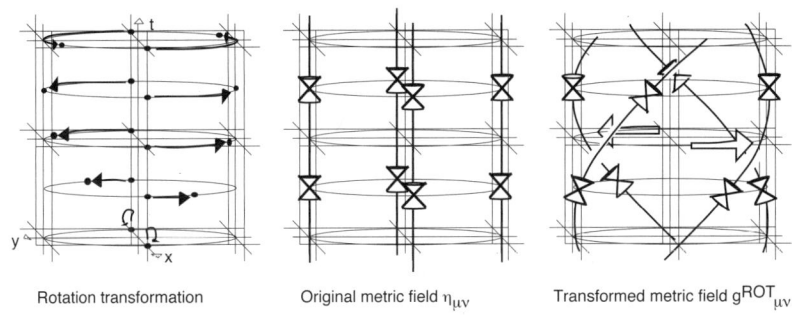

Rotation transformation   Original metric field $\eta_{\mu\nu}$ Transformed metric field $g^{ROT}{}_{\mu\nu}$

Fig. 5.8. Rotation transformation (5.12) creates a rotation field

The metric $\eta_{\mu\nu}$ has constant components. So we know without calculation that it is admissible in the coordinate system x_α^{LIM}—it satisfies both the source free field equations $T_{il}^x = 0$ and the condition that picks out x_α^{LIM}, the coordinate condition (5.11). We remove the metric field $\eta_{\mu\nu}$ and seek to apply the rotation field $g_{\mu\nu}^{ROT}$ to the bare coordinate system x_α^{LIM} as shown in Figure 5.9. Will x_α^{LIM} admit the rotation field, $g_{\mu\nu}^{ROT}$? Since the field equations $T_{il}^{x} = 0$ are covariant under transformation (5.12), the rotation field $g_{\mu\nu}^{ROT}$ satisfies it. But to be admissible in x_{α}^{LIM}, $g_{\mu\nu}^{ROT}$ must also satisfy the coordinate condition (5.11). A short calculation shows that it does not. We find:

$$\sum_{\kappa} \frac{\partial_{\gamma} ROT_{\kappa\alpha}}{\partial x_{\kappa}} = (\omega^2 x, -\omega^2 y, 0, 0) \neq 0,$$

so that (5.11) fails and the rotation field $g_{\mu\nu}^{ROT}$ is inadmissible in the coordinate system x_{α}^{LIM}. This is the most direct way to arrive at the result. There is another indirect path. The coordinate condition (5.11) is satisfied by the metric $\eta_{\mu\nu}$. It will also be satisfied by the rotation field $g_{\mu\nu}^{ROT}$ if coordinate condition (5.11) is covariant under the rotation transformation (5.12). This transformation is non-linear and unimodular. So an alternate calculation is to test the covariance of coordinate condition (5.11) under non-linear, unimodular transformation, just as Einstein does on the facing page p. 43/3 6 22L.

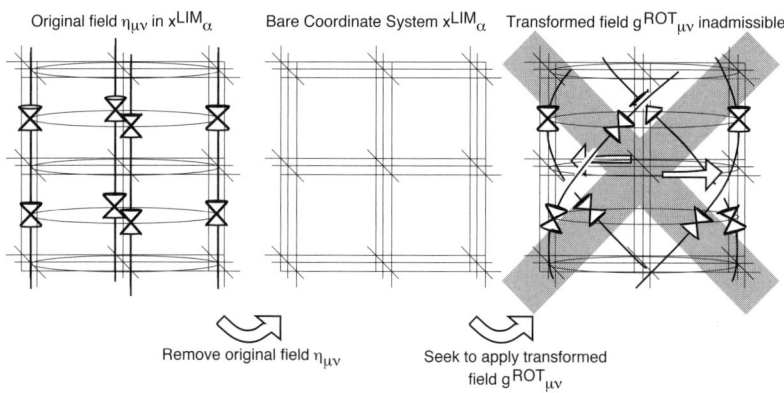

Original field $\eta_{\mu\nu}$ in $x^{LIM}{}_{\alpha}$ Bare Coordinate System $x^{LIM}{}_{\alpha}$ Transformed field $g^{ROT}{}_{\mu\nu}$ inadmissible

Remove original field $\eta_{\mu\nu}$ Seek to apply transformed field $g^{ROT}{}_{\mu\nu}$

Fig. 5.9. The coordinate system x_{α}^{LIM} will not admit the rotation field g_{α}^{ROT}

During the years of his "Entwurf" theory, Einstein never recognized that his hole argument depended upon the perilous presumption of the independent reality of the coordinate systems. It is an essential part of the present conjecture that Einstein was unaware, correspondingly, that his manipulations depend upon the presumption of the independent reality of the coordinate system x_{α}^{LIM}. Again, Einstein was so hasty in the earlier presentations of his hole argument that it was unclear whether they used active or passive transformations. Presumably this reflected a lack of attention in distinguishing the two types of transformations. We suppose a similar lack of attention in deciding whether transformations of (5.11) should be understood actively or passively.

5.6.2 The Anathema of Absolute Coordinate Systems

The fundamental goal of Einstein's work was to find a theory free of the iner-tial systems of special relativity, which were absolute in their failure to admit the rotation field (5.13).

The failure of x_α^{LIM} to admit the rotation field $g_{\mu\nu}^{\mathrm{ROT}}$ would have been of the most serious concern to Einstein. For it showed him that his theory harbored coordinate systems whose properties were routinely decried by him. The coordinate systems x_α^{LIM} would admit the special relativistic metric $\eta_{\mu\nu}$ but it would not admit the rotation field $g_{\mu\nu}^{\mathrm{ROT}}$. That is, the coordinate systems x_α^{LIM} behaved just like the inertial systems of special relativity that Einstein was so determined to eradicate. As he explained at the time of the "Entwurf" theory (Einstein 1914a, 176; translation from Beck 1996, 282):[26]

> The theory presently called "the theory of relativity" [special relativity] is based on the assumption that there are somehow preexisting "privileged" ref-erence systems K with respect to which the laws of nature take on an espe-cially simple form, even though one raises in vain the question of what could bring about the privilegings of these reference systems K as compared with other (e.g., "rotating") reference systems K'. This constitutes, in my opinion, a serious deficiency of this theory.

These inertial systems, as Einstein explained in his text (Einstein 1922, p. 55) supplied special relativity with the absolute elements that he would seek to eliminate in the general theory of relativity.[27]

> [Special relativity is based on] . . . the assumption that all inertial systems are . . . preferred, for the formulation of the laws of nature, to spaces of reference in a different state of motion . . . this preference for definite states of motion . . . must be regarded as an independent property of the space-time continuum. The principle of inertia, in particular, seems to compel us to ascribe physically objective properties to the space-time continuum . . . from the standpoint of the special theory of relativity we must say, *continuum spatii et temporis est absolutum.*

In 1913 it would appear to Einstein that the inertial systems of special relativity and now also the coordinate systems x_α^{LIM} endow their spacetimes with certain preferred or absolute motions. These are defined by the natural[28] straights of either coordinate system, as shown in Figure 5.10. The trajectories of free bodies are preordained to follow these straights; they may not be curved. Once x_α^{LIM} is admitted into the theory, its spacetime will not admit a rotation field $g_{\mu\nu}^{\mathrm{ROT}}$ in relation to which x_α^{LIM} would take on the character of a rotating reference system.

Promising as the gravitation tensor T_{il}^x seemed, it appeared inadmissible if Ein-stein's program of the elimination of absolutes was to succeed. For if the theory built around this gravitation tensor was to yield the correct Newtonian limit, it was at the cost of introducing exactly such an absolute element in the form of the limiting coor-dinate systems x_α^{LIM}.[29]

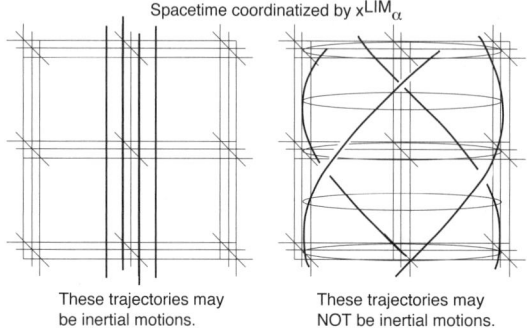

Spacetime coordinatized by x^{LIM}_α

These trajectories may be inertial motions.

These trajectories may NOT be inertial motions.

Fig. 5.10. The coordinate system x^{LIM}_α endows spacetime with absolute properties

5.6.3 The Theta Condition

This conjecture explains why Einstein's next step in the Zurich Notebook is to seek to replace the coordinate condiion (5.11) by another stated as a covariance condition contrived to admit rotation transformations (5.12).

The conjecture now also explains the calculations to which Einstein turns on the page following, p. 45/ 3 6 23L. Having just been thwarted by a coordinate condition of insufficient covariance, he decided to prevent another such failure by defining the coordinate condition from the start as a covariance requirement that had sufficient covariance for his purposes. So he stipulated a restriction to a class of coordinate systems within which the quantity $\theta_{i\kappa\lambda}$ transforms as a tensor, where:

$$\theta_{i\kappa\lambda} = \frac{1}{2} \left(\frac{\partial g_{i\kappa}}{\partial x_\lambda} + \frac{\partial g_{\kappa\lambda}}{\partial x_i} + \frac{\partial g_{\lambda i}}{\partial x_\kappa} \right).$$

His goal is clearly to replace the coordinate condition (5.11); he writes in the middle of his calculations that "[condition (5.11)] is not needed." It appears from calculations on other pages that Einstein designed his new coordinate condition to embrace the rotation transformations (5.12). He failed in this last goal, but only just.[30]

If Einstein believed that the covariance of his theory is restricted to that of the coordinate condition he imposes for recovery of the Newtonian limit, then he gains nothing in limiting the use of the coordinate condition to that special case of the Newtonian limit. He might as well impose the condition universally. That is, he might as well use it as what we have described as a "coordinate restriction" in Section 5.4.2. His gravitation tensor, to be used universally, will then be whatever remains of the tensor he starts with, after the coordinate restriction has been applied. This seems to be Einstein's purpose. On p. 45/ 3 6 23L, he takes the gravitation tensor T^x_{il} and adds and subtracts terms in $\theta_{i\kappa\lambda}$ until he arrives at a gravitation tensor of the form required by (5.8) and which is also by construction a tensor under unimodular transformation for which $\theta_{i\kappa\lambda}$ transforms as a tensor. That he freely adds these terms shows that his interest lies just in the final result and its covariance, for that final result will not be a quantity to which T_{il}^x reduces in the restricted set of coordinate systems.

If these last transformations had included the rotation transformations (5.12) then Einstein would have succeeded where he had failed on the previous page, in finding a gravitational field equation, covariant under rotation transformations and of form (5.8). But they did not include them and, apparently for this reason, the proposal of the theta condition was abandoned. Nonetheless, the introduction of this theta condition on p. 45/ 3 6 23L is an ingenious response to the difficulties Einstein believed he encountered with the gravitation tensor T_{il}^x on the preceding page p. 44/ 3 6 22R.

5.6.4 The Structure and Fate of the Entwurf Theory

The conjecture explains why Einstein was uninterested in finding the generally covariant gravitational field equations that reduce to his "Entwurf" equations. It also suggests that recognition of the admissibility of the gravitation tensor T_{il}^x and rejection of the hole argument could come at the same time since they are based on the same error.

The conjecture explains why Einstein set up and developed the Entwurf theory as he did and illuminates his return to general covariance. It suggests something quite general about the way Einstein would have sought to build his gravitation theory. According to the conjecture, as noted in Section 5.6.3, the covariance of the theory as a whole is limited to the covariance of the coordinate condition used to recover the Newtonian limit. The coordinate condition asserts the existence of coordinate systems x_α^{LIM} which in turn attribute absolute properties to spacetime, whether we are in the domain of the Newtonian limit or not. Thus Einstein purchases no additional covariance for his theory if he considers his gravitational field equations before they are reduced by the coordinate condition used to recover the Newtonian limit. He may as well work with the field equations after they have been reduced to the form (5.8).

This turns out to be just what Einstein does. The gravitational field equations published in the "Entwurf" theory have the form (5.8). From remarks in several places, for example (Einstein 1914a, 177–178), we know that he was sure that the "Entwurf" gravitational field equations were reduced forms of some unknown generally covariant equations, but he dismissed efforts to discover them as "premature" in the "Entwurf" paper (Einstein and Grossmann 1913, I. Sect. 5). (His attitude had hardened after he found the hole argument; then he dismisses these efforts as "of no special interest." (Einstein 1914a, 179).) That these efforts should be dismissed so quickly right from the first publication of the "Entwurf" theory is inexplicable in the modern view. For finding these equations would immediately dispel the uncertainty surrounding his theory: he did not know the extent of the covariance of the equation of the "Entwurf" theory. He could then use those generally covariant equations as his field equations and thereby present the world a theory that was manifestly generally covariant. Under the conjecture, however, his lack of interest is readily explicable. Finding those generally covariant equations would not allow him to add any covariance to his theory.

The conjecture allows us to see the connections between the events comprising Einstein's return to general covariance late in 1915. In our documentary records, that return began in earnest with a letter to Erwin Freundlich of September 30, 1915, (CPAE, vol. 8A, Doc. 123) in which an alarmed and weary Einstein reported his horror at discovering that his "Entwurf" equations were not covariant under the rotation transformation (5.12). In his communication of November 4 to the Prussian Academy, Einstein (1915) reports his return to the search for generally covariant gravitational field equations and that his choice of gravitation tensor is T_{il}^x.

We can now readily see how they could be connected. Their common feature is rotational covariance, that is, covariance under (5.12): Einstein had just found that his "Entwurf" equations lack it; he had rejected T_{il}^x because the associated coordinate condition (5.11) lacked it. We can guess many scenarios that lead from the discovery of the lack of covariance of the "Entwurf" equations to the readmission of the gravitation tensor T_{il}^x. For example,[31] Einstein was shocked to find that even the "Entwurf" gravitational field equations lacked covariance under rotation transformations (5.12). That he mistakenly thought these equations unique made the problem all the more acute. It would be natural in that circumstance to review the other candidate gravitation tensors from his earlier investigations that were covariant under transformation (5.12). They were the Ricci tensor and T_{il}^x. The Ricci tensor remained inadmissible because of its incompatibility with the flatness of weak, static fields. The tensor T_{il}^x did have the requisite covariance; it failed only when the associated coordinate condition (5.11) was considered. A devastated Einstein, now willing to think things through once again from the start, might well now see that his reasons for rejecting T_{il}^x were based on the error of according the coordinate systems x_α^{LIM} an existence independent of the metric field. The result would be his November 4 communication of the gravitation tensor T_{il}^x to the Prussian Academy. His final choice of the Ricci tensor and then Einstein tensor would only come in later communications that month after he recognized his other error of requiring the spatial flatness of weak, static fields.

But what of the hole argument? By November 4, Einstein could not have thought it succeeded in showing that a generally covariant theory was physically uninteresting for he was urging acceptance of a theory of near general covariance. One possibility is that Einstein had merely decided it must be flawed and that he would seek that flaw once the more pressing problem of finding generally covariant gravitational field equations had been solved. The conjecture suggests another possibility. According to it, the error of the hole argument and the error of the rejection of T_{il}^x are the same—improperly according a reality to coordinate systems independently of the metric fields defined on them. So once he located the error in one he had automatically found the error in the other. We might well understand that he would delay formulating a polished, public statement of the error of the hole argument until after November 1915. The real work was the completion of the theory by finding generally covariant equations, not drawing further attention to his earlier errors.[32]

There is scant evidence directly connecting the rejection of the hole argument and Einstein's discovery of the "Entwurf" theory's lack of rotational covariance. Most striking is a remark made to de Sitter in a letter of January 23, 1917 (CPAE, vol. 8A, Doc. 290). He reflected on two errors in his review, (Einstein 1914): the hole argument and another defective consideration. "I noticed my mistakes from that time," he recalled, " when I calculated directly that my field equations of that time were *not* satisfied in a rotating system in a Galilean space." Without the conjecture of this paper, it is hard to see why this calculation in a rotating system would have any direct bearing on the hole argument. With the conjecture, the connection is direct.

5.7 Conclusion

The present conjecture resolves outstanding puzzles in our history without the need to conjecture a new error by Einstein. The hole argument proves to capture precisely the essential obstacle separating Einstein from general covariance, although prior to 1915 he misidentified the obstacle it revealed.

The case for the conjecture is necessarily indirect. Unlike Einstein's error concerning the spatial flatness of weak, static fields, we do not have a direct, written admission by Einstein that he committed it. However some such error must be conjectured to complete our account of Einstein's search for his gravitational field equations. The other candidate explanation is the supposition that Einstein was just unaware of the modern use of coordinate conditions, even though he had the mathematical manipulations associated with them in his notebook. I do not believe he had this unawareness for reasons sketched in Section 5.4.2. The final decision depends considerably on a question of plausibility. Do we lean towards an obtuse Einstein, who persistently overlooks the obvious? Or do we prefer an Einstein able to commit an error of Byzantine sophistication? In the absence of good evidence for the former error, I choose the latter. The resulting account just takes the one other fundamental error that Einstein later freely admitted, the error of the hole argument. It asks after the consequences if that error were committed also months earlier in another context, that of the recovery of the Newtonian limit from candidate gravitational field equations.

The result is a compelling account of how Einstein came to abandon the search for generally covariant gravitational field equations in 1913. It was not just an oversight on Einstein's part. Very formidable obstacles separated him from the final, generally covariant gravitational field equations of 1915. He had to abandon his presumption of the spatial flatness of weak, static fields. Yet he had multiple items of independent evidence for it: it was suggested by his principle of equivalence, by the equations of motion of a particle in free fall and by the simplest form naturally taken by the gravitational equations in the weak field. Even if he could have seen past this problem, I now conjecture that a deeper misconception assured his failure. It lay buried beneath his conscious awareness, but misdirected fatally his attempts to use coordinate conditions. As long as he tacitly attributed an independent reality to spacetime coordinate systems, he must demand that the covariance of his theory be limited to the covariance of the coordinate condition used to recover the Newtonian limit from his gravitational field equations. Not even Einstein could be expected to find gravitational field equations that were otherwise admissible and associated with a coordinate condition of sufficiently broad covariance to support a generalized principle of relativity.

These were obstacles worthy of an Einstein and able to delay him for over two years in his struggle with his general theory of relativity. The hole argument proves to be more than an afterthought used to explain a decision already taken for other reasons. This argument, which Einstein repeatedly offered to explain the inadmissibility of generally covariant gravitational field equations, turns out to depend essentially on one of the two major obstacles recounted here—although Einstein misdiagnosed the import of the argument prior to 1915. We now see that it does not force us to abandon general covariance; rather it shows us we must abandon the notion that coordinate systems have a reality independent of the metric fields defined on them. Until Einstein did that, his quest for a generally covariant theory could only fail.

References

Beck, Anna (1996). *Translation. The Collected Papers of Albert Einstein. Vol. 4 The Swiss Years: Writings, 1912–1914*. Don Howard, consultant. Princeton University Press, Princeton.

Earman, John and Janssen, Michel (1993). Einstein's Explanation of the Motion of Mercury's Perihelion. In: *The Attraction of Gravitation: New Studies in the History of General Relativity: Einstein Studies, Volume 5*. John Earman, Michel Janssen and John D. Norton, eds., Birkhäuser Boston.

Einstein, Albert (1907). Über das Relativitätsprinzip und die aus demselben gezogenen Folgerungen, *Jahrbuch der Radioaktivität und Elektroni* **4**, 411–462; **5**, 98–99.

Einstein, Albert (1911): Über den Einfluss der Schwerkraft auf die Ausbreitung des Lichtes. *Annalen der Physik* **35**, 898–908; translated as "On the Influence of Gravitation on the Propagation of Light." In *The Principle of Relativity*. H.A. Lorentz et al., eds. Dover, 1952, 99–108.

Einstein, Albert (1912a): Lichtgeschwindigkeit und Statik des Gravitationsfeldes, *Annalen der Physik* **38**, 355–69.

Einstein, Albert (1912b): Zur Theorie des Statischen Gravitationsfeldes, *Annalen der Physik* **38**, 443–58.

Einstein, Albert (1913). Zum gegenwärtigen Stande des Gravitationsproblems, *Physikalische Zeitschrift* **14**, 1249–1262.

Einstein, Albert (1914): Die formale Grundlage der allgemeinen Relativitätstheorie, *Königlich Preussische Akademie der Wissenschaften* (Berlin). *Sitzungsberichte*, 1030–1085.

Einstein, Albert (1914a). Prinzipielles zur verallgemeinerten Relativistätstheorie. *Physikalische Zeitschrift* **15**, 176–80.

Einstein, Albert (1915). Zur allgemeinen Relativitätstheorie. *Königlich Preussische Akademie der Wissenschaften* (Berlin). *Sitzungsberichte*. 1915, 778–86.

Einstein, Albert (1915a). Erklärung der Perihelbewegung des Merkur aus der allgemeinen Relativitätstheorie. *Königlich Preussische Akademie der Wissenschaften* (Berlin). *Sitzungsberichte*, 831–839.

Einstein, A. (1916). Die Grundlage der allgemeinen Relativitätstheorie. *Annalen der Physik* **49**, 769–822; translated without p. 769 as "The Foundation of the General

Theory of Relativity." In *The Principle of Relativity*. H.A. Lorentz et al., eds. Dover, 1952, 111–164.

Einstein, Albert (1922). *The Meaning of Relativity*. 5th ed. Princeton University Press, Princeton, 1974.

Einstein, Albert (1954). *Relativity: the Special and the General Theory*. Trans. R. W. Lawson. 15th rev. ed. Methuen, London, 1977.

Einstein, Albert and Grossmann Marcel (1913): *Entwurf einer verallgemeinerten Relativitätstheorie und einer Theorie der Gravitation*. B.G. Teubner, Leipzig (separatum); with addendum by Einstein In *Zeitschrift für Mathematik und Physik* **63** (1914), 225–61. (CPAE, vol. 4)

Einstein, Albert and Rosen Nathan (1935). The Particle Problem in the General Theory of Relativity. *Physical Review* **48**, 73–77.

Howard Don and Norton John D. (1993). Out of the Labyrinth? Einstein, Hertz and the Göttingen Answer to the Hole Argument. In *The Attraction of Gravitation: New Studies in History of General Relativiy*. John Earman, Michel Janssen, John D. Norton, eds. Birkhäuser Boston, 30–62.

Janssen, Michel (1999). Rotation as the Nemesis of Einstein's 'Entwurf' Theory. In *The Expanding Worlds of General Relativity. Einstein Studies, Volume 7*. Hubert Goenner, et al. eds. Birkhäuser Boston, 127–157.

Janssen, Michel and Renn Jürgen (forthcoming). "Untying the knot: How Einstein found his way back to field equations discarded in the Zurich notebook, in *The Genesis of General Relativity*, Vol. II. General Relativity in the Making: Einstein's Zurich Notebook, Commentary and Essays. Springer.

Klein, Martin J. et al. eds. (1993). *The Collected Papers of Albert Einstein. Volume 5 . The Swiss Years: Correspondence, 1902–1014*. Princeton University Press, Princeton. (CPAE, vol. 5).

Klein, Martin J. et al., eds. (1995). *The Collected Papers of Albert Einstein. Volume 4. The Swiss Years: Writing, 1912–1914*. Princeton University Press, Princeton. (CPAE, vol.4)

Kretschmann, Erich (1915). Über die prinzipielle Bestimmbarkeit der berechtigten Bezugssystemen beliebiger Relativitätstheorien." *Annalen der Physik* **48**, 907–982.

Norton, John D. (1984). How Einstein found his Field Equations: 1912–1915. *Historical Studies in the Physical Sciences* **14**, 253–316; reprinted in *Einstein and the History of General Relativity: Einstein Studies, Volume 1*. Don Howard and John Stachel, eds. Birkhäuser Boston, 1989, 101–159.

Norton, John D. (1985a). What was Einstein's Principle of Equivalence? *Studies in History and Philosophy of Science* **16**, 203–246; reprinted in *Einstein and the History of General Relativity: Einstein Studies, Volume 1*. Don Howard and John Stachel, eds. Birkhäuser Boston, 1989, 5–47.

Norton, John D. (1985b). Einstein's Struggle with General Covariance. *Rivista di Storia della Scienza* **2**, 191–205.

Norton, John D. (1986). Einstein's Struggle with General Covariance. In *Proceedings of the Fourth Marcel Grossmann Meeting on General Relativity*. R. Ruffini, ed. Elsevier Science Publishers, 1837–1848.

Norton, John D. (1987). Einstein, the Hole Argument and the Reality of Space. In *Measurement, Realism and Objectivity*. John Forg, ed. Dordrecht, Reidel, 153–188.

Norton, John D. (1989). Coordinates and Covariance: Einstein's view of spacetime and the modern view. *Foundations of Physics* **19**, 1215–1263.

Norton, John D. (1992). Einstein, Nordström and the Early Demise of Scalar, Lorentz-Covariant Theories of Gravitation. *Archive for the History of Exact Sciences* **45**, 17–94.

Norton, John D. (1992a). The Physical Content of General Covariance. In *Studies in the History of General Relativity: Einstein Studies, Volume 3*. Jean Eisenstaedt and Anne Kox, eds. Birkhäuser Boston.

Norton, John D. (1993). General Covariance and the Foundations of General Relativity: Eight Decades of Disute. *Reports on Progress in Physics* **56**, 791–858.

Norton, John D (forthcoming). What Was Einstein's 'Fateful Prejudice'? In *The Genesis of General Relativity*. Vol. II. Jürgen Renn, ed., General Relativity in the Making: Einstein's Zurich Notebook, Commentary and Essays. Springer.

Renn, Jürgen ed. (forthcoming). *The Genesis of General Relativity*, Vol. I. General Relativity in the Making: Einstein's Zurich Notebook, Introduction and Source. Springer.

Renn, Jürgen (manuscript). Progress in a Loop; General Relativity as a Triumph of the 'Entwurf' Theory.

Renn, Jürgen, Sauer Tilman, Janssen Michel, Norton John, and Stachel John (forthcoming). Commentary in *The Genesis of General Relativity*, Vol. II. General Relativity in the Making: Einstein's Zurich Notebook, Commentary and Essays, Jürgen Renn, ed. Springer.

Renn, Jürgen and Sauer Tilman (1999). Heuristics and Mathematical Representation in Einstein's Search for a Gravitational Field Equation. In *The Expanding Worlds of General Relativity: Einstein Studies, Volume 7*. Hubert Goenner et al, eds. Birkhäuser Boston, 87–125.

Renn, Jürgen and Sauer Tilman (1996). Einstein's Zuricher Notizbuch. *Physikalische Blätter* **52**, 865–872.

Renn, Jürgen and Sauer Tilman et al. (forthcoming). A Commentary on Einstein's Zurich Notebook.

Schulmann, Robert et al. eds. (1998). *The Collected Papers of Albert Einstein. Volume 8. The Berlin Years: Correspondence, 1914–1918. Part A: 1914–1917*. Princeton University Press, Princeton. (CPAE, vol. 8A).

Speziali, Pierre ed. (1972). *Albert Einstein–Michele Besso: Correspondance 1903–1955*. Hermann, Paris.

Stachel, John (1980). Einstein's Search for General Covariance. Paper read at the Ninth International Conference on General Relativity and Gravitation, Jena. In *Einstein and the History of General Relativity: Einstein Studies, Volume 1*. Don Howard and John Stachel, eds. Birkhäuser Boston, 1989, 63–100.

Notes

[1] (Einstein and Grossmann 1913)

[2] For an entry into this extensive literature see (Stachel 1980), (Norton 1984, 1985b, 1986), (Earman and Janssen 1993), (Howard and Norton 1993), Editorial Notes in (Klein et al. 1995), (Janssen 1999), (Renn and Sauer 1996; 1999).

[3] Presented with commentary and annotation as Document 10 in (CPAE, vol. 4) and in (Renn, forthcoming).

[4] The group was founded in 1991 under the direction of Peter Damerow and Jürgen Renn as the Working Group Albert Einstein, funded by the Senate of Berlin and affiliated with the Center for Development and Socialization, headed by Wolfgang Edelstein at the Max Planck Institute for Human Development in Berlin. It was continued after 1995 under the direction of Jürgen Renn as part of the project of studies of the integration and disintegration of knowledge in modern science at the Max Planck Institute for the History of Science in Berlin. Its members include Michel Janssen, John D. Norton, Jürgen Renn, Tilman Sauer and John Stachel who are the coauthors of (Renn and Sauer et al., forthcoming). I am grateful to all members of this group for their contributions to and stimulating discussion of the material in this paper.

[5] In 1913, Einstein and Grossmann did not use the summation convention in their publications. The indices μ and ν range over 1, 2, 3 and 4.

[6] For Einstein's abbreviated version of the calculation that follows, see (Einstein 1913, Sect. 8). He later explains to Michele Besso in a letter of December 21, 1915, that this result was misleading (CPAE, vol. 8A, Doc. 168).

[7] I continue to follow the notational conventions of Einstein and Grossmann's "Entwurf" paper. With the exception of the Christoffel symbols, all indices are written "downstairs." The contravariant form of the metric $g_{\mu\nu}$ is written with the corresponding Greek letter as $\gamma_{\mu\nu}$. Commas denote coordinate differentiation.

[8] Einstein was aware in 1912 that the spatial geometry associated with acceleration need not be Euclidean. As he remarked in his (Einstein 1912a, Sect. 1), the geometry fails to be Euclidean in the space association with uniform rotation. For further discussion of Einstein's use of the principle of equivalence, see (Norton 1985a).

[9] That is, the trajectories of reference bodies of the inertial frame are given by the timelike curves in spacetime picked out by constant values of the coordinates X, Y and Z. The trajectories of the reference bodies of the accelerated frame are given by the timelike curves in spacetime picked out by constant values of the coordinates x, y and z.

[10] The term $g_{44} = a^2 x^2$ cannot be interpreted directly as a Newtonian potential since we are no longer dealing with the case of a weak field.

[11] (Einstein and Grossmann 1913, I Sect. 5). An even simpler choice for the first term would have been $\sum_{\alpha,\beta} \gamma_{\alpha\beta} \left(\partial^2 / \partial x_\alpha \partial x_\beta \right) \gamma_{\mu\nu}$. Einstein's choice of the term in

(5.8) does not affect the outcome since the two agree in first order quantities in the weak field.

[12]To ensure spatial flatness, Einstein does need an assumption of comparable strength. With it, the result of spatial flatness everywhere is quickly recovered. If Laplace's equation $\Delta\Psi = 0$ holds everywhere in a sphere of radius R, then a lemma asserts that the value of Ψ at the center is just the integrated average of the value of Ψ on the sphere's surface. Pick some arbitrary point in space and consider a family of spheres centered on it that extend to spatial infinity. If Ψ is to approach the same constant value Ψ_∞ in all directions at spatial infinity, then we must have $\Psi = \Psi_\infty$ at the center, if the lemma is to hold for all the spheres. Replace Ψ successively by each value of $\gamma_{\mu\nu}$ in (5.8b) and we conclude that each has Minkowskian values throughout the spacetime.

[13]Pick one component γ_{ik}. For a sphere centered on the sun and for a radial coordinate r, Gauss' theorem tells us that $\int_A \partial\gamma_{ik}/\partial r \; dS = \int_V \Delta\gamma_{ik}dV = \int_V T_{ik}dV = 0$ where A is the area of a sphere and V its volume. Therefore $\partial\gamma_{ik}/\partial r = 0$ so that, allowing for spherical symmetry, γ_{ik} is constant.

[14]These coordinates were then called "isothermal" and are now commonly called "harmonic" since the coordinate condition (5.10) is equivalent to the one that has the form of a wave equation $\Box x_\mu = 0$.

[15]"p. 37" refers to the pagination of the Zurich notebook introduced in CPAE, vol. 4). "3 6 19L" uses the system of designation associated with the control numbers in the Einstein Archive. It refers to the left-hand side of page 19 of the document 3-6, which is the Zurich notebook.

[16]That Einstein's public endorsement of them in 1915 was brief is readily explained by his recognition over the weeks following that weak, static fields need not be spatially flat, so that the Einstein tensor became admissible as a gravitation tensor and was quickly chosen by him.

[17]The form given is quoted directly from the Zurich notebook and the annotations on the terms are Einstein's.

[18]There are similar remarks in the paper (Einstein 1915, 1056).

[19]The supposed unawareness is incompatible with Einstein's labeling of terms on p. 44/ 3 6 22R. He introduces the decomposition of the Ricci tensor apparently aware in advance that one part, the quantity T_{il}^x, will reduce to the Newtonian form (5.8) under imposition of the requirement (5.11). If (5.11) is not being used as a coordinate condition, his gravitation tensor is whatever T_{il}^x reduces to *after* imposition of (5.11). Yet Einstein carefully and clearly labels T_{il}^x as "presumed gravitation tensor"—just the appropriate labeling if (5.11) is being used as a coordinate condition.

[20]For an account of the vacillations see (Norton 1984, Sect. 5). For further discussion see also (Stachel 1980, Sect. 3–4), (Norton 1987).

[21]The simplification is that I consider a matter free metrical field, whereas Einstein considered a source matter distribution in which the hole is a matter free region.

[22]Einstein's formulae $G(x)$ and $G'(x)$ correspond to $g_{\mu\nu}(x_\alpha)$ and $g'_{\mu\nu}(x_\alpha)$ respectively. Einstein sent a very similar explanation to Paul Ehrenfest in a letter of December 26, 1915 (CPAE, vol. 8a, Doc. 173).

[23]Einstein's formula (1) is the line element $ds^2 = dx_1^2 + dx_2^2 + dx_3^2 - dx_4^2$ of special relativity. In continuing to explain how general relativity regards a space with this line element, he repeats what for present purposes is the key insight learned in Einstein's 1915 rejection of the hole argument: "[...] the coordinate system used [...] in itself has no objective significance [...]" (p. 155).

[24]Under a linear transformation from coordinate system x_α to x'_β, the coefficients $p_\alpha^\beta = \partial x'_\beta / \partial x_\alpha$ and $\pi_\beta^\alpha = \partial x_\alpha / \partial x'_\beta$ are constants and this constancy is all that is needed to secure the covariance of (5.11). If condition (5.11) holds in unprimed coordinates $\sum_k \partial \gamma_{\kappa\alpha} / \partial x_\kappa = 0$, then so does condition (5.11) in primed coordinates since

$$\sum_\mu \frac{\partial \gamma'_{\mu\beta}}{\partial x'_\mu} = \sum_{\mu\alpha} \pi_\mu^\lambda \frac{\partial}{\partial x_\lambda}(p_\kappa^\mu p_\alpha^\beta \gamma_{\kappa\alpha}) = \sum_{\mu\alpha} \pi_\mu^\lambda p_\kappa^\mu p_\alpha^\beta \frac{\partial \gamma_{\kappa\alpha}}{\partial x_\lambda} = \sum_\alpha p_\alpha^\beta \sum_\kappa \frac{\partial \gamma_{\kappa\alpha}}{\partial x_\kappa} = 0.$$

[25]While this is no longer the practice in relativity theory, Einstein then considered special relativity not just as the case of a Minkowski spacetime, but as a Minkowski spacetime in the inertial coordinate system associated with (5.2). The rotation field is by modern lights just another presentation of Minkowski spacetime, but Einstein treated it as a different case. This is not the place to debate whether this approach is viable. Our concern is to understand Einstein's reasoning at that time. For discussion of the rationale underlying Einstein's approach see (Norton 1989, 1992a, 1993).

[26]See also (Einstein 1913, 1260).

[27]I have used ellipses liberally in the quote to bring to the fore the aspects of present importance. The complete passage reads: "All of the previous considerations have been based upon the assumption that all inertial systems are equivalent for the description of physical phenomena, but that they are preferred, for the formulation of the laws of nature, to spaces of reference in a different state of motion. We can think of no cause for this preference for definite states of motion to all others, according to our previous considerations, either in the perceptible bodies or in the concept of motion; on the contrary, it must be regarded as an independent property of the space–time continuum. The principle of inertia, in particular, seems to compel us to ascribe physically objective properties to the space–time continuum. Just as it was consistent from the Newtonian standpoint to make both the statements *tempus est absolutum spatium est absolutum* so from the standpoint of the special theory of relativity we must say, *continuum spatii et temporis est absolutum*. In this latter statement, *absolutum* means not only 'physically real,' but also 'independent in its physical properties, having a physical effect, but not itself influenced by physical conditions.'"

[28]In a coordinate system (x_1, x_2, x_3, x_4), these are defined as the curves that satisfy $x_4 = A_i x_i + B_i$ for constants A_i and B_i where $i = 1, 2, 3$, where $A_1^2 + A_2^2 + A_1^2 + A_3^2 < 1$.

[29]For completeness I note how this conclusion would err in Einstein's later view. In the construction shown in Figure 5.9, we incorrectly suggest that we seek to apply the two fields $\eta_{\mu\nu}$ and $g_{\mu\nu}^{\text{ROT}}$ to the *same* coordinate system. The construction fails because of the illicit intermediate stage in which a bare coordinate system is still supposed to label the same events. There are coordinate systems x_α^{LIM} compatible with the rotation field g_α^{ROT}, picked out by (5.11). But their x_4 axes would appear helical if drawn in Figure 5.9 just like the free fall trajectories of particles in $g_{\mu\nu}^{\text{ROT}}$. Indeed one of these coordinate systems would be the image under rotation transformation (5.12) of the coordinate system x_α^{LIM} associated with $\eta_{\mu\nu}$. The coordinate systems x_α^{LIM} are able to induce absolute properties onto a spacetime of events only as long as we suppose that they are capable of labeling events independently of the metric fields defined on them. The erroneous view requires that it makes sense to assert counterfactual claims like: "This trajectory designated by this coordinate axis could have been a non-inertial motion if there were a different metric field." If removal of the metric field deprives a coordinate system of its ability to designate these trajectories, then the counterfactual loses its meaning.

[30]If the coordinate condition admits these rotation transformations, then it must admit transformations between the special relativistic metric $\eta_{\mu\nu}$ and the rotation field $g_{\mu\nu}^{\text{ROT}}$. Since $\eta_{\mu\nu}$ has constant coefficients, we have $\theta_{i\kappa\lambda} = 0$ for it. If $\theta_{i\kappa\lambda}$ transforms tensorially under (5.12), then we must find that $\theta_{i\kappa\lambda} = 0$ also for $g_{\mu\nu}^{\text{ROT}}$. On pp. 7-8/3 6 42L-42R Einstein is apparently checking this expectation when he seeks all fields compatible with the conditions $\theta_{i\kappa\lambda} = 0$, with metrics of unit determinant and $\partial g_{ik}/\partial x_4 = 0$. (These last two conditions are satisfied by $g_{\mu\nu}^{\text{ROT}}$.) His expectations are almost vindicated. The solution class includes a metric whose coefficients in the *covariant* form equal the coefficients of the metric $g_{\mu\nu}^{\text{ROT}}$ in its *contravariant* form. This is close, but it is not the metric $g_{\mu\nu}^{\text{ROT}}$.

[31]With the repertoire supplied by the conjecture, finding other scenarios is merely a challenge to one's ingenuity. Einstein may instead have begun by deciding that he must return to general covariance and so reappraised the hole argument, his public objection to such a return. With the error of that argument found, the readmission of tensor T_{il}^x is now possible since its rejection was based on that same error.

[32]Einstein appears to have delayed informing his correspondents of the error of the hole argument and avoided mentioning the argument directly in print thereafter. In the surviving correspondence, the first explanation comes in the letter to Ehrenfest of December 26, 1915 (CPAE, vol. 8a, Doc. 173), which advances the point-coincidence argument. That argument is published in his review article the following year (Einstein 1916, Sect. 3), but the argument is presented as one favoring general covariance without any indication that it is his own response to the hole argument. See (Howard and Norton 1993, Sect. 7) for the suggestion that Einstein may have chosen to formulate his response to the hole argument in terms of point-coincidences upon the unacknowledged inspiration of a paper by (Kretschmann 1915) from December 1915.

Einstein and the Principle of General Relativity, 1916–1921

Christoph Lehner

Max-Planck-Institut für Wissenschaftsgeschichte, Wilhelmstraße 44, 10117 Berlin, Germany;
lehner@mpiwg-berlin.mpg.de

In many of his expositions of general relativity, Einstein offered two principles as the foundation of the theory. The first was the well-known principle of equivalence that expressed the core idea of general relativity, that gravitation has to be seen as an effect of the space-time structure itself. The second principle, however, underwent dramatic changes in its expression and role within the formulation of the theory. Its basic idea is the generalization of the special principle of relativity, demanding that all reference systems, not only inertial ones, should be treated on equal footing. Hence, it is frequently called the principle of general relativity. In this paper, I will consider the somewhat puzzling story of how Einstein expressed this idea during the years 1916–1921, after he had published his seminal papers on general relativity, but during a period when he was still exploring the meaning and implications of the theory and when he had to explain and defend it on many occasions.[1]

Since the early days of Einstein's work on general relativity, Einstein offers a well-known argument for the relativity of inertia from the relativity of motion. He credits this argument to Ernst Mach who criticized the Newtonian concept of absolute space with the following observation: The physical distinction between inertial and accelerated motion cannot be based on the fact that a system moves against absolute space since this is a completely unobservable entity, but it has to be explained by the motion against other physical bodies. Einstein uses this argument to demand that inertial forces be caused not by acceleration against absolute space, but by acceleration against other masses, and calls this postulate the "hypothesis of the relativity of inertia." [See, for example, (Einstein 1913, 1260–1261)]. Einstein's formulation of the principle in the following years was influenced by his struggle with the question whether general relativity could be formulated in a generally covariant way. In 1915, he recognized that one of the arguments he believed to hold against general covariance, the famous hole argument, was flawed.[2] This recognition was expressed in the point-coincidence argument: Two solutions of the field equations that can be transformed into each other by a continuous mapping are equivalent, since all observable physical facts can be expressed by purely local statements (e.g., coincidences of space time events). With the achievement of general covariance, Einstein believed that he had satisfied the demand for a generalization of relativity in a maximal way: If *all* coordinate systems are equivalent for the formulation of the theory, then especially there cannot be a difference between coordinate systems that represent frames in accelerated motion and ones that represent frames in inertial motion. Consequently, Einstein formulated the principle of general relativity thus: "The laws of physics have to be constituted such that they are valid in reference systems in any state of motion." (Einstein 1916, 772) and assumed that this also implied the relativity of inertia.[3]

However, already in 1917 he realized that this is not necessarily the case since even though general relativity is generally covariant, the structure of space-time does not merely depend on the distribution of matter in the universe but also on the boundary conditions of the field equations at infinity. Hence, even in an empty universe, there will be a distinction between inertial and accelerated motion imposed by the boundary conditions. He attempted to circumvent this problem by postulating a finite closed universe in which there was no need for such boundary conditions (Einstein 1917). However, the criticism that Willem de Sitter leveled against this model forced him to acknowledge that relativity of inertia was not an essential part of general relativity.[4]

Therefore he decided to split off a "Mach's principle" from the principle of general relativity in (Einstein 1918), the only place where he offers three principles as the foundation of general relativity. Mach's principle appears as the third principle, stated concisely as: "The G-field [metric field] is completely determined by the masses of bodies." This obviously addresses the role of boundary conditions in the determination of the metric field. Einstein adds that he hasn't properly distinguished this principle from the principle of general relativity in the past, but that not all his colleagues share his opinion that the principle is a necessary part of general relativity.

The second principle is the principle of equivalence. As the first principle, however, Einstein proposes a completely new formulation of the principle of relativity: "Nature's laws are merely statements about temporal-spatial coincidences: therefore, they find their only natural expression in generally covariant equations." This is obviously a reformulation of the point-coincidence argument, now used as an argument for general covariance. General covariance has taken the place of general relativity? Although this is logically permissible under the argument mentioned above, that general covariance implies general relativity in the old sense, it is somewhat surprising: On the face of it, general covariance seems a formal property of a theory, as opposed to a principle about the real world. Einstein addresses just this complaint, made recently by Ernst Kretschmann (1917): General covariance is merely a formal requirement and the point-coincidence argument itself leads to the conclusion that any physical theory can be brought into general covariant form. Einstein does not contradict very strongly, but rather points out that general covariance has great heuristic value taken together with the demand for theoretical simplicity. Although it might be mathematically possible to bring non-covariant theories like Newtonian mechanics or special relativity into generally covariant form, the resulting theories would be so cumbersome that they would never be taken serious as physical theories.

As John Norton shows, many readers found this response unconvincing, especially since already in the 1920s covariant formulations of Newtonian mechanics and special relativity were developed (Norton 1993, 817–829). However, Norton also points out an issue that neither Einstein nor Kretschmann addressed (Norton 1992) that the physical content of the new relativity principle lies in its premise. The claim that physical laws are only statements about spatiotemporal coincidences is not at all trivial. One might add, that it is not even plausible without the fundamental reinterpretation of space-time properties that general relativity produced. Newton's bucket is—on the face of it—a striking counterexample to this claim.

It is striking that this equation of general relativity and general covariance does not reappear in Einstein's writings of the following years. One possible reading of his writings during this period is that Einstein is becoming disenchanted with the use of these principles, especially the principle of general relativity. When Einstein mentions them later, he does so for their historical and heuristic role, but not as an essential part of the foundation of the theory. In 1920, on this reading, a new view of the foundations of general relativity emerges in the paper "Ether and Relativity" (Einstein 1920). There is no talk about the theory being derived from principles; rather, Einstein postulates the metric field as the fundamental entity of his theory. He calls it the "New Ether" that determines both the metric and inertio-gravitational properties of space-time and that is inextricably bound up with the existence of space-time. The characteristics of this geometrodynamical field form the mathematically sound foundations of general relativity, not any ambiguous principles.

On the other hand, it has to be acknowledged that during this period Einstein keeps using his principles not only in historical accounts, but also for the systematic foundation of general relativity and, most strikingly, in his defenses of the theory against its critics. While it would be possible to read this as a somewhat disingenuous use of arguments that Einstein himself had already abandoned for their polemical value, I will propose that Einstein still was convinced that general relativity could and should be based on physical principles even though their content had to be modified in the light of the criticism by Kretschmann and de Sitter.

The two most extensive expositions of general relativity written in these years were the unpublished paper "Fundamental Ideas and Methods of the Theory of Relativity, Presented in Their Development"[5] and the Princeton lectures (Einstein 1922). In these documents, Einstein offers a new formulation and argument for general relativity which is considerably less known than his statements from 1918. I will argue, however, that it replaces the point-coincidence argument as the fundamental justification for the principle of general relativity.

In section 15 of "Fundamental Ideas . . . ," with the title "The Fundamental Idea of General Relativity in its original Form," Einstein first introduces the principle of equivalence and compares the relation of gravitation and inertia it postulates to the fact that in special relativity electric and magnetic fields have only relative existence, depending on the state of motion of the observer. Then he continues:

> The empirical fact of the independence of gravitational acceleration from the material is therefore a powerful argument for *extending the postulate of relativity* to coordinate systems moving against each other at a nonuniform velocity. [My italics] (Einstein 2002, 265).

Einstein continues talking about the role of the principle of equivalence as a heuristic instrument for exploring properties of the gravitational field. Then he states:

This is based on the hypothesis that the principle of relativity is also valid for coordinate systems accelerated relative to each other. (Einstein 2002, 266).

After a discussion of the equality of inertial and relative mass, Einstein begins a new section (16), "General Reasons for a General Postulate of Relativity." It is only here that he introduces the Machian argument against absolute acceleration. Similarly, in the Princeton lectures, Einstein writes after considering the equivalence of inertia and gravitation:

We call the assumption of the full physical justification of this conception the "principle of equivalence"; this is obviously suggested by the law of equality of inertial and gravitational mass and indicates an *extension of the principle of relativity* to coordinate systems moving non-uniformly against each other. [My italics] (Einstein 1922, 37).

Mach's principle is discussed in detail only at the end of the lectures, in the context of cosmological considerations.

As these quotes show, Einstein offers two arguments for a general principle of relativity. The second one is the Machian argument previously discussed, but the first argument that is not based on general considerations about the nature of space. Rather, it is based on the principle of equivalence, which requires as a necessary condition a generalized principle of general relativity. The reasoning behind this is not hard to understand:

Generally, in any theory of motion we will have a distinction between inertial and non-inertial motion, i.e., motion without or with external forces. In Newtonian mechanics, this distinction coincides with the distinction of non-accelerated and accelerated motion in absolute space. The principle of equivalence requires that there is no physical difference between inertial motion without a gravitational field and free fall in a gravitational field. If this equivalence is to be a fundamental principle rather than a coincidence in physical phenomena, there can be no structure like Newtonian absolute space that distinguishes a priori which motions are inertial and which are not. This means that all states of motion are equivalent in principle, before a specific distribution of matter in the universe is specified. This statement I read as the mature formulation of Einstein's principle of general relativity. It is necessary for the interpretation of the principle of equivalence as a fundamental principle of physics. This necessity provides a sufficient argument for general relativity even without the Machian argument.

If the principle of general relativity is understood this way, the direct link between general relativity and general covariance that Einstein postulated in 1918 is broken: general relativity as a physical principle is quite independent of general covariance as a formal property of the theory since it addresses the physical distinction between inertial and non-inertial states of motion. Just as one could formulate Newtonian mechanics or special relativity in generally covariant coordinates, so it is possible to formulate general relativity in a preferred coordinate system. Einstein does not argue that it is not possible to do so, but merely that such a formulation imposes a formal structure without physical relevance. This formulation of the principle of general relativity is obviously quite close to Einstein's original formulation in terms of accelerated and non-accelerated frames of reference, which he used from 1907 until his discovery of the point-coincidence argument in 1915. Nevertheless, it acknowledges the abandonment of frames of reference that a generally covariant theory requires. It is therefore a formulation suitable for use in a generally covariant theory.

Rather, the principle can be seen as an elaboration on the physical premise of the point-coincidence argument. It is a necessary condition for this premise (that all natural laws are statements about point coincidences) that there be no a priori preferred states of motion, because they certainly could not be described by purely topological laws. Einstein's new formulation of the principle of general relativity can therefore be seen as a weaker antecedent than the premise of the point-coincidence argument. Moreover, it is a less abstract claim than the latter, which might also have induced Einstein to prefer it over the point-coincidence principle. After all, Einstein's principles (compare the case of special relativity) were meant to function as an *empirical* base for his theories (Einstein 1919).

References

Einstein, Albert (1913). Zum gegenwärtigen Stande des Gravitationsproblems. *Physikalische Zeitschrift* **14**, 1249–1262.

— (1916). Die Grundlage der allgemeinen Relativitätstheorie. *Annalen der Physik* **49**, 769–822.

— (1917). Kosmologische Betrachtungen zur allgemeinen Relativitätstheorie. *Königlich Preußische Akademie der Wissenschaften* (Berlin). *Sitzungsberichte*, 142–152.

— (1918). Prinzipielles zur allgemeinen Relativitätstheorie. *Annalen der Physik* **55**, 241–244.

— (1919). Time, Space, and Gravitation. *The Times* (London), 28 November 1919, pp. 13–14.

— (1920). *Äther und Relativitätstheorie. Rede, gehalten am 5. Mai 1920 an der Reichs-Universität zu Leiden.* Springer, Berlin.

— (1922). *Vier Vorlesungen über Relativitätstheorie, gehalten im Mai 1921 an der Universität Princeton.* Friedrich Vieweg, Braunschweig.

— (1998). *The Collected Papers of Albert Einstein.* Vol. 8. Robert Schulmann, Anne J. Kox, et al., eds. Princeton University Press, Princeton.

— (2002). *The Collected Papers of Albert Einstein.* Vol. 7. Michel Janssen, Robert Schulmann, et al., eds. Princeton University Press, Princeton.

Kretschmann, Erich (1917). Über den physikalischen Sinn der Relativitätspostulate. A. Einsteins neue und seine ursprüngliche Relativitätstheorie. *Annalen der Physik* **53**, 575–614.

Norton, John (1992). The Physical Content of General Covariance. In *Studies in the History of General Relativity.* Jean Eisenstaedt, Anne J. Kox, eds. Birkhäuser Boston, Cambridge, 281–315.

— (1993). General Covariance and the Foundations of General Relativity: Eight Decades of Dispute. *Reports on Progress in Physics* **56**, 791–858.

Stachel, John (1989). Einstein's Search for General Covariance, 1912–1915. In *Einstein and the History of General Relativity*. Don Howard, John Stachel, eds. Birkhäuser Boston, Cambridge, 63–100.

Notes

[1] This is not coincidentally very close to the period covered by Vol. 7 of the Collected Papers of Albert Einstein (Einstein 2002). The present paper grew out of the discussions I had mainly with Michel Janssen over the annotation for this volume. I would like to thank him in this place for his invaluable help on many occasions in setting my head straight on general relativity. I am afraid he cannot be blamed for the claims made in this paper, however.

[2] For an account of the hole and point-coincidence arguments, see (Stachel 1989) and (Norton 1992, 1993).

[3] This belief is documented in a letter to Michele Besso from July 31, 1916 (Einstein 1998, 324–325).

[4] The debate between de Sitter and Einstein is discussed in an editorial note in (Einstein 1998, 351–357).

[5] The manuscript is in the Pierpont Morgan Library, New York. It has been published in (Einstein 2002, 245–281)

7

Einstein and the Problem of Motion: A Small Clue

Daniel Kennefick

Physics Department, University of Arkansas, Fayetteville, Arkansas 72701, U.S.A.;
danielk@uark.edu

7.1 Introduction

Einstein's work on the problem of motion constitutes the largest body of his work on general relativity (GR) after 1918. His research on this topic played an influential and sometimes controversial role in research on the problem of gravitational radiation from self-gravitating systems after the Second World War. The initials EIH, referring to the initials of Einstein and his coauthors of a 1938 paper, Leopold Infeld and Banesh Hoffman, are still strongly associated with the first-post-Newtonian order equations of motion in GR. Fittingly his research on the problem of motion has received some attention from historians of science, especially from Peter Havas, who has written several papers addressing this subject, in particular one appearing in the first volume of the Einstein Studies Series to which this paper is a sort of addendum (Havas 1989). Havas was himself an important figure in the history of this problem in the period after Einstein's death. I would like to stress that this paper is not intended to add any new material to the history of the problem of motion in GR, beyond pointing out the existence of a single interesting line in an undated, but dateable manuscript in Einstein's hand, and commenting on some of the lessons which might be drawn from this discovery. I also take the opportunity to discuss some recent commentary on EIH and Einstein's approach to the problem of motion, including one article appearing in a physics journal which may be unfamiliar to some in the History of GR community.

Einstein's work on the problem of motion was heavily influenced by his continually evolving attitudes towards other subjects, especially unified field theories and the role of singularities in field theories. I will rely heavily for the singularities topic on John Earman and Jean Eisenstaedt's paper (1999) which discusses Einstein's use of singularities in his work on the problem of motion from 1927 on. It is well known, and widely reported by his collaborators, such as Leopold Infeld (1941), that Einstein was strongly motivated to work on the problem of motion in GR because of his belief that it might help his quest for a unified field theory of gravity. Of course the search for such a unified theory was the principle goal of his research from 1920 on. A little background on the meaning of unification for Einstein in this period is thus in order.

Abraham Pais has identified three aims in Einstein's unification program (Pais 1982). The best known is the unification of the gravitational and electromagnetic fields, and about as well known is his hope that unification would lead to an explanation of quantum phenomena in terms of some underlying causal theory. But an equally important part of the unification program, for Einstein and other theorists, was the desire to end the duality of matter and field which underpinned the physics of the post-Maxwell era. This branch of the program sought to explain matter as a phenomenon arising out of the field concept. Fields would ultimately become their own sources. It was an essential part of this program, for Einstein, that matter be described by continuous functions of the field. One concrete attempt to achieve this goal, within the context of GR, was the Geon, an invention of John Wheeler's. The Geon was a wave packet which held itself together by its own gravitational attraction (Wheeler 1955). It was viewed for a while as a possible prototype of elementary particles constructed out of pure field, although nothing much came of that hope in the end. Nevertheless the idea of reducing matter to a construction of the field was a key element in Einstein's conception of a successful unified field theory. In his autobiography Wheeler describes his conversations with Einstein on his Geon work, shortly before the latter's death, and we learn that Einstein said he had considered similar entities but rejected them as both "unnatural" and liable to instability (which proved to be the case) (Wheeler 1998, 237–238).

It seems that most of Einstein's work on GR after 1918 was motivated by needs arising out of his unification program. We know that he continued to harbor hopes that physical phenomena which were considered characteristic of the quantum would arise in his work on GR, thus illustrating how quantum effects might prove to be ultimately classical in origin. Infeld, again in his autobiography, claims that it was only in the course of their work on the EIH paper that he was able to convince Einstein that the classical problem of motion in GR would not yield any glimmer of hope in this direction.[1] But Einstein had clung tenaciously to such hopes since his earlier work on the problem of motion with Jakob Grommer in 1927, as a number of contemporaneous letters make clear, especially letters to Hermann Weyl and Arnold Sommerfeld. In the interests of brevity I refer the reader to (Earman and Eisenstaedt 1999) and (Havas 1989) for citations and quotations from this correspondence.

The principle result arising out of Einstein's work on the problem of motion for which he was given credit by later physicists is that in general relativity the equations of motion follow from the field equations of the theory itself, known as the Einstein equations. In fact, as Havas has pointed out, it was realized very early on in the history of GR, and by a number of different theorists, that the field equations constrain the motion of massive bodies. That this should be so is a little surprising, since in many other theories, such as Maxwellian electrodynamics and classical Newtonian mechanics, equations of motion must be stated separately from equations of the field (or other force law, such as Newtonian gravitation). In his early accounts of GR, as for instance in his book based on the Princeton lectures, Einstein (1922) always introduced the equations of motion as an additional axiom or assumption. This fact led Havas to surmise that Einstein did not initially realize that the equations of motion do not have to be assumed, but can be derived from the field equations. In remarking on Einstein's statement in Einstein and Grommer that "the law of motion is a consequence of the field law" (translated and quoted in Havas 1989, p. 240), Havas says,

> This is a rather astonishing statement, considering that Einstein had given no indication in the decade since he had postulated the geodesic law that this postulate [that particles move along geodesics] might be superfluous.

This statement is correct. There are no statements by Einstein in print, post 1915, which indicate that the geodesic law might be redundant. But there are two places before 1927 in which Einstein refers to this matter which, in my view, only makes our astonishment greater at his neglecting to draw attention to this fact in print after 1915. The first reference comes before the birth of GR, in Einstein's paper of 1913 with Marcel Grossman which introduced his "Draft" or *Entwurf* theory of gravitation (Einstein and Grossman 1913). This theory was the immediate precursor of GR and contains many features of the final theory. In this paper Einstein sketched how the equations of motion for at least one (idealized) type of matter (that of pressureless dust) might be derived from the field equations of the *Entwurf* theory. As we shall see, this is interesting because in early accounts of GR he consistently introduced the equations of motion of matter independently from the field equations, even though the argument made for the 1913 theory carries over to the later theory.

The new piece of evidence comes next in the story. In a manuscript apparently dating from 1921 Einstein included a line of text in notes for a lecture or publication which says:

Diese Gleichung enth. schon Div. Gl. und damit Bew. Ges. des mat. Punktes. Expanded and translated, this reads,

> This equation [the field equations are referred to, as they are the last equation at the bottom of the preceding page] already contains the divergence equation and with it the laws of motion of material points.

This is a clear statement which indicates, if the document's dating can be relied upon, that Einstein did understand, well before he raised the topic in print in 1927 that the equations of motion followed from the field equations. It even appears that he intended to make this argument in print six years before Einstein and Grommer. But he seems to have decided against doing so. The central question addressed in this paper is why he chose not to do so. To emphasize Havas' point once more, it seems astonishing that he would not do so, all the more so if it was a deliberate omission and not the result of a failure of comprehension.

Einstein's refusal to rid his theory of a superfluous axiom (that particles move along geodesics) is worth noting, given his characteristic axiomatic approach to physics. One might have expected him to move heaven and earth to remove an unnecessary assumption. I will argue that he took this course because he could not make his argument general enough to satisfy himself. Because his axiomatic approach preferred to base itself on soundly argued principles with global applicability, he could not eliminate one of his axioms unless he felt he could make a very general argument for doing so. But as mentioned above, the argument in its nascent form of 1913 depended critically on the nature of one's model of matter and therefore could be applied strictly speaking only on a case-by-case basis for each model or type of matter. One way around this was to make use of the singularity model of matter, as proposed by Eddington and others, but Einstein throughout most of his career was violently opposed to this approach. That the use of singularities became, in the end, a characteristic component of his attack on the problem of motion suggests that he regarded this as a critical issue that justified extraordinary measures. In particular Don Howard has suggested that the timing of the Einstein and Grommer paper may be connected with the crisis in Einstein's response to the new quantum theory, which took place around the same time (Don Howard 1990). At the same time, I shall also argue that Einstein's determination to pursue the problem of motion via the use of singularities, in spite of his own aversion to such a course, did prove fertile because he developed a clever method of excising the singularities from his spacetime. It has been argued (Anderson 1997) that Einstein's approach to this problem could have been very influential.

7.2 Dating of the Manuscript

The manuscript containing the line about the derivation of the equations of motion, which is titled "On the Special and General Theory of Relativity" ("Über die spezielle und allgemeine Relativitätstheorie"), consists of 9 unnumbered loose-leaf pages and is undated. It is (EA 2 085) in the Einstein Archive and it appears in Volume 7 of the Collected Papers (Einstein 2002) as Doc 63. The line quoted above appears on p. 453 of that volume. It has been dated for the collected papers edition to late 1921, on the basis that it represents a draft for the published version of the Princeton lectures, *The Meaning of Relativity* (Einstein 1922). Einstein delivered five lectures at Princeton on May 9–13, 1921 during his first trip to the United States. The first two were popular accounts of special and general relativity. In the subsequent three lectures he covered the same ground at a much more technical level for an audience of scientists. It had originally been planned to publish a book based on typescripts of the actual lectures, but this seems to have proved impractical, so in September 1921 Einstein was obliged to write what amounted to a text-book on relativity to satisfy his agreement with the Princeton University Press. Einstein's handwritten manuscript of the book survives, in which he divided the material into five lectures, without giving them much relation to the actual lectures delivered (since the book is clearly at the level of the three technical lectures and not at the level of the two popular ones). However two of the lectures, or chapters, were collapsed together before publication so that the book appeared under the somewhat confusing subtitle "Four lectures given at Princeton University."[2]

There are several reasons for thinking that this manuscript is an abortive draft of the Princeton book. It begins with a heading "1. Vorlesung." followed by three pages of text on special relativity, which breaks off in mid-page. Then there is a further heading, "3. Vorlesung. Allg Rel. Theorie." where the 3 is overwritten on a 4 (or conceivably vice-versa). This section covers General Relativity. It corresponds roughly to the material presented in the fourth and fifth lectures actually given at Princeton and to the third and fourth chapters of the book. Therefore the manuscript reflects the uncertainty in numbering which seems to have existed concerning the lectures and the chapters in the book. The title of the manuscript says nothing about Princeton, but then neither does the title of the handwritten draft of the Princeton book which is titled, in the draft, "Fünf Vorlesungen über Relativitätstheorie." That it is a manuscript for a book, rather than a draft of a series of lectures, is suggested by the first section, which is fully written out (a style Einstein seems to have hardly ever used in giving lectures) and includes marginal topical descriptors. Admittedly the second section, the one on GR, reads like lecture notes and is certainly not a draft of a book. It could be no more than a sketch at best.

This somewhat contradictory evidence may be reconcilable in various ways. That the two fragments belong together is suggested only by their being placed together in the Einstein archive. That there was reason for doing so originally may be hinted by the fact, clearly visible in photocopies of the first page, that there was once a paper clip binding this page to others. The whole 9 pages may have been preserved together. In the current original Einstein archive in Jerusalem the pages have been individually laminated for preservation purposes. The paperclip is no longer available as a witness. If we assume that the fragments go together, then we can argue either that what began as a draft was quickly aborted and turned into a sketch of the technically complex second half of the book, or an abortive draft may have been kept together with what were originally notes for the fourth and fifth lectures at Princeton as a reference while writing the full draft.

Even if one were to argue that the part of the manuscript of interest to us, the section on general relativity, represents lecture notes, it still seems highly likely that they refer to the Princeton lectures and therefore date to 1921. I base this statement on the content of the notes, especially on the last part of the GR section, which ends with the lines

> Bemerkung über Elektron. Poincarés Druck P
> $(K_x - \partial P/\partial_x) = 0$
> $\dots \dots \dots$
> $\dots \dots \dots$
> Kosmischer Druck mit unbekanntem Nullpunkt.

This is a sketch for a discussion of cosmology on the basis of the "Poincaré pressure." In a paper of 1919 Einstein dealt with the idea that the cosmological constant term in his field equations of 1916 could be interpreted as a negative universal pressure which, in turn, he identified with the pressure much earlier proposed by Poincaré (1906) as a means of explaining the stability of electrons against their own electrostatic field (Einstein 1919). As late as 1923, in a paper with Grommer on the Kaluza five-dimensional unified theory, he proposed that the Poincaré pressure (the more usual term is "Poincaré stresses") might be associated with the mysterious "corner term" in the Kaluza field equations (Einstein and Grommer 1923). Therefore I would see this reference as dating the manuscript to the first half of the 1920s. In particular it is worth noting that both the typescript made from the actual fifth lecture at Princeton, and the final chapter of the published book and its manuscript, conclude with a discussion of the Poincaré pressure as a universal pressure. In general the presentation of GR in our fragment follows that of the Princeton book.

7.3 The Problem of Motion

In General Relativity the field equations take the form of an equality between two tensors, one of which, known as the Einstein tensor, is constructed out of the metric tensor and the Riemann curvature tensor and therefore contains information about the geometry of spacetime itself. This spacetime curvature may be thought of as an expression of the gravitational field. The other side of the equation, the right-hand-side, is a tensor known as the stress-energy tensor which contains information about the source of the field, that is to say, matter or other forms of mass. The form taken by the stress-energy tensor naturally depends on the model one uses for the matter in question. Einstein declared himself profoundly dissatisfied with this state of affairs more than once. In 1936 he stated

> [General Relativity] is sufficient—as far as we know—for the representation of the observed facts of celestial mechanics. But it is similar to a building, one wing of which is made of fine marble (left part of the equation), but the other wing of which is built of low-grade wood (right side of equation). The phenomenological representation of matter is, in fact, only a crude substitute for a representation which would do justice to all known properties of matter. (Einstein 1936, p. 370)

It is in the light of the dislike which Einstein bore for the right-hand side of the field equations that we must understand his approach to the problem of motion in relativity. It was clearly understood by several early relativists, in particular Weyl and Eddington, that the vanishing of the divergence of the stress-energy tensor implied the equations of motion of bodies in GR, as has been pointed out by Havas (1989). This observation lay at the heart of statements in at least half-a-dozen textbooks of 1921 (or within a year or two of that date) which, as Havas describes, discuss this point. I contend that Einstein considered including an account in his textbook of this period, but decided not to.

The line of argument runs, in general, as follows. Take the covariant derivative (or we may say, the divergence) of both sides of the field equation. The contracted Bianchi identities then require that the left-hand-side of the new set of equations (the covariant derivative of the Einstein tensor) be zero.[3] One then has the result that the covariant derivative of the stress-energy tensor is zero, from which expression one may derive integral conservation laws of energy, mass and momentum. From this equation, assuming some form of the stress-energy tensor, and therefore some model of matter, one can also derive equations of motion for material bodies. In the case of particles moving freely in an external gravitational field, these equations of motion will describe motion along geodesics of the spacetime expressed by that field (by a geodesic we mean the generalization of the concept of a straight line to curved spacetime).

It is important to note that other physicists besides Einstein naturally wished to show that the derivation of the equations of motion from the field equations was independent of the particular form of the stress-energy tensor chosen. The best known approach was that of Weyl, as discussed for instance in volume 5 of his famous textbook (Weyl 1923, p. 267), in which he solves the differential form of the conservation laws (recall the divergence of the stress-energy tensor shown to vanish in the preceding paragraph) inside the "world tubes" of matter in motion and matches this solution onto the solution of the field equations in the exterior spacetime outside of the world tubes in such a way as to remain independent of the particular model of matter assumed within. Einstein certainly was made aware of Weyl's method when Weyl wrote to him following the publication of Einstein and Grommer, as Havas shows (Havas 1989, p. 245–248). He still objected that Weyl's method appeared to him to place restrictions on the model of matter employed, at least in certain cases, such as that of accelerating charged particles (not, as Havas observes, a case which his own method dealt with). Without giving a proper account of Weyl's work, which has remained very influential to this day, I only wish to stress that Einstein, reasonably or not, still continued to insist that this general line of argument, even in its most ingenious manifestation, did not suit him because it was insufficiently independent of assumptions on the internal structure of matter.

Now we do know that Einstein had realized as early as 1913 that the covariant derivative of the stress-energy tensor for pressureless dust leads to the geodesic equation, because of the statement to this effect in Einstein and Grossman 1913. As it happens, the left-hand-side of that theory's field equations, although very different from that of GR, also had the property that its divergence vanished. Therefore from the start, as it were, even before the final theory was discovered, Einstein knew of this route to the equations of motion. What our manuscript shows is that he had not forgotten it in the period between 1913 and 1927, which is what one might easily assume from his complete silence on the subject. On the contrary, he apparently considered including some discussion of this matter in the closest thing to a textbook on GR that he ever wrote, but decided against it.

Both in the actual lectures[4] and in the eventual book the equations of motion are introduced by stating as an additional assumption that bodies move along geodesics of the spacetime. For Einstein to decide to skip discussion of such an important topic, we have to conclude that he had very compelling reasons. It seems obvious that his distrust of the stress-energy tensor's role in the field equations was a major factor. In the paper with Grommer, Einstein (it seems reasonable to assume that the introduction to the paper represents a summary of Einstein's views of the subject, rather than Grommer's) discusses three ways of addressing the problem (Einstein and Grommer 1927, 2–3). The first is the axiomatic approach,[5] the second is the "right-hand-side" approach, and the third is to use the vacuum Einstein equations, with the left-hand-side equal to zero and the sources represented as singularities of the field. As I will discuss in the next section, Einstein also harbored grave doubts about the validity of this third approach.

7.4 Singularities

So it seems clear that Einstein did not discuss the problem of motion before 1927 because he was dissatisfied with an approach which, by proceeding on the basis of the stress-energy tensor, lacked generality. Any results could be said to apply only for the particular state or model of matter treated. In such a situation any claims about being able to drop the geodesic motion assumption as an axiom would not be forceful. Perhaps more importantly, since quite general claims can be made using this approach with due concern, as in Weyl's method, Einstein seemed to feel that the vacuum field equations offered a truer guide to the form of an eventual unified field theory, and so his unified field theory program dictated his preference for proceeding on the basis of the left hand side of the equations alone. One obvious way to proceed in this case was to treat all matter as singularities of the field, in the style of Newtonian point masses. In this case the right-hand side of the field equations could be done away with altogether. But we know that Einstein was averse to this approach. In *The Meaning of Relativity* in 1921 he inveighs against this use of singularities, in the context of electromagnetism, in a footnote

> It has been attempted to ... consider ... charged particles as proper singularities. But in my opinion this means giving up a real understanding of the structure of matter. (Einstein 1922, p. 33)

Therefore Einstein's objections to this approach have something in common with his objections to the right-hand-side approach. He would like to come up with a "unified" conception of the structure of matter and doesn't wish to sidestep the issue by a tactical maneuver. But it seems that when push came to shove he preferred to operate on the basis of the vacuum field equations.

Einstein's attitude to the role of singularities in relativity is a complex topic covering a great span of time with many subtleties. There is little point in discussing the ramifications here. I take as my starting point Earman and Eisenstaedt's (1999) view that Einstein frequently ignored distinctions between apparent and real singularities because what was of more importance to him was whether or not a singularity could be said to correspond to a mass point. They show how, over time, Einstein's aversion to singularities softened, but essentially only in so far as it permitted him to adopt the singularity model of matter. As the quote above suggests, in 1921 he was still hoping to avoid that contingency. But in 1927 comes an admission of defeat, in the paper with Grommer. The context here is his enumeration of the three possible ways of tackling the problem of motion.

> All attempts of the last few years to explain the elementary particles of matter by continuous fields have failed. The suspicion that this may not be the right way for understanding material particles has become very strong in us ... We are thus led to a third approach, which ... assumes singular world lines. (Einstein and Grommer 1927, p. 4) Havas' (1989, p. 240) translation

Einstein continued to resist a complete conversion to the use of singularities. He held out the hope that in the final version of a unified theory of fields, singularities would not only be unnecessary but would in fact be excluded and in addition, he drew comfort from the venerability of the practice of treating masses as some form of singularity.

> In Newtonian mechanics, particles are represented as singularities of a scalar field f which satisfies Laplace's equations everywhere outside the singularities (Einstein and Infeld 1949, p. 209)

So Einstein came to accept the use of singularities as a practical necessity, though as we shall see, he found a way of having his cake and eating it too, by means of a very clever stratagem which allowed him to excise his sources from the rest of his spacetime. But we may certainly ask why, having waited so long, he capitulated in 1927. A possible answer is suggested by Don Howard (1990), who suspects that the timing is connected with the birth of the new quantum mechanics around this time. In early 1927 Einstein worked on a paper which tried to show that there was a causal explanation of Schroedinger's wave mechanics based on over-determination of classical variables. At about this time he wrote to Max Born that he had hoped to

> show that one can attribute quite definite movements to Schroedinger's wave mechanics, without any statistical interpretation. — letter to Born, undated (Born 1969, p. 136)

That there is a connection between this abortive approach to the quantum problem and his work on the problem of motion in the same period is clear. We have the testimony of Einstein himself, when informing Hermann Weyl (who, as we have seen, had already made important contributions to this problem) of his work with Grommer on the problem of motion.

> I attach so much value to the whole business because [I want] ... to know whether or not the field equations are disproved by the established facts about the quanta. — Einstein to Weyl, 26/4/27 (quoted and translated in Havas, 1989, p. 247).

Thus a crisis situation with dire implications for underlying principles of physics, such as causality, which Einstein held dear, may have determined him to attack the problem of motion in GR with any tools available, in the hope that it would shed some light on difficulties elsewhere.

He did not return to the matter again until the late 1930s. Then he approached the problem of motion more ambitiously. Instead of dealing with test particles moving along geodesics of an external gravitational field, as in the 1927 papers, he addressed the N-body problem of an ensemble of masses moving and interacting only with each other. This problem, although limited to an approximation of slow motions of the sources and weak gravitational fields, is a much more practical one, applying, for instance, to the case of the solar system. And indeed the EIH solutions are said to form the basis for ephemerides used in satellite tracking today.

What is especially interesting about EIH, from our point of view, is how Einstein managed to carry through a very clever method of removing his singularities from the problem, while still retaining them as an aid to calculation. When he concluded that singularities represented the best way forward he did not just swallow hard and accept them as a necessary evil. He realized that Stoke's Law permitted him to convert the volume integrals over the source, which characterize the typical approach to this problem, into surface integrals over a surface enclosing that volume. This meant that the integration did not go over the singular point. Thus he avoided the obvious technical challenge of integrating over such a singularity and at the same time he hid away the singularity from the rest of spacetime. It can be argued that it does not matter what is inside the surface of integration, as long as it approximates to a point source which produces the same field at the surface.

Interestingly although EIH is a very famous paper in the history of relativity, and although advocates of the rival fast motion approximation, such as Havas, decry its influence on the subsequent development of the field, this surface integral method was not widely copied. EIH's general approach, based on a "slow-motion" expansion of the equations of motion, continued to be very relevant, especially in connection with the problem of emission of gravitational radiation by systems like binary stars. This radiation problem can be tackled as an extension of the problem of motion to higher order than was attempted by Einstein, Infeld and Hoffman (to the fifth power of the small velocity parameter, as opposed to the second power). For several decades EIH figured fairly prominently in the rhetoric of what became known as the "quadrupole formula controversy," after the radiation formula first derived in Einstein's 1918 paper on gravitational waves. Then oddly enough, at the end of this controversy, in the 1980s, it was argued that Einstein's surface integral trick contained the solution not only to some difficulties that relativists encountered in the problem of gravitational wave emission, but also to some very famous problems which bedeviled classical radiation theory throughout the 20th century.

The relativist James Anderson (1987) employed the surface integral method of EIH, together with later innovations (at least in so far as their introduction into GR is concerned), such as matched asymptotic expansions to produce a very clear derivation of the quadrupole formula based on the slow motion approximation applied to a binary star system. In the process of conducting a historical study of this controversy (Kennefick 1999) Anderson's work was several times cited to me as one of those decisive contributions which clarified matters in the last stages of the controversy. Since then Anderson has argued that EIH has much broader implications for the radiation problem in field theories. It has been a longstanding problem in classical electrodynamics, since the turn of the 19th century, that the radiation problem as applied to point charges results in divergent integrals and runaway solutions, which must be dealt with by various more or less ad hoc forms of maneuver. Anderson argues that the updated EIH approach, as we may call his version with its matching techniques, eliminates all of these problems. The electromagnetic radiation problem is attacked from within the framework of GR and the Einstein equations, the field equations determine what Anderson calls "conditions of motion" (a more rigorous version of the EIH equations of motion), and the surface integral method appears to do away with all of the divergent integrals and other undesirable features of the traditional approach. Therefore Anderson's work suggests that Einstein's take on the use of singularities is not just a case of sweeping them under the rug. The characteristic difficulties that are encountered in the use of point masses in any field theory may conceivably be eliminated by Einstein's clever approach.[6] Anderson laments that this aspect of EIH was not sufficiently widely appreciated in the physics community. Certainly while EIH has had its detractors, most relativists today would view it as Einstein's most important contribution to their subject after 1920.

7.5 Conclusions

It seems that Einstein often failed to mention in print something he knew well if, for some reason, he did not regard the issue as cut and dried. Examples include the argument that the existence of the coriolis force and the equivalence principle imply that gravity had to be a tensorial theory (not scalar like Nordstrom's) attested only by a letter from Ehrenfest, (see Einstein 1993, Doc. 380) in which Ehrenfest states that Einstein had discussed this with him, acknowledging Einstein's priority for the idea. Another case is that of the "ghost-field" in which Einstein envisaged a wave-like guide field which could explain interference and other wave-like effects exhibited by corpuscular light-quanta, and which was only known at the time, as Lorentz put it, via "verbal communication" and "hearsay," (Lorentz 1927) because Einstein never put these ideas into print. At the same time they enjoyed wide circulation, as Stachel tells us (Stachel 1986).

The story of Einstein's interest in the general relativistic problem of motion, only a small sidelight of which has been addressed here, offers a useful insight into Einstein's approach to physics. On the one hand, it shows us his dogged insistence on certain points of principle and his refusal to let go of what he saw as the fundamentals of physics. On the other hand, it gives us a glimpse of his opportunism, which has been much remarked on in recent Einstein scholarship. We may ask, how does Einstein's outlook differ from that of other physicists? Opportunism, I would say, is a characteristic quality of physicists. Perhaps it is the *defining* quality of physicists. Incidentally this opportunism can be viewed in an entirely positive light. Physicists like to solve problems and are willing to experiment and innovate in a way not open to professionals in more practical disciplines, such as engineers. In physics, a conservative character like Max Planck can become a revolutionary not in spite of himself, but by following his physicist's impulse to crack the problem at hand, no matter where it leads him. And indeed Einstein's own experience with quantum mechanics, in the long run, was that of a reluctant revolutionary. But perhaps we can see the following difference between Einstein's opportunism and that of other physicists. In their paper discussing Einstein's struggles with singularities, Earman and Eisenstaedt begin by saying "Creative scientists often succeed by ignoring or pushing aside foundations problems. This was not Einstein's way." A page later they remark "but always an opportunist, Einstein attempted to employ singularities in general relativity in order to treat the problem of motion of a test body in a gravitational field." Einstein was an opportunist, but he preferred not to opportunistically ignore foundational issues in physics, which he generally regarded as indispensable to his approach. Instead he preferred to ignore the problem itself, rather than compromise the fundamentals. This is the precise inverse of the typical physicist's method.[7] It also seems a hopeless way to make progress, and certainly no other physicist would have refused to address the problem of motion in GR for a decade, knowing as much as Einstein seems to have done in 1921 and presumably before. But Einstein preferred to throw out or disregard the problem when faced with a conflict between progress in tackling it and adherence to the principles. If the question does not make sense in the light of what was "known" about the foundations of physics, then it must be the wrong question. Only later when a crisis situation urged action on him did he return to this issue. In the end he found an ingenious way, via the surface integrals, of preserving his principles. He made use of singularities, thus permitting him to solve the vacuum field equations and deal with matter in the abstract without loss of generality. His method of surface integrals hides or "effaces" the singularities, so that in principle their presence or absence makes no difference.

It may seem paradoxical to say that Einstein was at once principled and opportunistic. But I think that Einstein's opportunism has its own peculiar flavor that is well illustrated by the story I've presented. Earman and Eisenstaedt (1999, 229-230) discuss Einstein's "principled opportunism" and mention Einstein's reference to the "unscrupulous opportunism" of physicists in "Reply to Criticisms" in Schilpp (1949, p. 684). Thus we can argue, with them, that Einstein was a believer in what Irish-American politicians used to call "honest graft," that is to say reprehensible or underhanded behavior that proceeds from relatively pure motives.[8] The only remark of Earman and Eisenstaedt's which I would not endorse is where they say, in reference to the use of singularities in EIH, "this piece of opportunism has to be judged, in retrospect, as conceptually ill-founded." I have argued that in fact Einstein's method in dealing with these singularities in the problem of motion was a classic moment of Einsteinian slight of hand that any physicists would appreciate, even if it has been suggested that not enough of them did, hardly surprising given the isolation of general relativity within mid-20th century physics.

What would most physicists do when faced with the choice Einstein outlines at the beginning of Einstein and Grommer? These choices were, to introduce geodesic motion as an axiom of the theory, to proceed via the right-hand side of the field equations and the differential conservation laws, and finally to proceed via the vacuum field equations and the singularity model of matter. Clearly very many physicists of the day took option number two, the stress-energy tensor route. Those who did not would hardly blanch at the prospect of route number three. Einstein's choice can hardly be upheld as superior to the other options. It is not necessarily a sound route at all, as Havas has argued (see note 4). It completely ducks one of the central problems of the theory. But I suspect that Einstein felt that a statement of principle is always to be preferred to a messy calculation backed up with even messier conjecture. A few years down the road, times had changed and he had stronger motives propelling him, so he tackled the problem. Nevertheless he ultimately came up with a typically brilliant way of preserving his principles. He would use the vacuum field equations, but he would circumvent the dreaded singularities by a clever stratagem. Whatever may be said about Einstein's priority in the problem of motion, this was a masterly Einstein moment, and Anderson has gone so far as to say that these papers [EIH and subsequent papers with Infeld on the problem of motion] contain what is arguably one of Einstein's greatest contributions to physics. (Anderson 1997, p. 4676)

Readers of the Einstein Studies literature will be a little surprised by Anderson's statement. Not much praise has been lavished on EIH in historical studies of Einstein's work over the years. Pais devotes only a paragraph to it, and is at pains to point out that "the same or nearly the same results were obtained much earlier by [others]." Havas, the principle authority on this topic (and Pais' source for that quote), gives a very thorough and balanced account of Einstein's work in this area, and of the work of many others, especially before the war. But Havas was, in general, a noted critic of the slow-motion approach and EIH in particular and therefore not disposed to sing its praises. In addition it was part of his role to recover much other work that had been completely forgotten or overlooked, and EIH was bound to suffer in the telling of that story. Earman and Eisenstaedt, as we saw, are uncomfortable with EIH's use of singularities. So it is perhaps worth stressing that EIH retains an important position in the historical lore of today's relativists, remains one of Einstein's more often cited papers, and is a paper which some physicists feel should have received even more attention than it did. This is not bad work for a man who had once hesitated even to mention the problem it addressed in print.

Acknowledgements. I would like to thank my colleague Michel Janssen, with whom I collaborated on the annotation of document 63 of volume 7 of the *Collected Papers of Albert Einstein*. I also owe him a debt of gratitude for his encouragement in writing this paper. Another colleague at the Einstein Papers Project Christoph Lehner gave me a number of useful suggestions, especially in bringing to my notice occasions on which Einstein failed to mention in print ideas that he had discussed privately with his colleagues. I should also thank Don Howard, who provided some very helpful comments and insights shortly before I gave this talk at HGR6 and John Stachel, who made a number of useful remarks after the talk. I would also like to thank Tilman Sauer and Thibault Damour for valuable comments and suggestions after reading the manuscript.

References

Anderson, James L. (1987). Gravitational Radiation Damping in Systems with Compact Components. *Physical Review D* **36**, 2301–2313.

Anderson, James L. (1997). Asymptotic conditions of motion for radiating charged particles. *Physical Review D* **56**, 4675–4688.

Born, Max (1969). *Albert Einstein Max Born Briefwechsel 1916–1955*. Nymphenburger Verlagshandlung, München.

Earman, John and Eisenstaedt, Jean (1999). Einstein and Singularities. *Studies in the History and Philosophy of Modern Physics* **30**, 185–235.

Einstein, Albert (1918). Über Gravitationswellen. *Königlich Preussische Akademie der Wissenschaften (Berlin). Sitzungsberichte*, 154–167.

Einstein, Albert (1919). Spielen Gravitationsfelder im Aufbau der materiallen Elementarteilchen eine wesentliche Rolle? *Königlich Preussische Akademie der Wissenschaften (Berlin). Sitzungsberichte*, 349–356.

Einstein, Albert (1922). *The Meaning of Relativity: Four Lectures delivered at Princeton University May, 1921*. Methuen, London.

Einstein, Albert (1936). Physics and Reality. *Journal of the Franklin Institute* **221**, 349–382. Translation of Physik und Realität. *Journal of the Franklin Institute* **221**, 313–347 (1938). Reprinted in Albert Einstein (1938) *Ideas and Opinions*. Dell, New York, 283–315.

Einstein, Albert (1993). *The Collected Papers of Albert Einstein, Vol. 5*. Martin J. Klein, A. J. Kox and Robert Schulmann eds. Princeton University Press, Princeton, N.J.

Einstein, Albert (1998). *The Collected Papers of Albert Einstein, Vol. 8*. Robert Schulmann, A. J. Kox, Michel Janssen and Jószef Illy eds. Princeton University Press, Princeton, N.J.

Einstein, Albert (2002). *The Collected Papers of Albert Einstein, Vol. 7*. Michel Janssen, Robert Schulmann, Jószef Illy, Christoph Lehner and Diana Kormos Buchwald eds. Princeton University Press, Princeton, N.J.

Einstein, Albert and Grommer, Jakob (1923). Beweis der Nichtexistenz eine "uberall regulären zentrisch symmetrischen Feldes nach der Feld-theorie von Th. Kaluza." In *Scripta Universitatis atque Bibliotecae Hierosolymitanarum: Mathematics et Physica*, Vol. 1, pp. 1–5.

Einstein, Albert and Grommer, Jakob (1927). Allgemeine Relativitätstheorie und Bewegungsgesetz. *Königlich Preussische Akademie der Wissenschaften (Berlin). Sitzungsberichte*, 2–13.

Einstein, Albert and Grossman, Marcel (1913). *Entwurf einer verallgemeinerten relativitätstheorie und einer Theorie der Gravitation*. Teubner, Leipzig. Reprinted in *Zeitschrift für Mathematik und Physik* **62**, 225–259 (1914).

Einstein, Albert, Infeld, Leopold and Hoffman, Banesh (1938). The Gravitational Equations and the Problem of Motion. *Annals of Mathematics* **39**, 65–100.

Einstein, Albert and Infeld, Leopold (1949). Motions of Particles in General Relativity Theory. *Canadian Journal of Mathematics* **3**, 209–241.

Havas, Peter (1989). The Early History of the 'Problem of Motion' in General Relativity. In *Einstein and the History of General Relativity*. Don Howard and John Stachel eds. Birkhäuser Boston, 234–276.

Howard, Don (1990). Nicht Sein Kann was Nicht Sein Darf, or the Prehistory of EPR: 1909-1935: Einstein's early worries about the quantum mechanics of composite systems. In *Sixty-two years of uncertainty: Historical, Philosophical and Physical Inquiries into the Foundations of Quantum Mechanics*. Arthur I Miller ed. Plenum, New York, 61–111.

Infeld, Leopold (1941). *Quest: The Evolution of a Physicist*. Doubleday, New York.

Kennefick, Daniel (1999). Controversies in the History of the Radiation Reaction Problem in General Relativity. In *The Expanding Worlds of General Relativity*. H. Goenner, J. Renn, J. Ritter, T. Sauer eds. Birkhäuser Boston, 207–234.

Lorentz, Hendrik Antoon (1927). *Problems of Modern Physics.* Ginn and Co., Boston. Reprinted in 1967, Dover, New York.

Pais, Abraham (1982). *Subtle is the Lord . . . : The Science and the Life of Albert Einstein.* Oxford University Press, Oxford, 288–291.

Poincaré, Henri (1906). Sur la Dynamique de l'électron. *Circolo Mathematico di Palermo. Rendiconti* **21**, 129–175.

Riordan, William L. (1905). *Plunkitt of Tammany Hall.* McClure, Phillips and Co., New York.

Rowe, David (2002). Einstein's Gravitational Field Equations and the Bianchi Identities. *Mathematical Intelligencer* **24**, 57–66.

Schilpp, Paul Arthur (1949). *Albert Einstein, Philosopher-Scientist.* Cambridge University Press, London.

Stachel, John (1986). Einstein and the Quantum: Fifty years of Struggle. In *From Quarks to Quasars: Philosophical Problems of Modern Physics.* R. Colodny ed. University of Pittsburgh Press, Pittsburgh. Reprinted in *Einstein from 'B' to 'Z'.* Birkhäuser Boston, 367–402.

Weyl, Hermann (1923). *Raum-Zeit-Materie.* Springer, Berlin.

Wheeler, John (1955). Geons. *Physical Review* **97**, 511–536.

Wheeler, John (1998). *Geons, Black Holes and Quantum Foam: A life in Physics.* Norton, New York, 237–238.

Notes

[1] "My contribution concerned one essential aspect only. I furnished the proof that the problem of motion can throw no light on the quantum theory. Here my skepticism won . . . the proof held. He [Einstein] said: 'Yes, I am now convinced that we cannot obtain quantum restrictions for motion from the gravitational equations'" (Infeld 1941, p. 260).

[2] The book was translated and appeared first in its English edition, because for contractual reasons the German edition could not precede the edition published by Princeton University Press.

[3] Admittedly the Bianchi identities were not known generally to physicists in the first couple of years after 1916, but they were presumably known to Einstein by 1921 since he had received two letters concerning their existence in 1918 [Rudolf Förster to Einstein, 16/2/18 (Einstein 1998, Doc. 463) and Friedrich Kottler to Einstein, 30/3/18 (Einstein 1998, Doc. 495); see Rowe (2002) for a full account of the multiple rediscoveries of these important identities.]

[4] Typescripts of the two popular lectures are preserved in the Einstein Archive as (EA 4 016) and (EA 4 017), published as Appendix C to Einstein 2002. Abstracts, in English (the lectures were all given in German), of the three technical lectures are preserved as (EA 5 027). See annotations to Doc. 71 of Einstein 2002 for citations of newspaper accounts, including synopses, of the lectures.

[5]Havas (1989) has observed that Einstein was unaware that, strictly speaking, the axiomatic approach is not valid because even in the linearized version of GR, the field equations do impose constraints on the motion of bodies. Einstein always believed that the motion constraints were entirely a consequence of the non-linearity of the theory and could be ignored when dealing with linearized gravity.

[6]In his autobiography Infeld says that all of the basic ideas for the paper were Einstein's, see quote in note 1 above.

[7]Don Howard has emphasized how Einstein's opposition to quantum mechanics was a "highly principled" one (Howard 1990 86-91). It may be that Einstein's brand of opportunism played into his rejection of quantum mechanics. It certainly seems to have surprised contemporaries that a theory which built on much of Einstein's own work should have been rejected by him on what seemed very dogmatic grounds, especially since the new theory had the marvelous property that it actually worked and greatly expanded the number of calculations which theorists could accomplish. Einstein himself seems to have expected originally that quantum theory would have to modify theories like general relativity. But by the early 1920s he seems to have already begun harboring the hope that his brilliant success with general relativity would light the way to a classical explanation of quantum phenomena. So his opposition to quantum mechanics also had an opportunistic, as well as a principled flavor. He felt "ashamed" of successes built on unprincipled opportunism, his own physicist's opportunism being of an untypically principled variety.

[8]George Washington Plunkitt (1842–1924), the man who left us the immortal phrase "I seen my opportunities and I took 'em," which summed up his definition of honest graft, was a Tammany Hall politician in New York whose political philosophy is recorded in (Riordan, 1905, pp. 3–10).

8

A Note on General Relativity, Energy Conservation, and Noether's Theorems

Katherine Brading

Department of Philosophy, University of Notre Dame, Notre Dame, IN, 46556, U.S.A.;
kbrading@nd.edu

The variational problem posed by Emmy Noether in her seminal 1918 paper leads to three theorems, two of which she presents in that paper and the third of which is due to F. Klein, also in 1918.[1] The origins of these theorems lie in the discussions of Klein, Noether, D. Hilbert and A. Einstein over the status of energy conservation in generally covariant theories such as General Relativity. In this paper I will outline one thread of this discussion and show how the three theorems of Noether and Klein can be brought to bear. The particular thread of interest begins with Klein's observation (in his response to Hilbert's (1916) first note on the foundations of physics) that the energy conservation law associated with Hilbert's energy vector is a mathematical identity, in contrast to the familiar energy conservation laws of mechanics which are not identities.[2] These two aspects—the claim that energy conservation is an identity, and the claim that this marks a contrast with other theories—are picked up by Hilbert and by Einstein, and are the subject of this note.

8.1 Historical Background

Klein's 1917 response to Hilbert[3] includes a section specifically on Einstein's theory (Klein 1917, 476–477, comment 9) in which he considers the energy conservation law found in Einstein's 1916 paper "Die Grundlagen der allgemeinen Relativitätstheorie," consisting of the vanishing of the divergence of two terms:

$$\partial_\nu(T_\sigma^\nu + t_\sigma^\nu) = 0 \tag{8.1}$$

where T_σ^ν and t_σ^ν are the so-called energy components associated with the electromagnetic and gravitational fields, respectively. Using the field equations, the "energy," can be re-written as:

$$T_\sigma^\nu + t_\sigma^\nu = -\partial_\rho \left(\frac{\partial G^*}{\partial g_{,\rho}^{\mu\sigma}} g^{\mu\nu} \right) \tag{8.2}$$

where G^* is the gravitational Lagrangian depending on the $g^{\mu\nu}$ up to first derivatives only (see Einstein 1916). Einstein and Klein agree that the divergence of the right-hand side of (8.2) vanishes as an identity:

$$\partial_\rho \partial_\nu \left(\frac{\partial G^*}{\partial g^{\mu\sigma}_{,\rho}} g^{\mu\nu} \right) \equiv 0. \tag{8.3}$$

However, for Klein this further implies that (8.1) holds as an identity. He points out the relationship between Einstein's terms T^ν_σ and T^ν_σ and those appearing in his own treatment, and concludes that Einstein's energy conservation law is therefore an "identity."

In 1917 Klein had begun a correspondence with Einstein,[4] and on 13 March 1918 Einstein writes to Klein beginning his letter as follows:[5]

> Highly esteemed Colleague,
> It was with great pleasure that I read your extremely clear and elegant explanations on Hilbert's first note. However, I do not find your remark about my formulation of the conservation laws appropriate. For equation (8.1) is by no means an identity, no more so than (8.2); only (8.3) is an identity.

Klein replies to Einstein immediately (20 March 1918),[6] attempting to clarify his point, the essence of which is that Einstein's conservation law (8.1) can be re-expressed as the divergence of two terms: a term which itself vanishes via the field equations (hence the vanishing of the divergence of this term is "physically meaningless"), and a term whose divergence vanishes identically. Hence the taking of the divergence does not have physical significance. Einstein replies on 24 March 1918,[7] writing that he "does not concede" that either Klein's relations or his (the relations (8.1)) are "devoid of content." Rather, he says, "What they contain is *a part* of the content of the field equations."[8]

After this letter from Einstein the correspondence on this issue slows down, but Klein continues to work on it with the assistance of Noether. In 1918 Noether and Klein each publish papers that together contain three theorems, the result of work that they had been doing together.[9] On 15 July, Klein writes to Einstein with the reasoning found in his 1918 paper that in essence leads to the Boundary theorem (see below). Further details and discussion of the Klein–Einstein correspondence during this period leading to the Noether and Klein papers, and of the crucial role played by Noether, can be found in (Rowe 1999, see especially pp. 212–28). The content of these papers enables us to resolve both aspects of the story mentioned above, but first let us mention the historical background to the second aspect.

In his reply to Klein, Hilbert (Klein 1917, 477–482) agrees with Klein,[10] and goes further, postulating that conservation of energy holding "identically" is *characteristic* of any generally covariant theory. He writes:[11]

"With your considerations on the energy theorem I am in full factual agreement: with Emmy Noether, whose help I called upon for clarification of questions pertaining to the analytical treatment of my energy theorem more than a year ago, I found accordingly that the energy components set up by me, just as those of Einstein, can be formally transformed by means of the Lagrangian differential equations ... of my first contribution, into expressions whose divergence *identically*, that is without reference to the Lagrangian equations [...] vanishes.

"Since on the other hand the energy equations of classical mechanics, of the theory of elasticity, and of electrodynamics, are fulfilled only as a consequence of the Lagrangian differential equations of these problems, then it is justified if you accordingly do not recognise in my energy equations the analogues of those of your theory. Certainly I maintain that for *general* relativity, that is, in the case of general invariance of the *Hamiltonian* function, [such] energy equations ... in general do not exist ... I might designate this circumstance as a characteristic trait of the general theory of relativity. For my assertion, mathematical proof should be adduced."

Once again, Einstein is in disagreement with Hilbert and Klein. In his letter to Klein of 13 March 1918, Einstein insists that

"The relations here are exactly analogous to those for nonrelativistic theories."

As we shall see below, Noether's 1918 paper is explicitly concerned with giving the mathematical proof that Hilbert sought for his claim.

8.2 Discussion[12]

We now turn our attention to how to resolve these two related disagreements between Einstein, Hilbert and Klein, using results based on the 1918 papers of Klein and Noether entitled "On the differential laws for conservation of momentum and energy in Einstein's theory of gravitation" and "Invariant variation problems," respectively.

Klein's paper is concerned with results that follow for generally covariant theories, and in particular General Relativity.[13] The diffeomorphism freedom of General Relativity is a local symmetry in the sense that the symmetry depends on arbitrary functions of space and time. In its generalised form (i.e., applying to all Lagrangian theories that have a local symmetry), we call the theorem contained in Klein's paper the "Boundary theorem" for reasons to do with how it is derived (see Brading and Brown, 2003b). We can state this theorem as follows.

8.2.1 Boundary Theorem

If a continuous group of transformations depending smoothly on ρ arbitrary functions of time and space $p_k(x)(k = 1, 2, \ldots, \rho)$ and their first derivatives is a Noether symmetry[14] group of the Euler–Lagrange equations associated with $L(\varphi_i, \partial_\mu \varphi_i, x^\mu)$, then the following three sets of ρ relations are satisfied, one for every parameter on which the symmetry group depends:

$$\sum_i \partial_\mu \left\{ \left(\frac{\partial L}{\partial \varphi_i} - \partial_\nu \frac{\partial L}{\partial(\partial_\nu \varphi_i)} \right) b_{ki}^\mu \right\} = \partial_\mu j_{k(\text{Noether})}^\mu \qquad (8.4)$$

$$\sum_i \left(\frac{\partial L}{\partial \varphi_i} - \partial_\nu \frac{\partial L}{\partial (\partial_\nu \varphi_i)} \right) b_{ki}^\mu = j_{k(\text{Noether})}^\mu - \sum_i \left\{ \partial_\nu \left(\frac{\partial L}{\partial (\partial_\nu \varphi_i)} b_{ki}^\mu - \frac{\partial (\Delta \Lambda^\mu)}{\partial (\partial_\nu \Delta p_k)} \right) \right\},$$

(8.5)

$$\sum_i \left\{ \left(\frac{\partial L}{\partial ((\partial_\mu \varphi_i)} b_{ki}^\nu - \frac{\partial (\Delta \Lambda^\nu)}{\partial (\partial_\mu \Delta p_k)} \right) - \left(\frac{\partial L}{\partial (\partial_\nu \varphi_i)} b_{ki}^\mu - \frac{\partial (\Delta \Lambda^\mu)}{\partial (\partial_\nu \Delta p_k)} \right) \right\} = 0 \quad (8.6)$$

where the infinitesimal transformation $\delta_0 \varphi_i$ is given by

$$\delta_0 \varphi_i = \sum_k \left\{ a_{ki}(\varphi_i, \partial_\mu \varphi_i, x) \Delta p_k(x) + b_{ki}^\nu(\varphi_i, \partial_u \varphi_i, x) \partial_\nu \Delta p_k(x) \right\}, \quad (8.7)$$

Δp_k indicating that we are considering infinitesimal transformations, the a_{ki} and b_{ki}^μ depending on the particular transformation in question, and $j_{k(\text{Noether})}^\mu$ is the "Noether current"[15] associated with the kth arbitrary function:

$$j_{k(\text{Noether})}^\mu := -\sum_i \left\{ \frac{\partial L}{\partial (\partial_\mu \varphi_i)} \frac{\partial (\delta_0 \varphi_i)}{\partial (\Delta p_k)} + L \frac{\partial (\delta x^\mu)}{\partial (\Delta p_k)} - \frac{\partial (\Delta \Lambda^\mu)}{\partial (\Delta p_k)} \right\}. \quad (8.8)$$

The terms in Λ^μ occur when the action associated with the Lagrangian L is not strictly invariant under the transformations being considered, instead picking up a divergence term. This is the case for the so-called Einstein $\Gamma\Gamma$ action, for example.[16] The above three identities (8.4)–(8.6), along with that of Noether's second theorem (see below), are not independent of one another, but we present all four here since that is how they emerged historically.

Rearranging the first identity of the Boundary theorem, equation (8.4), we get:

$$\partial_\mu \left\{ j_{k(\text{Noether})}^\mu - \left(\frac{\partial L}{\partial \varphi_i} - \partial_\nu \frac{\partial L}{\partial (\partial_\nu \varphi_i)} \right) b_{ki}^\mu \right\} = 0. \quad (8.9)$$

Hence, defining

$$\Theta_k^\mu := j_{k(\text{Noether})}^\mu - \left(\frac{\partial L}{\partial \varphi_i} - \partial_\nu \frac{\partial L}{\partial (\partial_\nu \varphi_i)} \right) b_{ki}^\mu, \quad (8.10)$$

we have that

$$\partial_\mu \Theta_k^\mu = 0 \quad (8.11)$$

holds identically. From this, we infer the existence of the so-called "superpotentials" $U_k^{\mu\nu}$, such that

$$\Theta_k^\mu = \partial_\nu U_k^{\mu\nu}, \quad (8.12)$$

where

$$\partial_\mu \partial_\nu U_k^{\mu\nu} = 0 \quad (8.13)$$

holds identically. These mathematical manipulations allow us to re-write the Noether current in the following form:

$$j^{\mu}_{k(\text{Noether})} = \left(\frac{\partial L}{\partial \varphi_i} - \partial v \frac{\partial L}{\partial (\partial_v \varphi_i)} \right) b^{\mu}_{ki} + \partial_v U^{\mu v}_k. \tag{8.14}$$

In other words, the Noether current can be expressed as consisting of a term which vanishes when the field equations are satisfied,

$$\frac{\partial L}{\partial \varphi_i} - \partial_{\mu} \frac{\partial L}{\partial (\partial_{\mu} \varphi_i)} = 0, \tag{8.15}$$

and a term whose divergence vanishes identically.

Now consider the conservation law[17]

$$\partial_{\mu} j^{\mu}_{k(\text{Noether})} = 0. \tag{8.16}$$

Given that the Noether current can be re-written in the form (8.14), we see that (8.16) can be understood as the vanishing of the divergence of two contributions. The first contribution vanishes when the field equations are satisfied without any need to take the divergence; the divergence of the second contribution vanishes identically. We can therefore re-express Klein's concern over the status of Einstein's conservation law as a more general point about conservation laws for Noether currents associated with local symmetries (i.e., where the k subscript relates to an arbitrary function of space and time p_k). The Kleinian claim is that because we can re-write the Noether current in the above form, the taking of the divergence does not lead to a physically significant result; the conservation law (8.16) therefore lacks physical significance.

At least a part of Einstein's response seems to be that (8.16) holds only when the field equations are satisfied, and that we are therefore making use of physically significant information in order to move from (8.14) to (8.16). This is true, but it doesn't address the full weight of the problem: the term of the Noether current involving the Euler–Lagrange equations vanishes on-shell *without* any need to take the divergence of the Noether current. Taking the divergence plays a role only with respect to the second term, and there the divergence vanishes identically. We are back to the question: wherein lies the physical content in taking the divergence of the Noether current and finding that the resulting expression vanishes?

I think that the right thing to say at this point is as follows. We have shown that whenever we have a local symmetry, the associated Noether current can be re-written in the form (8.14) such that when the field equations are satisfied

$$j^{\mu}_{k(\text{Noether})} = \partial_v U^{\mu v}_k. \tag{8.17}$$

Part of the Kleinian worry is that the associated continuity equation for $j^{\mu}_{k(\text{Noether})}$ lacks physical content because of (8.13). But notice: while it is true that we can always write an expression of the form (8.17) when the field equations are satisfied, there remains the question of whether, and if so when, this equation expresses a physically significant relation. So far in doing the re-writing all we have done is mathematics, and only mathematics. The relation (8.17) gains *physical* significance only when it holds "not as an identity or definition, but as a field equation postulated to relate two separate systems" (Deser 1972, p. 1082). Consider, for example, the Maxwell field equations

$$J^\mu = \partial_\nu F^{\mu\nu}. \tag{8.18}$$

These equations are of the form (8.17), and

$$\partial_\mu \partial_\nu F^{\mu\nu} = 0 \tag{8.19}$$

holds simply in virtue of the antisymmetry of $F^{\mu\nu}$. Nevertheless, we do not say that conservation of electric charge is a mathematical identity without physical significance. This is because the equations (8.18) are not a mere mathematical re-expression of the current J^μ; they express a physically significant relation between two different types of field: on the left-hand side we have a current, J^μ, depending on the matter fields carrying the electric charge, and on the right-hand side we have an expression depending on the electromagnetic fields, $F^{\mu\nu}$. Thus, the current conservation law follows via (8.18) and (8.19), and since (8.18) is physically significant so is the current conservation law.

Similarly in the case of General Relativity, the re-expression of energy-momentum through a relation of the form (8.17) has physical content because it gives a relation between the behaviour of the metric and the matter fields, it is a field equation with physical content, and hence the conservation law that follows from it (via an identity for the right-hand side) also has physical content.

This is, I believe, how we should understand the first aspect of the story, concerning the claim that energy conservation is an identity. Turning now to the second aspect, the contrast with other theories alleged by Hilbert and disputed by Einstein, we need to look at Noether's paper (Noether 1918). In that paper Noether proved two theorems, the first holding with respect to the global symmetries of a theory, and the second holding with respect to local symmetries. We may state these two theorems as follows.[18]

Noether's First Theorem

If a continuous group of transformations depending smoothly on ρ constant parameters ω_k $(k = 1, 2, \ldots, \rho)$ is a Noether symmetry group of the Euler–Lagrange equations associated with $L(\varphi_i, \partial_\mu \varphi_i, x^\mu)$, then the following ρ relations are satisfied, one for every parameter on which the symmetry group depends:

$$\sum_i \left(\frac{\partial L}{\partial \varphi_i} - \partial_\mu \frac{\partial L}{\partial(\partial_\mu \varphi_i)} \right) \frac{\partial(\delta_0 \varphi_i)}{\partial(\Delta \omega_k)} = \partial_\mu j^\mu_{k(\text{Noether})}, \tag{8.20}$$

where $\Delta \omega_k$ indicates that we are taking infinitesimal symmetry transformations,

$$\delta_0 \varphi_i = \frac{\partial(\delta_0 \varphi_i)}{\partial(\Delta \omega_k)} \Delta \omega_k, \tag{8.21}$$

and where $j^{\mu}_{k(\text{Noether})}$ is the Noether current (8.8), the arbitrary functions p_k replaced by the arbitrary parameters ω_k.

Noether's first theorem is widely known for the general connection it makes between symmetries and conservation laws. When the left-hand side of (8.20) vanishes (for example via the field equations, but see Brown and Brading (2002), for a more detailed discussion) we arrive at a conservation law (8.16). This was not the main purpose of her paper, however. Rather, Noether was providing the proof that Hilbert has asked for concerning his conjecture, and for that we need also her second theorem.

Noether's Second Theorem

If a continuous group of transformations depending smoothly on ρ arbitrary functions of time and space $p_k(x)$ $(k = 1, 2, \ldots, \rho)$ and their first derivatives is a Noether symmetry group of the Euler–Lagrange equations associated with $L(\varphi_i, \partial_\mu \varphi_i, x^\mu)$, then the following ρ relations are satisfied, one for every parameter on which the symmetry group depends:

$$\sum_i \left(\frac{\partial L}{\partial \varphi_i} - \partial_\mu \frac{\partial L}{\partial(\partial_\mu \varphi_i)} \right) a_{ki} = \sum_i \partial_\nu \left\{ b^\nu_{ki} \left(\frac{\partial L}{\partial \varphi_i} - \partial_\mu \frac{\partial L}{\partial(\partial_\mu \varphi_i)} \right) \right\} \qquad (8.22)$$

where the infinitesimal transformation $\delta_0 \varphi_i$ is given by (8.7), above.

As we saw in Section 8.1 above, Hilbert's conjecture was that the difference between generally covariant theories such as General Relativity, and earlier theories such as classical mechanics, can be characterised by the differing status of energy conservation: in generally covariant theories the energy conservation law can be re-written, using the Euler–Lagrange equations, such that it holds "identically." The final section of Noether's paper concerns this "Hilbertian assertion" quoted above (see section 1). She writes:[19]

> From the foregoing we finally obtain the proof of a Hilbertian assertion concerning the connection between the lack of proper energy theorems and "general relativity," and this even in a generalized group-theoretic version.

Where Hilbert uses the term "identically," we shall mean that the current conservation law can be re-written in the form (8.14), this being what we concluded above based on the clarifications made by Klein. The proof then proceeds as follows. In theories that *do not* admit a local symmetry group, only Noether's first theorem (and not her second) can be obtained. In such theories, we apply Noether's first theorem to a global symmetry and obtain a corresponding relation of the form (8.20) from which we may proceed to a current conservation law.[20] However, in theories that admit a local symmetry group we can do two things: the first theorem can be applied to the global subgroup, from which we may proceed to conservation laws, and since the second theorem also applies we can combine it with the first theorem to arrive at what Earman has called "Noether's third theorem."[21] We equate the left-hand sides of the equations of the first and second theorem—and the consequence is just the first identity of the Boundary theorem (8.4). In other words, *only* when the global symmetry group is a subgroup of a local symmetry group can we re-write the Noether current in the form discussed by Klein, i.e., in the form (8.14). In classical mechanics (for example), the global space and time symmetry group is *not* a subgroup of a local symmetry group, so the energy conservation law (associated with global time translations) cannot be re-written in the form (8.14). The form (8.14) is indeed characteristic of generally covariant theories, or indeed of any theory with a local symmetry structure. In this way, Noether proved Hilbert's conjecture, and generalised it beyond the case of general covariance and energy conservation to all continuous global and local symmetry groups.[22]

8.3 Conclusions

The subject of this note has been a small historical thread in the long and complex story of the status of energy conservation in General Relativity, concerning two related claims made by Klein and Hilbert: that the energy conservation law is an identity in generally covariant theories, and that this marks a contrast with other (earlier) theories. Both these claims were disputed by Einstein. We have seen how three theorems proved by Noether and Klein can be brought to bear on this disagreement, showing that:

(1) Klein's worry over the physical significance of the energy conservation law in General Relativity was perhaps not adequately addressed by Einstein, even though in the end we side with Einstein against Klein, and

(2) the possibility of re-writing the energy conservation law in the form that so worried Klein does indeed depend upon the *local* symmetry structure of General Relativity.

Acknowledgements. This note is part of joint research projects with Harvey Brown (on Noether's theorems and the Boundary theorem) and Tom Ryckman (on the historical background to these theorems and their relationship to the work of Hilbert). I am most grateful to them both for long and detailed discussions, and to Tom Ryckman for his patient assistance in translating various passages. I am also grateful to Michel Janssen and Tilman Sauer for ongoing discussions of issues relating to this paper. Finally, I would like to thank Wolfson College, Oxford, for their support during the period that this paper was written.

References

Barbashov, B. M., and Nesterenko, V. V. (1983). Continuous Symmetries in Field Theory. *Fortschritte. der Physik* **31**, 535–567.

Brading, K., and Brown, H. R. (2003a). Symmetries and Noether's theorems. In *Symmetries in Physics: Philosophical Reflections*. Katherine Brading and E. Castellani, eds. Cambridge University Press.

Brading, K., and Brown, H. R. (2003b): Noether's Theorems, Gauge Symmetries, and General Relativity. Manuscript.

Brown, H. R. and Brading, K. (2002). General Covariance from the Perspective of Noether's Theorems. *Diálogos* **79**, 59–86.

Deser, S. (1972). Note on current conservation, charge, and flux integrals. *American Journal of Physics* **40**, 1082–1084.

Earman, J. (2003). Tracking down gauge: an ode to the constrained Hamiltonian formalism. In *Symmetries in Physics: Philosophical Reflections*. Katherine Brading and E. Castellani, eds. Cambridge University Press.

Einstein, A. (1916). Die Grundlagen der allgemeinen Relativitätstheorie. *Annalen der Physik* **49**, 769–822. Translated as The Foundation of the General Theory of Relativity in The Principle of Relativity. H. A. Lorentz et al. Eds. Dover, New York, 111–164.

Einstein, A. (1998). *The Collected Papers of Albert Einstein*, vol. 8, R. Schulmann et al. eds. Princeton University Press, Princeton, New Jersey.

Hilbert, D. (1916). Die Grundlagen der Physik. (Erste Mitteilung.). *Königliche Gesellschaft der Wissenschaften zu Göttingen. Mathematisch-physikalische Klasse. Nachrichten*, 395–407.

Klein, F. (1917). Zu Hilberts erster Note über die Grundlagen der Physik. *Königliche Gesellschaft der Wissenschaften zu Göttingen. Mathematisch-physikalische Klasse. Nachrichten*, 469-82.

Klein, F. (1918). Über die Differentialgesetze für die Erhaltung von Impuls und Energie in der Einsteinschen Gravitationstheorie. *Königliche Gesellschaft der Wissenschaften zu Göttingen. Mathematisch-physikalische Klasse. Nachrichten* 171–89. Translated by J. Barbour as "On the Differential Laws for Conservation of Momentum and Energy in Einstein's Theory of Gravitation," ms.

Noether, E. (1918). Invariante Variationsprobleme. *Königliche Gesellschaft der Wissenschaften zu Göttingen. Mathematisch-physikalische Klasse.* Nachrichten, 235–57. Translated by M. A. Tavel (1971) as "Noether's Theorem" In *Transport Theory and Statistical Physics*. **1**, 183–207. Page numbers refer to the English translation.

Rowe, D. (1999). The Göttingen Response to General Relativity and Emmy Noether's Theorems. In *The Symbolic Universe: Geometry and Physics 1890–1930*. J. Gray, ed. Oxford University Press, Oxford.

Sauer, T. (1999). The Relativity of Discovery: Hilbert's First Note on the Foundations of Physics. Archive for History of. Exact Sciences 53, 529–75.

Trautman, A. J. (1962). Conservation Laws in General Relativity. In *Gravitation: an Introduction to Current Research*. L. Witten, ed. Wiley, New York.

Notes

[1] For the variational problem and derivations of the theorems, see Brading and Brown (2003a) and (2003b).

[2] Klein (1918) p. 475.

[3] On Hilbert's first note on the foundations of physics, see Sauer (1999).

[4] See Rowe (1999), pp. 210–213.

[5] Einstein (1998), document 480, pp. 494–5 of the English translation. Equation numbers are ours; in Einstein's letter the numbers are (22), (23) and (24) and refer to the equations appearing in Klein's note.

[6] Einstein (1998), document 487, pp. 503–507 of the English translation.

[7] Einstein (1998), document 492, pp. 512–514 of the English translation.

[8] Einstein then goes on to give reasons in favour of his own version of the divergence relations rather than Klein's, but the difference between the two does not concern us here.

[9] Noether's paper was originally submitted to the Göttingen Society by Klein in January 1918. She continued to work on it, presenting it to the Society in July and finishing the paper by the end of September (see Rowe, 1999, p. 221).

[10] The friendly tone of this exchange masks the deep criticisms that Klein was making of Hilbert's work (see Rowe, 1999, p. 212).

[11] Hilbert's answer to Klein (1917), p. 477. Thanks to Tilman Sauer and to Tom Ryckman for translating this passage.

[12] The following discussion is reproduced in its essentials in Brading and Brown (2003a).

[13] Section 7 of Klein (1918) is about the relationship between Einstein's formulation of the conservation theorems and Klein's derivations.

[14] A "Noether symmetry" is a symmetry of the field equations that satisfies the requirement that the change in the action arising from an infinitesimal symmetry transformation is at most a surface term. See Brading and Brown (2003b).

[15] See Noether's first theorem, below.

[16] For further details and explanation, and for the derivation of the Boundary theorem, see Brading and Brown (2003b), where references to related results can also be found.

[17] More precisely, this is a continuity equation, and in physics (as opposed to mathematics) the term 'conservation law' is often reserved for expressions of the form $\frac{d}{dt}Q = 0$, where Q is here a conserved charge, these being obtained from continuity equations subject to certain conditions (see Brading and Brown, 2003b).

[18] For the derivations see Barbashov and Nesterenko (1983); Brading and Brown (2003b), and Trautman (1962).

[19] Noether 1918, p. 253–4, p. 201 of the English translation (Tavel) but amended translation (my thanks to Bjoern Sundt and Tom Ryckman).

[20] Note that there is no guarantee that the result is interesting—see Brading and Brown (2003b).

[21] See Earman (2003).

[22]Picking up on Hilbert's use of the term "proper" for the energy conservation laws in non-generally covariant theories, Noether terms such relations "improper." The origins and significance of this terminology in Hilbert's work is the subject of ongoing joint work with Tom Ryckman.

Weyl vs. Reichenbach on Lichtgeometrie

Robert Rynasiewicz

Department of Philosophy, Johns Hopkins University, Baltimore, MD 21218, U.S.A.;
ryno@jhu.edu

9.1 Introduction

In certain respects, what I have to offer is but a vignette of a professional relationship
gone sour. What makes it more than just a human interest story is the degree to which
the substantive issues involved appeal to the mathematical imagination. Apart from a
handful of toy examples, we are used to thinking in terms of global topologies other
than R^4 in relativity only in the transition from special to general theory. However, if
only light-cone structure is considered, the argument can be made that the most natural
arena for special relativity is the compactification of R^4 by the addition of a light-cone
at conformal infinity.

Weyl and Reichenbach were hardly of the same ilk, either philosophically or
mathematically.[1] But their common interest in the foundations of relativity theory
and geometry kept them in communication during the early 1920s. By mid-decade,
though, they had had it with one another, exchanging unkind words in print. Here I
shall explore their respective views on Lichtgeometrie and the physical foundations of
Minkowski geometry, disagreements over which can be seen to be largely responsible
for their falling out.

9.2 The Program of Reichenbach's "Bericht"

At the Deutsche Physikertag in Jena of 1921, held from the 18th through the 24th
of September, Reichenbach presented his initial sketch for a novel axiomatization of
relativity in a brief Vortrag, the record of which was published the following December in the *Physikalische Zeitschrift* with the title "Bericht über eine Axiomatik der
Einsteinschen Raum-Zeit-Lehre" (Reichenbach 1921). In keeping with its brevity, the
"Bericht," as I will henceforth refer to it, sets out only axioms and definitions without
indicating, even in outline, proofs of the considerable claims alleged to follow. For our
purposes, there are two notable features of Reichenbach's approach in the "Bericht."

First is the epistemological or "erkenntnislogische" style of the axiomatization.
The fundamental idea is to implement as axioms only propositions that make direct

contact, at least ideally so, with experimentally testable facts of experience. Thus, in the very second sentence he explains:

> Die Axiome enthalten dann diejenigen fundmaentalen Tatsachen, deren Bestehen der Theorie Existenzberechtigung verleiht; sie sind prinzipiel empirische Behauptungen, die also durch das Experiment nachgeprüft werden können.[2] (p. 683)

For the conceptual development of the theory, however, definitions need to be introduced, which, in contrast to the axioms, are "willkürlich Gedankenbildungen, die grundsätzlich weder empirisch bestätig noch widerlegt werden können."[3] (pp. 683–4) Reichenbach will later introduce the term of art, "constructive" axiomatizion, for this manner of proceeding (Reichenbach 1924). Its virtue (emphasized in Reichenbach's subsequent writings, but not in the "Bericht") is that it permits the separation of the factual content from the conventional components of the theory.[4]

The second notable feature is the particular grouping of the Axioms into two classes, "Lichtaxiome" and "Materialaxiome." The first class contains assertions solely about the physical properties of light, without making any reference to material objects. The second class expresses claims about the behavior of rigid rods and natural clocks. According to Reichenbach, it can be shown that a complete "Raum-Zeit-Lehre," can be constructed on the basis of the Lichtaxiome alone, that is, a pure "Lichtgeometrie." On the other hand, as is quite familiar in pre-relativistic physics, the behavior of rigid rods and natural clocks (together with some implicit criterion regarding distant-simultaneity) can also be used to underwrite a "Raum-Zeit-Lehre." The significance of the specific "Materialaxiome" chosen by Reichenbach is that, cumulatively, they entail the identity of the Raum-Zeit-Lehre which they give rise to with that of the "Lichtgeometrie" developed on the basis of the Lichtaxiome. The payoff, as Reichenbach sees it, is two-fold. The first is that such division of axioms is possible in the first place. The second is that, although the Lichtaxiome have been thoroughly confirmed by experiment, it has not yet been possible to completely confirm the full set of Materialaxiome.[5] Thus, although the theory of relativity can be understood to be a valid and complete theory insofar as it is founded on the Lichtaxiome, open issues remain as to the behavior of material structures in relativistic space-time. One lacuna Reichenbach mentions in the "Bericht" is confirmation of the transverse Doppler effect using Canal-rays. In addition, although unmentioned there, we might surmise that there are issues raised by Weyl's unified theory, specifically in terms of the adjustment [Einstellung] of material structures to the gravitational-gauge field.

I wish now to indulge your patience briefly in order to sketch the nature of the Lichtaxiome and the development of the claims made in the "Bericht" in connection with the resulting Lichtgeometrie, since this is crucial for the theme to be developed in the remainder. Reichenbach takes as his primitives two notions: first, the set of all world-lines of all possible observers (in whatever state of motion) in space-time, and second signaling between these world-lines by means of light-rays. The principal goal is to lay down conditions sufficient to define an inertial system of world-lines. The Lorentz transformations then emerge as the coordinate transformations that take one world-line from one inertial system to another.

In briefest outline, Reichenbach accomplishes this as follows.[6] A clock is defined to be any mechanism that induces an order preserving map from a given world-line to the real numbers. The first problem is to establish a system of world-lines possessing a common time-function. As a matter of definition, one calls a system of world-lines an "auf A bezogenes System," if, for each world-line B of the system, the time lapse of light transmission from A to B and back is always the same according to the arbitrarily adopted clock at A. Reichenbach then introduces the axiom that it is possible to choose a clock (in the generalized sense above) at A such that an "auf A bezogenes System" is also an "auf B bezogenes System" for the other world-lines B of the "auf A bezogenes System" (Axiom III). A system that satisfies this axiom is called a "Normalsystem," and the clock used to establish it as such is said to be a "Normaluhr."[7] At this point it is easy to see that, in Minkowski space-time, an inertial frame qualifies as a "Normalsystem." But so do other systems of world-lines, for example, systems expanding with respect to one another with a constant velocity, if one implements a "Normaluhr" that registers time as a logarithmic function of proper time. Hence further conditions need to be imposed.

To this end, Reichenbach first introduces as a definition the Einstein criterion of clock synchronization and imposes the "round-trip" axiom in order to guarantee that this method of clock synchronization is transitive (Axiom IV).[8] The final step is to define a criterion of equal distances between world-lines in terms of light signaling[9] and impose the axiom that it is possible to find a "Normalsystem" in which the resulting metric is Euclidean (Axiom V).

The punch line now, though stated as a definition, is that such a "Normalsystem" is to be called an "Inertialsystem," the points of which are to be said to be "zueinander ruhend," and the "Normalzeit" belonging to the system is to be called "gleichförmig." Of course (although I do not mean to intimate here that Reichenbach does not realise it), in order for this "definition" to have any significance for the project at hand, a representation theorem is in order to the effect that the only realizations of such "Inertialsysteme" in a space-time with Minkowski metric are classes of parallel time-like geodesics. As mentioned earlier, no proofs are given in the "Bericht." A footnote in the text refers the reader to "eine ausführliche Veröffentlichung der ganzen Untersuchung" which "wird später erfolgen." (Reichenbach 1922, 684, Note 1.) Such detailed presentation of the entire investigation did not appear until 1924 in the monograph *Axiomatik der relativistischen Raum-Zeit-Lehre* (Reichenbach 1924).

9.3 Weyl and Reichenbach, 1921–1925

In the introduction to his *Axiomatik*, there is a footnote that may provide a clue as to Reichenbach's inspiration to base his axiomatization on the construction of a Lichtgeometrie. In the main text, he pays special attention to the fact that in the presentation to come the space-time metric of special relativity is defined solely by means of light signals, and thus the latter alone suffice for the definition of simultaneity, as well as for those of the uniformity of time and the equality of spatial intervals.[10] The footnote refs the reader to the appendix of the fourth edition of Weyl's *Raum-Zeit-Materie* (Weyl 1921a) for a suggestion that this is indeed possible.[11]

There are in fact two appendices to the fourth edition, but it is clearly the first that Reichenbach has in mind. There we find:

> Um in der speziellen Relativitätstheorie die "normalen" Koordinatensysteme vor allen andern auszuzeichnen, ... , kann man nicht bloß der starren Körper, sondern auch der Uhren entraten.[12] (Weyl 1921a, 285)

Curiously, what Reichenbach does not mention is that the brief demonstration in the appendix takes advantage of not just light propagation, but also the inertial trajectories given by force-free point-masses. Perhaps the intended message is that Reichenbach has seen how to go even one-step further in eliminating the need for the latter.

In support of this clue, we know that a correspondence between Reichenbach and Weyl had commenced some nine months prior to the "Bericht" Vortrag of September, 1921, a correspondence that alerts Reichenbach in February of the existence of the just published fourth edition. The hypothesis that Reichenbach drew from there his inspiration for a Lichtgeometrie, however, would need to be reconciled with the report in the preface that the investigation began in the fall of 1920, i.e., prior to the appearance of Weyl's fourth edition.

In any event, it may appear remarkable that a correspondence developed at all. Apparently, Reichenbach initiated contact by sending Weyl a complimentary copy of his primarily philosophical *Relativitätstheorie und Erkenntnis Apriori* (Reichenbach 1920). But given that he had so serverely criticized Weyl in that work, it is difficult to gauge what his intentions might have been. Those criticisms concerned Weyl's views on the relation between mathematics and physics. In order to appreciate their vitriolic tone, they deserve to be quoted at length. As background, understand that in this work, Reichenbach attempts to carve out an eclectic position that combines various elements of Kantian philosophy with empiricism. The gist of the Kantianism is that, although a priori principles are needed, they are not to be viewed as epistemologically a priori, but only as provisionally adopted principles needed to make sense of an objective world. The major theme is that developments in physics, specifically the theory of relativity, may mandate thorough-going revision of the provisionally adopted principles. In the paragraph leading into his comments on Weyl, Reichenbach's discussion of the relation of mathematics to epistemology takes on more of an empiricist ring:

> Besonders zu beachten ist hier aber der Unterschied zwischen Physik und Mathematik. Der Mathematik ist die Anwendbarkeit ihrer Sätze auf Dinge der Wirklichkeit gleichgültig, und ihre Axiome enthalten lediglich ein System von Regeln, nach dem ihre Begriffe unter sich verknüpft werden. Die rein mathematische Axiomatik führt überhaupt nicht auf Prinzipien einer Theorie der Naturerkenntnis. Darum konnte auch die Axiomatik der Geometrie gar nichts über das erkenntnistheoretische Raumproblem aussagen. Erst eine physikalische Theorie konnte die Geltungsfrage des euklidischen Raumes beantworten, und gleichzeitig die dem Raum der Naturdinge zugrunde liegenden erkenntnistheoretischen Prinzipien aufdecken.[13] (Reichenbach 1920, 72-3.)

With the very next sentence, Reichenbach launches an attack on Weyl and his unified field theory:

> Ganz falsch ist es aber, wenn man daraus, wie z. B. Weyl und auch Haas, wieder den Schluß ziehen will, daß Mathematik und Physik zu einer einzigen Disziplin zusammenwachsen. Die Frage der Geltung von Axiomen für die Wirklichkeit und die Frage nach den möglichen Axiomen sind absolut zu trennen. Das ist ja gerade das Verdienst der Relativitätstheorie, daß sie die Frage der Geltung der Geometrie aus der Mathematik fortgenommen und der Physik überwiesen hat. Wenn man jetzt aus einer allgemeinen Geometrie wieder Sätze aufstellt und behauptet, daß sie Grundlage der Physik sein müßten, so begeht man nur den alten Fehler von neuem. Dieser Einwand muß der Weylschen Verallgemeinerung der Relativitätstheorie entgegengehalten werden, bei der Begriff einer feststehenden Länge für einen unendlich kleinen Maßstab überhaupt aufgegeben wird. Allerdings ist eine solche Verallgemeinerung möglich, aber ob sie mit der Wirklichkeit verträglich ist, hängt nicht von ihrer Bedeutung für eine allgemeine Nahegeometrie ab. Darum muß die Weylsche Verallgemeinerung vom Standpunkt einer physikalischen Theorie betrachtet werden, und ihre Kritik erfährt sie allein durch die Erfahrung. Die Physik ist eben keine "geometrische Notwendigkeit"; wer das behauptet, kehrt auf den vorkantischen Standpunkt zurück, wo sie eine vernuftgegebene Notwendigkeit war. Und die Principien der Physik kann ebensowenig eine allgemein-geometrische Überlegung lehren, wie sie die Kantische Analyse der Vernüpft lehren konnte, sondern das kann allein eine Analyse der physikalischen Erkenntnis.[14] (Reichenbach 1920, 73-4.)

Yet the ensuing correspondence, or at least that which survives, proceeded politely, if not at times outright amicably. Weyl's reply of February 2, 1921 opens with the words:

> Sehr geehrte Herr Kollege!
> Endlich nach vielen Wocken komme ich dazu, Ihnen zu danken für die freundliche Zusendung Ihrer Schrift "Relativitätstheorie und Erkenntnis a priori." Ich glaube, dass bei meiner abweichenden philosophischen Grundeinstellung eine Verständigung zwischen uns nur mühsam würde zu erzielen sein; aber ich verkenne darum nicht die grosse Ehrlichkeit Ihre Bemerkungen, das Erkenntnisproblem der realen Aussenwelt scharf zu erfassen. Aber ich darf mich wohl heute auf zwei Bemerkungen Beschränken, die weniger das Philosophische als das Physikalische betreffen.[15] (Document HR 015-68-04, Archives for Scientific Philosophy, University of Pittsburgh.)[16]

The first of the two remarks that follow addresses Reichenbach's treatment of the philosophical consequences of general relativity for a priori principles. Weyl points out that Reichenbach does not consider the position of one who would take the equality of inertial and gravitational mass to be simply a surd fact.

The second remark responds to Reichenbach's characterization of Weyl's views on the relation between mathematics and physics:

Es ist gewiss nicht wahr, dass für mich, wie Sie auf S. 73 sagen, Mathematik (!!, z.B. Theorie der ζ-Funktion?) und Physik zu einer einzigen Disziplin zusammenwachsen. Ich habe nur behauptet, dass die *Begriffe* in der *Geometrie* und der Feldphysik zum Zusammenfallen kommen. (HR 015-68-04)[17]

Weyl goes on in some detail, with numerous references to the fourth edition of *Raum-Zeit-Materie* (Weyl 1921a), as to why the general framework of his unified field theory makes no a priori commitments concerning the actual geometry of the world — in short, that depends on what action principle one chooses.

Reichenbach is satisfied at least to the extent that the following year, in a now frequently discussed article (Reichenbach 1922), he takes the opportunity to retract his earlier accusation:

Ich muß jedoch meinen früher erhobenen Einwand ([Reichenbach 1920], S. 73), daß Weyl die Physik aus der Vernüpft deduzieren will, zurücknehmen, nachdem Weyl dieses Mißverständnis aufgeklärt hat ([Weyl 1921d], S. 475).[18] (p. 367)

I have not been able to determine if Reichenbach had also sent Weyl a personal retraction. Perhaps he did so early in 1922 in a letter to Weyl that appears to be no longer extant. The letter had been forwarded from Zurich to Barcelona, where Weyl was giving his Catalonian Lectures (Weyl 1923). Despite continuing and deep differences of opinion on the foundations geometry and general relativity, Weyl's postcard in reply suggests a warming cordiality. Weyl begins playfully: "Ihren Brief von 18.1 habe ich erst jetzt-und-hier erhalten." (HR 015-68-03) After filling the postcard with remarks concerning a point of continued contention between them — the status of the principle of equivalence — Weyl finishes off punningly: "Auf alles weitere komme ich zurück, wenn ich wieder in Zürich sein werde." A final extant letter from Weyl to Reichenbach from May 20 of that year (HR 015-68-02), whose substance will concern us later, contains no hint of a decline in collegiality.

The relationship takes a markedly different course after the publication of Reichenbach's *Axiomatik* in 1924. Weyl reviewed it that November in less than enthusiastic terms:

Das Bestreben der Reichenbachschen Schrift geht offenbar dahin, beim Aufbau der Raum-Zeit-Lehre möglichst klar die Tatsachen von den auf sie gegründeten Festsetzungen zu scheiden. Insofern hat sie einen erkenntnistheoretischen Hintergrund; und der Verf. ist in der Tat mit Erfolg bemüht, die Voraussetzungen der Theorie nach allen Seiten zu beleuchten. Zur Hauptsache enthält die Schrift aber nicht eine philosophische, sondern eine rein mathematische Untersuchung, und sie muß sich daher auch eine Beurteilung nach mathematischen Gesichtspunkten gefallen lassen. In dieser Hinsicht aber ist sie wenig befriedigend, zu umständlich und zu undurchsichtig. Das eigentlich Wertvolle: die Aufstellung der Axiome a), b) c) und der Übergang von ihnen zur Raum-Zeit-Messung, zum Koordinatenraum und damit zur Möbiusschen Geometrie ließe sich bequem auf ein paar Seiten durchführen, wobei die Klarheit und Verständlichkeit nur gewinnen würde.[19] (Weyl 1924, 2128)

The axioms a) b) c) are not actually Reichenbach's, but those given by Weyl earlier in the review in order to more easily convey to the reader the gist of Reichenbach's more complicated system of light axioms. Since Reichenbach will take issue with the simplicity and clarity of these, they are worth a quick glance.

Axiom a) asserts the existence of a class of world-lines such that, if at any time light signals, sent simultaneously from a world-line A and reflected around two given closed "polygonal paths" formed by world-lines of the system, return to A simultaneously, then they will do so when sent from A around the same paths at any other time. Axiom b) is the same as Reichenbach's "round-trip" axiom generalized to arbitrary closed polygonal paths, viz., light signals sent simultaneously in opposite directions. One now defines the equality of distances between world-lines in terms of equal times of to-and-fro light transit. Axiom c), like Reichenbach's Axiom V, adds the final condition that the system of laws of Euclidean geometry hold for the induced spatial geometry.

A few months later (early 1925) Reichenbach responded to Weyl's review with a sense of indignity. First, he takes issue with the characterization of his work as a primarily mathematical investigation:

> Von mathematischer Seite ist über meine Untersuchungen kein anderes Urteil erlaubt als das Urteil "richtig" oder "falsch." Um mathematische Eleganz ist es mir hier nicht zu tun — die hat in der Relativitätstheorie genug Gelegenheit gehabt, sich auszuleben, und hat jedenfalls für die erkenntnistheoretische Klärung nur beschränkte Bedeutung gehabt.[20] (Reichenbach 1925, 37)

Next he takes up the issue whose presentation of the physical axioms is to be preferred:

> In den ersten Seiten seines Referates gibt Herr Weyl eine Darstellung meiner Axiomatik, die vermutlich ein Beispiel dafür sein soll, wie ich es "bequem auf ein paar Seiten" hätte besser machen können; ich überlasse es sehr gern dem Urteil der Leser, welche von beiden Darstellungen sie weniger "umständlich und undurchsichtig" finden. Ich für mein Teil have noch immer die Klarheit eines stufenweisen Aufbaus, der mit möglichst einfachen logischen Operationen auskommt, einem schillernden mathematischen Nebel vorgezogen, mit dem mancher seine Gedanken zu umgeben vorzieht. Der Plan meiner Untersuchung ist durch die Absicht diktiert, die Resultate der physikalischen Erfahrung möglichst deutlich aufzudecken und aus jedem neuen Erfahrungssatz so viel an ableitbaren Folgerungen herauszuholen, als irgend angeht. Wenn man mit einem Minimum von Begriffen arbeitet, wird dabei mancher Schritt umständlicher werden, als wenn man von vornherein mit der Gesamtheit aller verfügbaren Hilfsmittel beginnt.[21] (Reichenbach 1925, 37-8)

The informed reader may be genuinely puzzled as to just what these "Hilfsmittel" are supposed to be for, as we have seen, Weyl's Axioms a)–c) involve no notions other than those that Reichenbach allows himself.

Finally, Reichenbach censures Weyl for having missed the whole point of his work:

> Ich halte es aber für sehr bedauerlich, wenn ein Mathematiker von Herrn Weyls Rang den Zweck einer solchen erkenntnistheoretisch-logischen Untersuchung derart verkennt und mit seiner Autorität den Versuch zu unterdrücken sucht, der mathematisch und physikalisch so fruchtbar ausgebauten Relativitätstheorie jetzt endlich auch den logischen Unterbau zu geben, der letzten Endes allein die Gewähr ihrer Gültigkeit tragen kann.[22] (Reichenbach 1925, 38)

What could have triggered such a dramatic falling out between the two figures?

Recall the mention in Weyl's review of the "transition to Möbius' geometry," something not addressed in Reichenbach's reply.

9.4 Lichtgeometrie and Conformal Infinity

A peculiar feature of the printing in Reichenbach's *Axiomatik* is that there are occasional passages in the main text which appear in reduced font of the same size as the footnotes. No rationale for this is given. Concerning the production of the work, the preface, dated March 1924, mentions delays in publishing due to the difficult economic situation. It also tells us that the investigation was begun in the fall of 1920 and completed "in essesence" in March 1923. It is not inconsistent with this that these passages represent emendations at the stage of correcting the galley proofs.

However this may have been, the first such passage is unmistakably an insertion at some stage of editing. It occurs in the introduction shortly after the above mentioned sentence bearing the footnote to Weyl and represents a significant qualification to the claim that a complete and unique Lichtgeometrie can be constructed using only coordinative definitions that involve light.[22] The inserted passage reads:

> Die Festlegung des gleichförmigen Bewegungszustandes im Raume gelingt allerdings nur, wenn man es als zulässiges Kriterium ansieht, daß die entstehende Metrik an keiner Stelle des ganzen Raumes eine Singularität besitzt. Will man ein solches Kriterium nicht benutzen, so bleiben als lichtgeometrisch gleichberechtigt zwei Scharen von räumlichen Bezugssystemen bestehen, die gegeneinander beschleunigt bewegt sind und von denen jeweils das eine, gemessen am anderen, eine innere Dehnung (bzw. Kontraktion) erfährt, also nicht starr ist. Die Auszeichnung einer der beiden Scharen als starr und gleichförmig bewegt erfolgt dann mit Hilfe körperlicher Gebilde, entweder eines starren Stabes oder einer natürlichen Uhr. Diese körperlichen Gebilde werden in Definition 18 und 19 definiert; eine genaue Untersuchung der ganzen Frage gibt Sect. 16. Es bleibt aber bestehen, daß für jedes dieser beiden Scharensysteme die raum-zeitliche Metrik allein durch das Licht definiert wird; die Lichtgeometrie läßt also nur eine Zweideutigkeit der Schar offen, während sie innerhalb jeder Schar eindeutig alle Verhältnisse festlegt.[23] (Reichenbach 1924, 10)

It is important to get clear on the logic of this comment. Suppose that, for some reason or other, the criterion that there be no "singularities" is *not* permissible. Then the whole plan of the *Axiomatik* based on pure Lichtgeometrie threatens to collapse. Either one has to introduce coordinative definitions invoking material objects, in which case there really is no self-sufficient Lichtgeometrie as a standard against which claims about the behavior of material objects is completely factual rather than conventional, or else one has to understand the notion of Lichtgeometry in a significantly weakened sense, to wit, that it does not suffice to determine even the affine structure of space-time, in which case again it does not serve its purpose. Thus, the cogency of the program hangs by the thread of the "permissibility" of the criterion regarding "singularities."

It is remarkable that *Weyl had warned Reichenbach of this back in 1922!* On the second page of the letter dated May 20, he takes up discussion of Lichtgeometrie:

> Mit der Lichtgeometrie steht es doch so (das ist den Mathematiker längst bekannt, ich habe es auch im 1. Teil meines [Catalonian] Vortrags erwähnt): Diejenigen Abbildungen, welche Nullkegel in Nullkegel überführen, sind die Möbius'schen Kugelverwandtschaften. Beim Operieren im beschränkten Gebiet (was doch allein physikalisch vernünftig ist) genügen die Nullkegel also *nicht* zur Charakterisierung der Geometrie, wohl aber *Nullkegel* und *Gerade*. Benütz man freilich den unendlichen Raum, so wie ihn Euklid annimmt (nicht mit anderen Zusammenhangsverhältnissen), so genügen die Nullkegel, weil die ähnlichen Abbildungen die einzigen Kugelverwandtschaften sind, welche die unendlichferne "Kugel" in die unendlichferne überführen.[24]

The dilemma presented to Reichenbach is how, using only the sorts of axioms permissible in a constructive axiomatization, one could establish that, globally, space-time is topologically R^4.[25]

Mathematically, the problem arises as follows. As a warm-up exercise consider the Euclidean plane with a Cartesian coordinate chart adapted to the Euclidean metric. In this chart, the Möbius transformations take the form:

$$\xi = \frac{x}{x^2 + y^2},$$
$$\eta = \frac{y}{x^2 + y^2}.$$

Under them, circles centered at the origin are mapped to circles at the origin, the unit circle being the set of fixed points, and in general the mapping is angle-preserving. Since the transformations are undefined at the origin, any curve passing through the origin will have a singular point. However, by taking the one-point compactification of the plan by adding a point at infinity, the mapping becomes globally well-behaved. the null circle at the origin is sent to the infinite circle, whose circumference is the point at infinity. A circle whose circumference passes through the origin is mapped to an infinite circle represented by a straight line passing through infinity and returning on itself, and so on. Moreover, the mapping is it's own inverse. The counterpart of the above transformations in a Cartesian chart adapted a Minkowskie metric on R^4 are:

$$\xi = \frac{x}{s^2}; \eta = \frac{y}{s^2}; \zeta = \frac{z}{s^2}; \tau = \frac{t}{s^2},$$

where

$$s^2 = t^2 - x^2 - y^2 - z^2.$$

Again the transformations are undefined at the origin. But, since the metric is indefinite, they are also undefined for an entire three-dimensional submanifold, namely the set of all points on the null-cone passing through the origin. And again, the mapping can be made globally well defined by appropriately compactifying R^4, this time by the addition of a null-cone at infinity. The resulting conformal isometries are the "Möbius'schen Kugelverwandtschaften" to which Weyl refers. The analytic treatment of these becomes a bit tricky, since one has added to R^4, not one point, but an entire three-dimensional submanifold; the transformations as given in the Cartesian coordinate representation above don't indicate directly which point at infinity corresponds to which point on the null cone through the origin. Weyl's technique, borrowed from projective geometry, is to use a five-dimensional chart of homogenous coordinates. The first appendix to his Catalonian lectures gives a brief exposition. (Weyl 1923, 62–64). Since, however, this is not an appropriate occasion to probe the analytic details, I shall confine myself to some descriptive remarks concerning the behavior in conformal space-time of the two-fold classes of world-lines to which Reichenbach refers.

Think of one of these classes as the class of "really" inertial frames and introduce a Minkowski chart adapted to one of these inertial frames.[26] Starting at the $t = 0$ hypersurface, the zero-velocity curves eventually converge at future time-like infinity, and since this is the same point as past time-like infinity, return again after an infinite lapse of proper-time to the $t = 0$ hypersurface. Consider now the image of this frame under the Möbius tranformations. The effect on the $t = 0$ hypersurface is just the same as the Möbius transformations for Euclidean three-space. The coordinate origin is mapped to spatial infinity and the points of the hypersurface correspondingly inverted. If one follows the image points of the inertial frame toward the future, they trace out hyperbolic trajectories whose asymptotes are null curves. Thus, viewed in the Minkowksi chart, the image of the inertial frame expands radially from the $t = 0$ hypersurface and its world-lines never enter into the time-like future of the origin. (Similarly, by time-inversion symmetry, the image curves followed into the past remain space-like separated from the origin, appearing to contract in from infinity.) Continued indefinitely in the future direction, they intersect null infinity (each curve in a distinct point) and cross over into the causal past of the coordinate origin, pass through it, and proceed into the causal future, accelerating asymptotically to the speed of light. Eventually they cross null-infinity once again, emerge in the space-like past of the $t = 0$ hypersurface, and complete a round-trip at that hypersurface.

It may seem as though there is an asymmetry in global behavior here, since the world-lines of the inertial frame simply cross the image at infinity of the origin once, whereas their image curves cross null-infinity twice. But from the point of view of the image of the origin, null infinity is an ordinary light-cone and the points of the light cone emanating from the origin lie at null-infinity. And so, since each "really" inertial curve (excepting the one through the origin) passes once through the past-lobe and once through the future-lobe, it passes, from the point of view of the image of the origin, through null-infinity twice. On such a manifold, and a fortiori for any finite region, there is nothing about the behavior of light that breaks this symmetry and allows you to say which class of curves is "really" inertial, or equivalently which points lie "really" at infinity. Thus, one needs to introduce the notion of "straight line" (or inertial trajectory) as an additional primitive. Accordingly, in the first of his Catalonian lectures, Weyl's characterization of Minkowski space-time is based on these two primitives. In the style of the Erlangen Program, he identifies its automorphism group (i.e., the Poincaré group, although he does not name it as such) in terms of the automorphism groups for these respective components. In summary,

1. *Die Gruppe der ähnlichen Abbildungen ist der Durchschnitt der projektiven Gruppe und der Gruppe der Möbiusschen Kugelverwandtschaften.*
2. *Die erste läßt sich durch den Begriff der geraden Linie, die zweite durch den Begriff des Nullelementes kennzeichnen.*[27]

(Weyl 1923, 8)

Turning now to Reichenbach's promised "precise examination"[28] of the issue of the conformal underdetermination of inertial structure in Sect. 16 of the *Axiomatik* he tackles it in terms of the question, what set of transformations leave invariant the equation:

$$dx_1^2 + dx_2^2 + dx_3^2 - dx_4^2 = 0.$$

He continues — but *without any mention of Weyl*:

Die Frage dieser Transformationen ist von mathematischer Seite längst geklärt. Die allgemeinsten derartigen Transformationen sind neben den linearen noch die Kugelverwandtschaften, welche Kugeln in Kugeln überfurhen .... Jedoch haben alle diese nichtlinearen Transformationen eine Singularität, sie führen einen endlichen Punkt ins Unendliche über und umgekehrt. Dies bedeutet, daß es in den so erhaltenen Systemen K' einen Punkt gibt, für den die [Licht-]Axiome I bis V nicht gelten. Verlangt man deshalb, daß die Lichtaxiome für alle Punkte eines Systems gelten sollen, so ist nur noch die lineare Transformation möglich, d. h. der Bewegungszustand eines solchen Systems ist bis auf eine gleichförmige Translation festgelegt.[29] (Reichenbach 1924, 59–60)

Reichenbach freely admits that in a fixed finite region, light signaling cannot distinguish between uniformly moving frames and frames in "hyperbolic" motion. A few pages later, Reichenbach takes up the issue how one might be able to tell, on the basis of light signaling alone, whether one is in an inertial system or the system obtained from it by a Möbius transformation. He suggests a systematic method for searching for singular behavior in light signaling as one goes to larger finite domains. Unfortunately, Reichenbach does not realize that, in following out the method, whether a system displays singular behavior depends entirely on where one assumes infinity to lie in the initial parameterization of the world-lines. Finally, the crucial point raised in the passage inserted into the introduction — whether it is "permissible" to impose the criterion that the manifold admit a singularity free metric — is not substantively addressed.[30] At the end of the section, he merely mentions that if one forgoes imposing the constraint, material structures can be exploited. A footnote directs us to a third option, the use of test masses to determine inertial motion, as in the appendix to (Weyl, 1921a).

To sum up, we find in Reichenbach's *Axiomatik*, not only the insistence that a problem can be solved that Weyl has long argued cannot be solved, but also not even a mention that it was Weyl who had alerted him to the problem in the first place.

9.5 Weyl's *Axiomatik*

To my knowledge, there was no subsequent correspondence between Reichenbach and Weyl. Although we cannot conclude that there were not other contributing factors, their bitter disagreements over the treatment of Lichtgeometrie cannot be dismissed as a major, if not the primary cause. Indeed, the topic of Lichtgeometrie continued to occupy Weyl's attention. His five-part university lectures given at Göttingen for the Winter and Spring terms bear the title *Axiomatik*. Part II, sub-titled "Die Raum-Zeit-Lehre der speziellen Relativitästheorie" is devoted almost exclusively to the matter of Lichtgeometrie.[31] Although Reichenbach is not mentioned by name in the course of Part II, (Reichenbach 1924) is listed in the bibliography accompanying the entire set of lectures.

Weyl's light axioms here are essentially the same as given in his review (Weyl 1924) of Reichenbach's *Axiomatik*. And, as in the appendix to his Catalonian lectures, the analytic formulation uses homogenous coordinates, although the treatment here is far more detailed. Our concern is the final section, which reveals that the issues with Reichenbach are still on his mind.

He begins by rehearsing the fact that, if the global topology is assumed to be that of R^4, then the light-cone structure also fixes the affine structure. But, he goes on, this conception of the world-as-a-whole derives from our concept of congruence as abstracted from our experience with the rigid bodies. And, of course, congruence of rigid bodies has no place in a pure Lichtgeometie. Weyl conjectures that if it were really the case that only light signals were at our disposal, we would be led to a different picture of the world-as-a-whole, namely the four-dimensional indefinite Möbius "Kugelraum":

> Wir würden [die Welt als Ganzem] mit einem vierdimensionalen indefiniten MÖBIUSschen Kugelraum gleichsetzen, zu welchem der unendlichferne Nullkegel mit dazu gehört. Er ist übrigens ... nicht eine offene Mannigfaltigkeit. Sie ist nicht bloss räumlich, sondern auch zeitlich geschlossen: Im unendlichfernen Nullkegel ist die "unendlichferne Zukunft" an die "unendlich weit zurückliegende Vergangenheit" geknüüpft.[32] (II, 31)

At this point, a footnote to the text cites a series of precedents in the history of philosophy for the notion of closed time. These range from the ancient Greek Alkmaeon, to Laotze, to Nietzche (specifically, *Also sprach Zarathustra*).

The text continues with a further display of philosophical erudition:

> In diesem Medium sind natürlich alle MÖBIUStransformationen durchaus singularitätenfrei. *In einer reinen Lichtwelt gibt es kein Unendlichfernes*; der Aufblick zu "dem gestirnten Himmel über mir" belehrt mich über diese Wahrheit, welche die Menschenseele auf's tiefste zu erschüttern vermag. Aber wir sind keine reinen Lichtwesen; es liegt an anderen Naturgesetzen als an denen des Lichtes, denen wir auch untertan sind, dass ein bestimmter Nullkegel der Lichtwelt für uns zum unerreichbaren unendlichfernen Weltenraum wird.[33] (II, 32)

The passage in quotes "dem gestirnten Himmel über mir" is no doubt an allusion to a passage from Kant's, *Critique of Practical Reason*:

> Zwei Dinge erfüllen das Gemüt mit immer neuer und zunehmender Bewunderung und Ehrfurcht, je öfter und anhaltender sich das Nachdenken damit beschäftigt: der bestirnte Himmel über mir und das moralische Gesetz in mir.[34] (Kant 1990 [1788], 186.)

Weyl completes the line of reasoning:

> Wenn man also nicht von einer vorgefassten Meinung über die Welt als Ganzes ausgehen will, bedarf man weiterer Naturgesetze, um die gleichförmigen Translationen aus der Klasse aller einheitlichen Hyperbelbewegungen herauszuheben. Als solches eignet sich, wie wir sahen, das Galileische Trägheits - gesetz. Aber man kann sich auch, wenn man das vorzieht, auf das Verhalten der starren Masstäbe oder Uhren stützen.[35]

In a different context, these very same words might well have been directed at Reichenbach.

9.6 Conclusion

One would certainly like to know how Reichenbach responded to Weyl's letter of May 1922, informing him of the underdetermination of affine structure by conformal structure. It may be that further correspondence covered the same ground over and over. Reichenbach's attitude is not easy to fathom, given that his overriding concern was to separate out the conventional from the factual components of relativity. One wonders why he attempted so persistently to banish temporally closed topologies from the realm of Lichtgeometrie. One might wonder as well whether the topological questions behind their dispute over Lichtgeometrie did not have an eventual impact. Reichenbach's next major work, *Philosophie der Raum-Zei-Lehre* (Reichenbach 1928), is often viewed as primarily a popularization of his *Axiomatik* for less technical audiences.

However, one of the major advances in that work beyond the *Axiomatik* involved extending questions of conventionality to topology, in particular the relation between causal anomalies and non-Euclidean topology. Was there a nudge here from his experience with Weyl?

Acknowledgments. I would like to thank Brigitta Arden of the University of Pittsburgh, Brigitte Uhlemann of the University of Konstanz, and Momota Ganguli of the Institute for Advanced Study for their most expeditious and friendly assistance in locating and providing archival materials. Special thanks to Frank Döring for his help in deciphering much of Weyl's handwriting. Stephan Hartmann was useful in various other respects.

References

Bateman, H (1910). The Transformation of the Electrodynamical Equations. *Proceedings of the London Mathematical Society* **8**: 228–264.

Carathéodory, C (1924). Zur Axiomatik der speziellen Relativitätstheorie. *Sitzungsberichte der Berliner Adademie. Physikalisch-mathematischen Klass*: 12–27.

Cunningham, E (1910). The Principle of Relativity in Electrodynamics and an Extension Thereof. *Proceedings of the London Mathematical Society* **8**: 77–98.

Kant, Immanuel (1990 [1788]). *Kritik der praktischen Vernunft*. Felix Meiner, Hamburg.

Reichenbach, Hans (1920). *Relativitästheorie und Erkenntnis apriori*. Springer, Berlin. Translated as Reichenbach (1965). Original language edition reprinted in Reichenbach (1979), pp. 191–302.

— (1921). Bericht über eine Axiomatik der Einsteinschen Raum-Zeit-Lehre. *Physikalische Zeitschrschrift* **22**: 683–687.

— (1922). Der gegenwärtige Stand der Relativitätsdiskussion. *Logos* **10**: 316–378. Reprinted in Reichenbach (1979), pp. 342–404. Translated with omissions as "The Present State of the Discussion on Relativivity" by Maria Reichenbach in Reichenbach (1959), pp. 1–45.

— (1924). *Axiomatik der relativistischen Raum-Zeit-Lehre*. Vieweg, Braunschweig. Reprinted in Reichenbach (1979), pp. 13–171. Translated as Reichenbach (1969).

— (1925). Über die physikalischen Konsequenzen der relativistischen Axiomatik. *Zeitschrift für Physik* **35**: 32–48. Reprinted in Reichenbach (1979), pp. 172–183.

— (1928). *Philosophie der Raum-Zeit-Lehre*. Walter de Gruyter, Berlin and Leipzig.

— (1959). *Modern Philosophy of Science*. Translated and edited by Maria Reichenbach. Routledge & Kegan Paul, London.

— (1965). *The Theory of Relativity and A Priori Knowledge*. Translation of Reichenbach (1920) by Maria Reichenbach. University of California Press, 1965, Berkeley and Los Angeles.

— (1969). *Axiomatization of the Theory of Relativity* Translation of Reichenbach (1924) by Maria Reichenbach. University of California Press, Berkeley and Los Angeles.

— (1979). *Gesammelte Werke*, edited by Andreas Kamlah and Maria Reichenbach, vol. 3, *Die philosophische Bedeutung der Relativitätstheorie*. Friedr. Vieweg & Sohn, Braunschweig/Weisbaden, 1979.

Robb, Alfred A. (1914). *A Theory of Space and Time*. Cambridge University Press, Cambridge, U.K.

Ryckman, T. A. (1994). Weyl, Reichenbach and the Epistemology of Geometry. *Studies in the History and Philosophy of Science* **25**: 831–870.

— (1996). Einstein Agonists: Weyl and Reichenbach on Geometry and the General Theory of Relativity. In *Minnesosota Studies in the Philosophy of Science*, vol. XVI. *Origins of Logical Empiricism*. Ronald N. Giere and Alan W. Richardson, eds. University of Minnesota Press, Minneapolis, MN, 165–209.

Rynasiewicz, Robert (2003). Reichenbach's ε-Definition of Simultaneity in Historical and Philosophical Perspective. *The Vienna Circle and Logical Empiricism: Re-Evaluation and Future Perspectives*. Vienna Circle Institute Yearbook 10 [2002]. Kluwer, Dordrecht, Boston, and London.

Wald, Robert M. (1984). *General Relativity*. University of Chicago Press, Chicago and London.

Weyl, Hermann (1921a). *Raum-Zeit-Materie*. 4. Auflage. Springer, Berlin.

— (1921b). *Space-Time-Matter*. English translation of Weyl (1921a) by Henry L. Brose. W. P. Dutton, New York.

— (1921c). Das Raumproblem. *Jahresbericht der Deutsche Mathematikervereinigung* **30**: 92–108. Reprinted in Weyl (1969), pp. 212–228.

— (1921d). Über die physikalischen Grundlagen der erweiterten Relativitätstheorie. *Physikalische Zeitschrift* **22**: 473–480. Reprinted in Weyl (1969), pp. 229–236.

— (1921e). Feld und Materie. *Annalen der Physik* **65**: 541–563. Reprinted in Weyl (1969), pp. 237–259.

— (1922). Das Raumproblem. *Jahresbericht der Deutsche Mathematikervereinigung* **3**: 205–221. Reprinted in Weyl (1969), pp. 328–344.

— (1923). *Mathematische Analyse des Raumproblems: Vorlesungen gehalten in Barc - elona und Madrid*. Springer, Berlin.

— (1924). Review of Hans Reichenbach, *Axiomatik der relativistischen Raum-Zeit-Lehre. Deutsche Literaturzeitung* **30**: 2122–2128.

— (1968). *Gesammelte Abhändlungen*, Band II. Edited by K. Chandrasekhar. Springer-Verlag, Berlin.

Notes

[1] For previous studies on the interaction between Weyl and Reichenbach on the foundations of geometry and general relativity see (Ryckman 1994, 1996).

[2] "The axioms then contain those fundamental facts whose existence grant the theory its right to existence; in principle, they are empirical claims which thus can be tested by experiment."

[3]"[The definitions] are arbitrary conceptual elements, which at bottom are capable of neither empirical confirmation nor refutation.

[4]An obvious consideration, *inter alia* derives from Einstein's contention that the one-way speed of light is not empirically measurable, and hence is chosen by mere convention in his formulation of special relativity.

[5]Es wird gezeigt werden, daß allein auf diese [Licht-]Axiome eine vollständige Raum-Zeit-Lehre aufgebaut werden kann. Die Materialaxiome besagen die Identität der so entwickelten "Lichtgeometrie" mit der Raum-Zeit-Lehre der starren Maßstäbe und Uhren. Es darf als wichtigstes Resultat dieser Untersuchung aufgefaßt werden, daß also auch ohne die Geltung der Materialaxiome, deren empirische Bestätigung noch nicht restlos durchgeführt werden konnte, die Relativitätstheorie eine gültigige und vollständige physikalische Theorie ist. (Reichenbach 1922, 684)

[6]I ignore the preliminary issue of defining time order for a single world-line and take for granted Axiom (II), which asserts that light signals that depart simultaneously from world-line *A* to world-line *B* (in an arbitrary state of motion with respect to *A*) and are reflected back to *A* return concurrently.

[7]Although Reichenbach does not mention so, it presumably follows that the clocks implemented at all other world-lines in order for the system to qualify as a "Normalsystem" also have the status of "Normaluhr."

[8]Axiom IV states: "Werden von einem Punkt *A* eines Normalsystems zwei Lichtsignale um einen geschlossenen Dreiecksweg *ABCA* gleichzeitig in entgegengesetztem Sinne geschickt, so kehren sie gleichzeitig zurück."

"If two light signals are sent from a point *A* of a "normal" system around a closed triangle *ABCD* at the same time in opposite directions, then they return at the same time." (Reichenbach 1922, 684)

Reichenbach uses the term "Punkt" for what I have been calling a world-line. Note that the symmetry of clock synchronization follows from the Einstein definition alone without the assumption of the round-trip axiom.

[9]"Definition 6. Werden zwei Signale gleichzeitig von *A* längs der Wege *ABA* und *ACA* geschickt, und kehren sie gleichzeitig nach *A* zurück, so heiß *AB = AC*."

"If two signals are sent simultaneously from *A* along the paths *ABA* and *ACA* and return to *A* at the same time, [the spatial separation] *AB* is said to be the same as [the spatial separation] *AC*." (Rcichenbach 1922, 684)

[10]Es düfte von Interesse sein, daß in der vorliegenden Darstellung aus dem Bedürfnis heraus, mit einem Minimum von Axiomen auszukommen, die raum-zeitliche Metrik der speziellen Relativitätstheorie allein durch Lichtsignale definiert wird, also neben der Gleichzeitigkeit auch die Gleichförmigkeit und die Streckengleichheit. (Reichenbach 1924, 10)

[11]Ein Hinweis darauf, daß dies möglich ist, findet sich übrigens im Anhang der 4. Auflage von Weyl, "Raum-Zeit-Materie."

[12]To distinguish "normal" co-ordinate systems among all others in the special theory of relativity, . . . , we may dispense with not only rigid bodies but also with clocks. (Weyl 1921b, 313)

[13]"Of particular note here, however, is the difference between physics and mathematics. Mathematics is indifferent to the application of its laws to things in reality, and its axioms contain only a system of rules, according to which its concepts are related to one another. Pure mathematical axiomatization leads in no way to principles of a theory of natural knowledge. And thus the axiomatization of geometry cannot say anything at all about the epistemological problem of space. Only a physical theory can answer the question of the validity of Euclidean space, and likewise reveal the epistemological principles lying at the basis of the space of objects in nature."

[14] It is entirely false, however, if one then wants to draw the conclusion anew that mathematics and physics are merging into a single discipline, e.g. Weyl and also Haas. The question of the validity of axioms for reality and the question as to what axioms are possible are to be kept absolutely separate. That is indeed the very merit of the theory of relativity, that it takes away from mathematics the question of the validity of geometry and has referred it to physics. Now, if one sets forth further laws from a generalized geometry and maintains that they must be the foundations of physics, then one commits the old mistakes all over again. This objection must be leveled against Weyl's generalization of the theory of relativity, in which the concept of a fixed length for an infinitely small measuring rod is abandoned altogether. To be sure, such a generalization is possible, but whether it conforms with reality does not depend on its meaning for a generalized infinitesmal geometry (Nahegeometrie). For this, Weyl's generalization must be regarded from the standpoint of a physical theory, and it is subject to critique solely on the basis of experience. Physics is simply not a "geometric necessity;" whoever maintains this reverts to the pre-Kantian Standpoint, where it was a necessity given by Reason. A generalize-geometric consideration can teach the principles of physics just a little as the Kantian analysis of Reason could teach it. Rather only an analysis of physical experience can do this.

[15]"Finally, after many weeks, I've gotten around to thanking you for the friendly sending of your writing, "Relativitätstheorie und Erkenntnis a priori." I believe that, due to my divergent basic orientation in philosophy, an agreement between us would be reached only with difficulty; however, I do not thereby underestimate the great sincerity of your remarks to grasp sharply the problem of knowledge of the real external world. However, let me confine myself today to two remarks, which concern less the philosophical than the physical." (Quoted by permission of the University of Pittsburgh. All Rights reserved.)

[16]All further references to letters from Weyl to Reichenbach will state only the document number from this archive.

[17]Emphasis in original. "It is certainly not true, as you say on p. 73, that, for me, mathematics (!!, e. g. theory of the ζ-function?) and physics are growing together into a single discipline. I have claimed only that the *concepts* in geometry and field physics have come to coincide."

[18]"I must, however, take back my objection, raised earlier ([Reichenbach 1920], p. 73), that Weyl wants to deduce physics from Reason, since Weyl has cleared up this misunderstanding ([Weyl 1921d], p. 475)."

The passage from (Weyl 1921d), to which Reichenbach refers, begins: "Von vershiedenen Seiten is gegen meine Theorie eingewendet worden, daß in ihr aus reiner Spekulation Dinge a priori demonstriert würden, über welche nur die Erfahrung entscheiden kann. Das is ein Mißverständnis." Weyl goes on to explain that his "Prinzip der Relativität der Größe" does not entail the non-integrability of the length of a tangent vector around a closed loop anymore than does Einstein's general relativity entail the non-integrability of parallel displacement around a closed loop. Rather these theories only permit the possibility of non-integrability in the respective cases.

[19]"The aspiration of Reichenbach's book is obviously to distinguish as clearly as possible in the construction of the space-time theory the empirical facts from stipulations based upon them. To this extent it has an epistemological background; and the author is concerned, in fact with success, to illiminate on all sides the presuppositions of the theory. However, in the main, the book contains not a philosophical, but a purely mathematical investigation, and consequently it must be judged from a mathematical point of view. In this regard, though, it is not very satisfactory, too tedious and too obscure. Its sole value: the setting out of axioms a) b) c) and the transition from these to space-time measurement, to coordinate space and thereby to Möbius geometry, can be suitably carried out in a few pages, whereby clarity and comprehensibility would only have been gained."

[20]"From the mathematical side, no other judgment of my investigation is permissible other than the judgment "true" or "false." I am not concerned here with mathematical elegance — there has been plenty of opportunity to indulge in this in the theory of relativity, and it has had, in any event, only limited significance for epistemological clarification."

[21]"In the first few pages of his review, Herr Weyl gives a presentation of my axiomatization, which, likely, is supposed to be an example of how I could have made it 'suitably' better 'in [only] a few pages': I gladly leave it to the reader to judge which of the two presentations is found to be 'tedious and obscure'. I for one have always preferred the clarity of a step-by-step construction, which manages with the simplest possible logical operations, to a shimmering mathematical fog with which many prefer to surround their ideas. The plan of my investigation was dictated by the intention to lay bare the results of physical experience as clearly as possible and to extract from each new law of experience as many derivable consequences as are at all relevant. If one works with a minimum of concepts, then as a result many steps become more tedious, than if at the outset one begins with the totality of all available auxiliary resources."

[22]However, I think is is very regrettable, if a mathematician of Herrn Weyl's rank so misunderstands the goal of such a logico-epistemological investigation and with his authority seeks to suppress the attempt to give the theory of relativity, which has

been so fruitfully developed mathematically and physically, now at last also the logical foundation which alone can ultimately secure its validity.

[22] A point of clarification is in order here. The *relativistic* Lichtgeometry is not uniquely determined by the the light axioms. One can equally well introduce coordinative definitions that lead to the Lichtgeometrie of classical optics in Newtonian space-time. Indeed, this is Reichenbach's main lemma for separating the factual from the conventional components of space-time geometry. See (Rynasiewicz 2003) for how this relates to Reichenbach's thesis of the conventionality of simultaneity. The claim of uniqueness concerns the light axioms *together with* appropriately chosen coordinative definitions that involve only light signaling.

[23] "The establishment of the uniform states of motion in space succeeds, of course, only if one considers it to be a permissible criterion that the resulting metric possesses a singularity at no place in the entire space. If one does not want to use such a criterion, then there remain two classes of light-geometrically preferred spatial reference systems, which are accelerated with respect to one another and which are such that at any time the one, as measured from the other, experiences an internal expansion (respectively, contraction), and thus is not rigid. The selection of one of the two classes as rigid and uniformly moving is then carried out with the help of material objects, either rigid rods or natural clocks. These material structures are defined in Definition 18 and 19; 16 gives a precise investigation of the entire question. It nonetheless holds that, for each of these two classes of systems, the space-time metric is definable solely in terms of light; the light-geometry thus leaves open a two-fold multiplicity of classes, while it determines within each class all ratios uniquely."

[24] Emphasis in original. "As for the Light-geometry, this is in fact how matters stand (which has been well-known to mathematicians for a long time, and which I also talked about in Part 1 of my [Catalonian] lectures): Those mappings, which carry null-cones over to null-cones, are Moebius's 'sphere transformations'. When operating in a restricted region (which, of course, is the only physically reasonable case) the null-cones do *not* thus suffice to characterize the geometry, but rather *null-cones* and *straight lines*. Of course if one uses infinite space, just as Euclid assumed it (without additional [projective] properties), then the null-cones suffice, because the similarity mappings [i.e., congruences] are the only 'sphere transformations' which carry the infinitely distant "sphere" over to the infinitely distant 'sphere'." (HR 015-68-02)

[25] Other contemporary investigations of conformal invariance can be found in (Bateman 1910), (Carathéodory 1924), and (Cunningham 1910).

[26] The chart, of course, does not cover the entire manifold.

[27] Emphasis in original. "1. The group of similarity mappings is the intersection of the projective group and the group of Möbius sphere-transformations. 2. The first can be characterized by the concept of the straight line, the second by the concept of the null-element." Briefer but similar treatments are given in (Weyl 1921c, 1922). Interestingly enough, in all three places the axiomatization of (Robb 1914) is cited. In his *Axiomatik* Reichenbach nowhere mentions Robb's work.

[28]The reader might also find of interest Andreas Kamlah's explanatory note to this section (Reichenbach 1979, 460–464).

[29]The question of the transformations on the mathematical side was settled long ago. The most general such transformations are, besides the linear, just the sphere-transformations, which take spheres to spheres. However, all the non-linear transformations have a singularity, which takes a finite point to infinity and conversely. This means that in the system K' so obtained, there exists a point for which the [light] axioms I through V do not hold. Thus, if one requires that the light axioms should hold for all points of a system, only the linear transformations are possible, i.e., the state of motion of such a system is established up to a linear transformation.

[30]Although, there is no singularity-free *Minkowski* metric on the completion of R^4 by conformal infinity, it does admit a singularity-free Lorentz metric, since there is a conformal isometry of Minkowski space-time into an open region of the Einstein static universe. See (Wald 1984, 273).

[31]The typescript of these lectures are now at the Mathematics-Natural Sciences Library of the Institute for Advanced Study. Part II is thirty-three pages in length, not including two pages of diagrams. The other four parts are titled: I. Geometrie, III. Raum und Zahl, IV. Grundlegung von Algebra und Analysis, and V. Topologie.

[32]"We would have identified [the world as a whole] with a four-dimensional indefinite Möbius "Kugelraum", to which the infinitely distant null-cone belongs. It is, moreover, not an open manifold. It is not just spatially, but also temporally closed: At the infinitely distant null-cone is the "infinite future" to which "the past stretching infinitely backward" is joined."

[33]"In this medium, all Möbius transformations are naturally free of singularities. *In a pure Lichtwelt there is no infinitely distant*; looking upward at 'the star-filled heavens above me' reveals to me this truth, which is capable of moving the human soul to the greatest depths. However, we are not pure light-beings; there are laws of nature other than those of light to which we are subject and which establish that a particular null-cone of the Lichtwelt becomes for us the unreachable, infinitely distant outerspace."

[34]"Two things preoccupy the mind with ever newer and increasing wonder and awe, the more often and more persistently they are contemplated: the starry heavens above me and the moral law within me."

[35]Thus, if one does not want to start from a preconceived opinion concerning the world in its entirety, other laws of nature are needed in order to distinguish uniform translations from the class of all uniform hyperbolic motions. Suitable for this, as we have seen, is the Galilean law of inertia. However, if one prefers, one can also rely on the behavior of rigid rods or clocks.

Dingle and de Sitter Against the Metaphysicians, or Two Ways to Keep Modern Cosmology Physical

George Gale

Department of Philosophy, University of Missouri—Kansas City, Kansas City, MO 64110
U.S.A.; galeg@umkc.edu

Summary. It would be hard to find two more radically different personalities than the irascible Herbert Dingle and the courtly Willem de Sitter. Yet, when it came to their philosophy of science, these two otherwise-so-different men were united against a common enemy, those they both called the "metaphysicians." Right from 1917, de Sitter attempted always to keep cosmology tightly bound to real observations made upon a real world. In *Kosmos*, written near the end of his life, he re-affirms most strongly his principle that "there is nothing an orthodox physicist abhors more than metaphysics." Dingle, for his part, accepts early on the positivist use of the verifiability principle to eliminate metaphysics from science, and continuously wields the principle as a weapon against those errant cosmologists who would sacrifice science for a sort of mysticism. Both men reject the strict and literal use of the term "universe," and for the same reasons: there is no observation, no verification, of statements containing that term. Both men reject the "cosmological principle" as Milne and others use it, on the grounds, as de Sitter puts it, that "we have . . . no means of communicating with other observers, situated on faraway stars." Eddington, although always closely associated with de Sitter personally, comes in for his own fine share of criticism. After de Sitter's death, Dingle carried on the battle alone, always on the bases that he and de Sitter had earlier established. The two peaks in Dingle's long struggle were the notorious 1937 controversy in the pages of *Nature*, a nasty dogfight which managed to involve almost every single important physicist in Britain; thirteen years later, the long war with the metaphysicians ended with the pyrrhic victory of Dingle's Royal Astronomical Society Presidential Address' invective against the latest and greatest metaphysical creation, Bondi's steady state universe theory. In the end, however, it would be a mistake to believe that the campaign of de Sitter and Dingle accomplished nothing. On the contrary, it is clear that their critique succeeded magnificently in keeping the metaphysicians at least somewhat in check, and, more importantly, maintaining cosmology's connection to the real and observable world. As I will show, the common philosophical spirit of the two men grows out of precisely the same terrain: both men are exquisitely, excruciatingly, anchored in the rich empirical detail of observational astronomy. Unlike most of the other cosmologists, both men knew exactly what it took to construct data out of astronomical observations, both men knew exactly how hard is the subsequent task of interacting their hard-won data with theory, and it was this direct experience of real, genuine empirical science that they brought into the fray with the cosmological metaphysicians. And cosmology was the better for it.

10.0 Background

Modern cosmology got off to a slow start. In 1917, Einstein discovered that his general theory of relativity (GTR) could be applied to the universe as a whole, yielding a model of the universe that was dense with matter, stable in its geometry, and closed. Within a few short months, de Sitter found another solution to the GTR equations; his model was also closed, but it was stable just because it was empty of matter. The two solutions respectively came to be called "A" and "B," following de Sitter's terminology.

Although the two models were fascinating just because they existed, thereby providing something for theorists to work with, neither was particularly acceptable as a candidate for reality: the observed universe was neither everywhere dense, nor everywhere empty (de Sitter 1930, 481–482). Over the years, Einstein and de Sitter argued about the two models, each pointing out the strong points of his, and the weaknesses of the other's. Beyond this, however, not much progress was made. Friedmann published his own solution in 1922 and 1924; it differed from both A and B in that its geometry evolved over time. Although Einstein reviewed both of Friedmann's papers, he did not see that they offerred an acceptable alternative to A and B, most likely because their geometry was not static. Friedmann's work went unacknowledged. In 1927 LeMaître published his solution. Although he did not know Friedmann's solution, his model was similar to it because—unlike A and B—its geometry evolved over time as well. LeMaître's initial fate was identical to Friedmann's: no one, not even LeMaître's mentor Eddington, perceived the model as an acceptable alternative to A and B (de Sitter 1930, 482). In other words, just like A and B, theoretical cosmology itself was globally static and unevolving over the years, at least until 1930.

On the observational front, in contrast, progress was being made. First reports of observed nebular spectrum wavelength shifts (1917) doubly surprized astronomers: first, because shifted spectra seemed common, and secondly, because the great majority of the shifts were toward the red (de Sitter 1917, 27). Slipher, who had begun his spectral observations in 1912, made interim reports, and produced a final review of his observations in 1925 (North 1994, 523). Humason and Hubble continued Slipher's work, which culminated in Hubble's 1929 announcement of a direct relation between distance and degree of red shift of any given nebular spectrum.

A chance event immediately linked Hubble's announcement to theory. At the January 1930 monthly meeting of the Royal Astronomical Society, de Sitter happened to remark to Eddington that it was too bad that there were only two theoretical models. Eddington agreed (Deprit 1995, 363). Their remarks were published in the February *Observatory*, soon to be read by an astonished LeMaître, who immediately sent another copy of his 1927 paper to Eddington. Eddington, mortified, rushed a translation and review of LeMaître's work into print (McVittie 1987; Eddington 1930).

Now there were three models.

A year later, at a meeting of the British Association for the Advancement of Science during the Autumn of 1931, a gathering of the majority of practicing cosmologists certified LeMaître's model as the most likely correct description of the universe (de Sitter 1931b). As de Sitter remarked "There can be not the slightest doubt that LeMaître's theory is essentially true" (de Sitter 1931b, 707).

Unfortunately, things changed almost immediately. As de Sitter remarked early in 1932, Heckmann had discovered unforeseen freedom in the values of both 1 and the curvature of space (de Sitter 1933a, 159). Cosmologists, liberated by their unforeseen freedom, created dozens of new solutions to the GTR equations. And now, suddenly, where before there had been only three, there were now indenumberably many: instead of the problem being too few cosmological models, the problem now was that there were far, far too many. How to choose among them that one which best matched the universe in which we lived? How indeed. So begins our story.

10.1 Introduction

Reducing the number of the potential candidate-models was a plausible first step. Unfortunately, as de Sitter notes, the reduction could not be accomplished by observation:

> there is nothing in our observational data to guide us in making the choice . . .
> The observations give us *two* data, viz. the rate of expansion and the average density, and there are *three* unknowns: the value of λ the sign of the curvature, and the scale of the figure, i.e., the units of *R* and of the time (de Sitter 1932, 127).

A year later, de Sitter not only argued that "astronomical observations give us no means whatever to decide which of these possible solutions corresponds to the actual universe," but went far beyond, concluding that "the choice must, as Sir Arthur Eddington says, depend on aesthetical considerations."[1] (de Sitter 1933b, 630). In England, Eddington was himself trying mightily to craft a new approach to pruning down the number of candidate universes. He called his foray "Fundamental Theory;" it was a 'pure' theory, he noted, simply because it used no particular observational data in its development.

Eddington's efforts had been first announced to the cosmological world during the Autumn 1931 BAAS meeting (Eddington 1931); he developed his ideas during the following year's lectures in the U.S., which were soon published as *The Expanding Universe* (Eddington 1932). Eddington had initially noticed that when he constructed a range of various functions from the six so-called universal constants, the products of all the manipulations seemed to be roughly 10^{40}.[2] The constancy of this value told him that something was afoot here, something he took to be of importance to cosmology. De Sitter reacted to Eddington's efforts in characteristic fashion:

> Sir Arthur Eddington has recently published a remarkable formula, linking up the numerical data referring to the universe with those referring to the electron [which] at first sight it might seem to make the problem determinate by adding one more datum (de Sitter 1933a, 183–184).

De Sitter's hope, of course, is that Eddington's formula might fit in with the other two used to determine the nature of the Universe. But as we shall soon see, the Dutch astronomer quickly came to strongly dislike what his close friend was attempting with the 'fundamental theory,' and this because it violated de Sitter's own philosophical principles. In the end de Sitter's philosophical beliefs led him to attack Eddington's efforts.

But Eddington was not alone in de Sitter's gunsights. For the same philosophical reasons, de Sitter was flat-out opposed to another Englishman's attempts to pare down the proliferation of cosmological models.

Oxford's E.A. Milne had announced a program in May of 1932 to come up with a purely rational—eschewing, as with Eddington, concrete astronomical observations—cosmological theory, a theory which would clearly pick out just the universe which we find ourselves in. Milne summed up his program in the following way:

> The general object of the investigations in question was to determine the consequences of the assumption that the universe is, on the average, homeogeneous ... By an extreme application of the principle of the economy of thought I investigated the consequences without appealing to any empirical quantitative 'laws of Nature' whatsoever ... I did so because it early appeared that very much more could be deduced from it than was commonly recognized (Milne et al. 1937, 997).

If anything, de Sitter disliked Milne's program even more than Eddington's. In this dislike, he was soon joined by Herbert Dingle, well-known British astrophysicist, and budding philosopher of science. Together, de Sitter and Dingle rapidly became outspoken critics of the sort of effort Eddington, Milne, and others had mounted in hopes of constraining the burgeoning number of candidate universe models. Dingle and de Sitter's shared reasons for opposition unite the two men in what must be one of the odder philosophical-scientific pairings ever. We now turn to that story.

10.2 De Sitter: Empiricist All the Way Down

10.2.1 The beginnings

Willem de Sitter was born in Sneek, Friesland, on 6 May 1872.[3] He went to Groningen with the idea of becoming a mathematician. Early on he took a job in the Astronomical Laboratory, where Kapteyn was carrying out a long project of measurements on the Cape photographic survey. In 1896 Sir David Gill visited the lab, encountered de Sitter at the measurement machine and took a liking to him. Gill soon invited de Sitter to come to the Cape and become a computer. After his qualifying exams were successfully completed, de Sitter journeyed to the Cape, where Gill put him to work making observations with the heliometer, working on stellar parallaxes, and assisting with polar triangulations. He also did some of his own work on visual and photographic photometry.

It was during this time that Gill suggested a dissertation topic to de Sitter: reduction and discussion of a series of heliometer observations of Jupiter's satellites which had been made in 1891. Originally, the goal of the observations had been improvement of the tables of the motions of Jupiter's satellites, but for various reasons, this work had not been undertaken. De Sitter undertook it, and his great success in providing "strongly determined correction of considerable amount to the inclinations and nodes of all the satellites, and an accurate determination of the mass of Jupiter" forever linked him to the Jovian system: work on Jupiter and its family were to occupy him until his death thirty-five years later (Jones 1935, 344).[4]

His work at the Cape developed de Sitter's fundamental astronomical skills in two ways. First, he not only learned how to use all the usual equipment to make observations, he learned as well what counts as a good observation and what can be done with it. With this latter realization, de Sitter came to good knowledge of the genuine, actual, empirical basis of astronomical science. Secondly, reducing the observational data from the Jovian satellites practiced the young astronomer's mathematical skills in thorough and useful ways. Such a powerful combination of these two skills—empirical, observational skill plus theoretical, mathematical skill—was lacking in the colleagues with whom de Sitter would later practice cosmology—with the interesting exception of Dingle.

De Sitter soon came to the notice of the wider European astronomical community. In 1911 he published a paper in *Monthly Notices* which predicted minute deviations of lunar and planetary motions away from the Newtonian model's results and toward those of the special relativistic model. Then, in 1916 and 1917 he published three seminal papers on the general theory in *Monthly Notices*. These papers presented detailed examination of general relativity's mathematical core, and, even more importantly, predictions about some of its astronomical consequences. The Astronomer Royal, H. S. Jones, argues that "the British eclipse expeditions of 1919, which provided the first evidence in support of Einstein's conclusions as to the amount of deviation of rays of light in passing near the Sun, would probably not have been sent out had de Sitter's papers not appeared." (Jones 1935, 345). Eddington, of course, directed one of the expeditions, thereby forging another link in his friendship with de Sitter.

Throughout the rest of his career, de Sitter was to demonstrate at all times this wonderful balance of observational and theoretical talent. Yet, although his talents were balanced nicely, there was no such symmetry in his philosophical tenets.

10.2.2 Philosophical considerations

De Sitter's philosophy of science was decidedly empiricist. For example, in *Kosmos* he makes the following claim:

> Every physical theory must begin and end in observation. Its origin is the attempt to account for observed phenomena on a basis of reason, and consequently the final test necessarily is comparison with observations; no theory can survive which is not able successfully to stand this test (de Sitter 1932, 6).

The process described here evidently combines both induction and deduction. First, a set of observations is given, and an attempt is made to come up with an explanatory hypothesis for the observations. Then, secondly, the hypothesis is tested against

further observations, apparently via deductive predictions. Aside from the logic, however, what stands out most clearly is the tight interweaving of observation into the
whole process.

Einstein's general theory of relativity (GTR) is a prime example of how this process works. According to de Sitter, GTR "is a purely physical theory, invented to
explain empirical physical facts, especially the identity of gravitational and inertial
mass." (de Sitter 1933a, 147). Because of GTR's being embedded in the observational
process described above, there is "nothing metaphysical about its origin." (de Sitter
1932, 112). This claim is repeated, just as strongly, in another place: "it has nothing
metaphysical about it." (de Sitter 1933a, 147). Even though "it has, of course, largely
attracted the attention of philosophers," clearly, "that is not what it set out to do, that
is only a by-product." (de Sitter 1932, 112).

Distinguishing physics from metaphysics is important to de Sitter. He engages the
topic in a number of places. His statement in Kosmos couldn't be clearer: "Strictly
speaking every assertion about what has not been observed is outside physics and
belongs to metaphysics."[5] Indeed, "there is nothing an orthodox physicist abhors more
than metaphysics." (de Sitter 1932, 5). Sometimes, however, de Sitter simply can't
avoid weakening the boundary keeping physics and metaphysics separate. λ is a good
example of this difficulty.

In an article in late 1931, de Sitter described the origin of this term:

> Very soon after the completion of the theory Einstein was led, by certain con-
> siderations of a philosophical or metaphysical nature, to introduce into these
> equations a certain quantity, denoted by the Greek letter *lambda*, and called
> by him the "cosmological constant" ... We do not know what its exact phys-
> ical meaning is, and have as yet very little insight into its connection with
> other fundamental constants of nature. (de Sitter 1931b, 2).

λ clearly has metaphysical origins. For this reason alone de Sitter should oppose it.
But, although he is obviously sensitive to λ's transgression of the demarcation between
physics and metaphysics, he accepts the hypothesis nonetheless. The reasons he offers
for acceptance are interesting ones. First,

> it has such obvious advantages from many points of view, that it has been
> generally adopted, even before any phenomenon has been observed which
> could be explained by the equations containing *lambda*, but not without it.
> (de Sitter 1931b, 3).

Unfortunately, de Sitter does not elaborate at all on the "obvious advantages" of
λ so we are unable to evaluate them. Later on he hints that the reason for retaining
l lies in the mathematical equations and the role λ plays therein. In particular, "the
behaviour of *lambda* is not more strange or mysterious than that of the constant of
gravitation *kappa*, to say nothing of the quantum-constant h or the velocity of light
c." (de Sitter 1931b, 11). Thus, just insofar as we find ourselves able to accept other
constants of nature, so also should we find ourselves able to accept λ.

Another aspect of cosmology fraught with metaphysics is its inherent need to ex-
trapolate: "It should not be forgotten that all this talk about the universe involves a

tremendous extrapolation, which is a very dangerous operation." (de Sitter 1931a, 708). Again, the boundaries were quite clear: "We have a direct knowledge only of that part of the universe of which we can make observations;" to this region de Sitter gives a special name—"I have already called this 'our neighborhood.'" (de Sitter 1932, 113). Physics, strictly so-called, can only take place within our neighborhood: "All assertions regarding those portions of the universe which lie beyond our neighbourhood either in space or in time are pure extrapolations." But, being extrapolations, should these assertions be rejected? Apparently not. Cosmology demands at least a small bit of metaphysics, or, perhaps more neutrally stated in this case, a small bit of philosophy: "In making a theory of the universe we must, however, adopt some extrapolation, and we can choose it so as to suit our philosophical taste." Caution, however, is the watchword, since "we have no right to expect it to be confirmed by future observations extending to parts now outside our reach." (de Sitter 1931a, 708).

One feature of extrapolation mitigates against its being pure metaphysics. We have a "philosophical or aesthetical" conviction, "on which extrapolation is naturally based," that "the particular part of the universe in which we happen to be is in no way exceptional or privileged." (de Sitter 1932, 113). De Sitter's argument here is straightforward. Extrapolation has two components. First, we have observational knowledge, that is to say, science, about our neighborhood. Secondly, we have a philosophical— explicitly metaphysical, in fact—principle that "Our neighborhood is not exceptional," a principle that came later to be called "the Copernican Principle," for the obvious reason that Copernicus had moved earth from the privileged center to the unexceptional planetary zone. Together, the two components justify extrapolation beyond our neighborhood. Although the move contains a philosophical component, at bottom, it is founded upon observation. Hence, in de Sitter's eye, extrapolation is risky, but acceptable; it's not quite metaphysics.

Another saving grace of extrapolation will not become fully clear until later, when we examine de Sitter's criticisms of Eddington and Milne. A brief remark will suffice here. Until now, I have stressed observation as the main basis for de Sitter's boundary between physics and metaphysics: physics concerns the observable, metaphysics not. But there is an additional element underlying the boundary, beyond observation itself. Included implicitly in de Sitter's notion of "an observable" is the understanding that any general proposition concerning observables must originate inductively. That is, put most plainly, observables are constructed on the basis of inductive logic. As we shall see, both Eddington and Milne are chided because their work involves *hypotheses*,[6] a priori propositions originating in reason or imagination, rather than originating in induction over perceptibles. Pure hypotheses—Popperian 'conjectures'—are viewed with deep suspicion by de Sitter. On the other hand, because propositions about, say, spatial homogeniety on cosmological scales, are reached by extrapolation, and extrapolation ranges over neighborly observables, there is an inductive basis to such proposals. Which basis thereby justifies them in de Sitter's eyes. Not so the moves of Eddington and Milne, to which we now turn.

10.2.3 Against the metaphysicians

Eddington and de Sitter were fast friends.[7] Yet their friendship did not enjoin respect-
ful criticism, especially from de Sitter's corner. On the question of how to pare down
the vast number of candidate cosmologies, a question of great interest to both men, Eddington was by far the more creative. And de Sitter, although intrigued by Eddington's
creativity, was by no means uncritically accepting of it. As noted above, Eddington's
proposal involved his 'fundamental theory,' an effort to generate a cosmology without
use of any particular astronomical datum. De Sitter introduces Eddington's idea during
a discussion of the problem of too many models:

> It has already been pointed out that there is no observational evidence avail-
> able which would enable us to decide which of the several possible solutions
> represents the actual universe. This not because the data are not sufficiently
> accurate, but because they are deficient in number. (de Sitter 1933a, 182).

De Sitter turns to Eddington's proposal because, he notes, "at first sight it might
seem to make the problem determinate by adding one more datum." (de Sitter 1933a,
184). Of course, Eddington's proposal, de Sitter admits, is remarkable:

> Sir Arthur Eddington has recently published a remarkable formula, linking
> up the numerical data referring to the universe with those referring to the
> electron. As published by Eddington the formula reads

$$\frac{\sqrt{N}}{R} = \frac{mc^2}{e^2} = 3.54 \times 10^{12}$$

> where N is the number of H atoms; R is the radius of curvature of the Universe
> in de Sitter's solution B; and m, c, and e take their usual values (de Sitter
> 1933a, 183-184).

After rephrasing Eddington's formula slightly, de Sitter remarks the crux of the
matter: "If the formula were accepted, then the observed density and the coefficient of
expansion, or ρ and R_B would be sufficient to determine all required characteristics
of the universe." (de Sitter 1933a, 183). This, of course, is precisely the goal for both
men.

Unfortunately, in the end, Eddington's proposal is ultimately unacceptable to de
Sitter because it is metaphysical.

But at first glance, the situation had not looked entirely hopeless. Eddington, in a
personal communication, had told de Sitter that he believed there to be only a finite
number of distinguishable electrons in the universe. This pleased de Sitter because,
"if this were so there would be physical basis for the finiteness of the universe—
though not an *observational* basis, but one depending on the structure of our theory
of the electron" (de Sitter 1933a, 183). Here de Sitter makes a rather neat distinction:
observational basis vs. physical basis. On this view, we could accept the claim "there
is a finite number of distinguishable electrons in the universe" on either of two bases.
First, we could count all the electrons, which would give us an observational basis.

This method, of course, is implausible. So we turn to the second alternative instead: we could deduce the number from 'the structure of our theory of the electron.' Apparently, de Sitter believes that if we accept our theory of the electron, then consequences of that theory are acceptable as well.

But ultimately de Sitter decides that even this plausibility is denied Eddington's proposal. In the end, until we can demonstrate that our theory of the electron supports the consequence Eddington claims, "the assertion that the universe is finite is a pure a priori assumption, which can be based only on philosophical or metaphysical grounds." Eddington's proposal—which de Sitter's apologizes for having "dwelt rather long on"—even "if it be adopted, we can decide which of the several possible solutions represents our actual universe only by making an a priori hypothesis" (de Sitter 1933a, 184).

Clearly, to make an a priori hypothesis is to do metaphysics—which the de Sitterian physicist hates worst of all to do. Thus is Eddington's pure, fundamental theory rejected.

De Sitter gives Milne's equally rationalist—that is, a priori—proposal precisely the same short shrift.

Milne's attempt to construct a cosmology with a unique solution soon came to be called 'kinematic relativity' for the simple reason that it relied solely upon special relativity's kinematics. The main feature of Milne's theory, which Robertson deemed "the cosmological principle," was quite distinctive. Milne introduces it in this way:

> According to these very elementary considerations, which only involve the principles of the special theory of relativity, ...every observer can regard himself as the center of the universe by choosing his time axis so as to point away from the time-space origin...the world is then perfectly egocentric at all points, and the moving picture of the world as made by any one observer is identical with that made by any other observer (Milne 1932, 10).

True to the spirit of STR, Milne begins from the point of view of the observer: what does any given observer—situated anywhere in the cosmos—see when they look out upon the universe? Milne's answer is strictly STR: they must each get the same universe, just as Einstein had earlier answered a similar question, saying that co-moving observers must get the same physics.

De Sitter doesn't like Milne's move one single bit. His view, just like Robertson's, was that the principle of relativity should be *extrapolated* from observations made in our neighborhood, and not somehow imposed upon the theory, a priori, based on some wild idea about alien observers "situated on faraway stars." Milne's proposal is rejected with a scoff:

> It has even been proposed to replace the principle of invariance by other principles expressing this relativity, such as the so-called "cosmological principle," which asserts that statistically the world pictures of two different observers must be the same. We have, however, no means of communicating with other observers, situated on faraway stars, or moving with excessive velocities, such as that of a β-particle, and the building of world structures on the

supposed experiences of these fictitious observers is equivalent to the introduction in disguise of certain specific assumptions regarding the interpretation
of our own observations. (de Sitter 1934, 598).

As always, by 'introducing specific assumptions' de Sitter means nothing other
than "a priori."[8] Note the negative connotations of the adjectives in this passage: "so-
called," "faraway," "excessive," "supposed," "fictitious," "disguise." It is clear that de
Sitter is having nothing at all to do with Milne's cosmological principle, and for two
reasons.

First, it had no observable consequences: we have no way to communicate with
aliens. Given this inability, it is otiose and metaphysical to even think about them.

Secondly, Milne hypothesizes his principle a priori, he imposes it as an initial
assumption of the theory.

Thus, Milne's proposal violates both the logic and the content of what de Sitter
takes to be "physical" as opposed to "metaphysical."

It is useful to look at Robertson—from whom de Sitter got the name cosmological
principle—on Milne. It is easy to see that the two critics stand together. In regard to
the principle Robertson remarks:

> It must be said, however, that there exists an essential distinction in the meta-
> physical status of the uniformity postulate in the two theories; whereas in
> relativistic cosmology it is set up as an extrapolation of observation for the
> purpose of determining a suitable ideal structural background for the actual
> world, it is used in Milne's theory as an *a priori* principle which may even be
> applied to the construction of a theory of gravitation (Robertson 1933, 158).

Robertson, exactly as de Sitter, views the status of the invariance/uniformity postulate as acceptable when it is based upon extrapolation; but unacceptable when based
upon hypothesis. Hypothesis involves a priori metaphysics; extrapolation involves induction over observation. Only the latter is science.

Together, Eddington and Milne are unscientific in de Sitter's view just insofar
as they do not practice science as inductive observationalists. Indeed, the two men
are wrong on both counts: they start from hypotheses generated in their minds, not
their perceptions; and they use deductive logic after their generalizations, not inductive logic before. That, at least, is de Sitter's claim against the metaphysicians. But de
Sitter was not alone in his view, he had a loud and enthusiastic teammate in Herbert
Dingle.

10.3 Dingle: Against the Metaphysicians

10.3.1 Dingle's background

When Dingle was 14, he had to drop out of school and go to work as a clerk. Habits
of mind he learned on this most empirical of jobs stuck with him throughout his long
life. After 11 years clerking and home study, he was awarded a Royal Scholarship for

Physics at Imperial College. His practical skills, especially with experimental machinery, led to his being appointed Demonstrator in the Physics Department even before his graduation in 1918. During his undergraduate work, Dingle had become a great admirer of Alfred Fowler, the justly famous spectroscopist. He followed Fowler into this field, focussing especially upon spectroscopy's astronomical applications. Dingle was very good at his work, and was elected Fellow of the Royal Astronomical Society already by 1922.

Dingle's work in spectroscopy was pioneering. His first interest was revealing the spectra of ionized and doubly ionized fluorine, which he did with great success. Then, because there was a strong need by astronomers for spectra of iron in the infrared, Dingle worked out ways to use the newly available IR-sensitive photographic plates to capture the spectra. At all times his work showed a strong committment to getting the empirical data, even if it required the most elegant and delicate observational methods or pushing new and untried techniques beyond what could have been expected.

But not all Dingle's efforts were concentrated on the empirical side of astronomy. From very early on he showed an abiding interest in two related but quite distinct subjects: relativity and philosophy of science. In 1922 he wrote a popular book entitled *Relativity for All*. The book was quite a hit, revealing not only Dingle's grasp of a subject still considered arcane by many, but also his enormous communicational skills, skills which would stand him in good stead throughout his often controverted and frequently polemical career. In 1932 Dingle got a Rockefeller fellowship. He went to CalTech to work on theoretical cosmology, mostly with R.C. Tolman, who was then fully engaged writing his classic masterpiece *Relativity, Thermodynamics and Cosmology*.[9] Dingle worked out the basic formula for a very general metric, which Tolman included in the book. Tolman also strongly encouraged Dingle to investigate the stability of the spatial homogeniety of the universe, a problem which had originally been tinkered with by Eddington and his students (McCrea and McVittie 1930; McVittie 1931; McVittie 1932). Dingle's results were published in MN the following year. In 1940 Dingle published a textbook on relativity in Methuen's well-regarded monograph series. The book was for a long time taken to be one of the best introductions to the topic.

Relativity was favored by Dingle for several reasons, not the least of which was his high regard for Einstein. But most important was Dingle's regard for the 'operational' philosophical perspective he believed embodied in the theory. Long an admirer of Mach's sort of positivist empiricism, Dingle was taken with P.W. Bridgman's more worked out and explicit version, operationalism strictly so-called (Bridgman 1928). Dingle's mature philosophy of science held that "the essential basis of science was the rational correlation of elementary experiences of the natural world;" indeed, as his Lowell Lectures at Harvard in 1937 publicly illustrated, Dingle was a stalwart "in the now unfashionable tradition of British empiricism" (Whitrow 1980, 335). As we shall soon see, Dingle's sometimes heroic empiricism would lead him, alongside de Sitter, against the Metaphysicians.

History and philosophy of science in Britain owe much to Dingle. He was one of the founders of the British Society for the History of Science, and served as its president from 1955–1957. In 1948 he initiated a subgroup of BSHS, which called itself the Philosophy of Science Group. A few years later the group split off and became the British Society for the Philosophy of Science; at about the same time Dingle was instrumental in launching *The British Journal for the Philosophy of Science*, which unto today remains one of the mainstays of the field worldwide. In the midst of all this, much to his genuine surprize, he was elected to the presidency of the Royal Astronomical Society. His 1953 Presidential Address was devoted almost entirely to a brilliantly witty and wickedly polemical assault on that dreaded spawn of the Metaphysicians, Hermann Bondi and his steady-state universe (Dingle 1953b).

It is now time to directly inspect what Dingle had against the Metaphysicians.

10.3.2 Esse est percipi

Dingle rejected metaphysics strenuously whenever and wherever he found it. What he took to be metaphysics was quite straightforward: he agreed with Carnap, "perhaps the leading exponent of the most active of modern schools of philosophy, the so-called 'logical positivism,'" to wit " 'I will call *metaphysical*' Carnap writes, 'all those propositions which claim to present knowledge about something which is over and above all experience.'" (Dingle is quoting from *Philosophy and Logical Syntax*, p. 15) But Carnap is not tough enough for Dingle, who goes on to say "I make this more rigorous: only that which is *practically observable*—that is, only that which would be observable if we were able to use known means of observation to the known limits of their possibilities—*is significant*." (Dingle 1938, 25). Metaphysics—claims about that which is not practically observable—is not just not science for Dingle, it is *not significant*, or, in the strongest positivist language, it is talk about nothing. For Dingle, to be is to be observed.

Dingle quite favors the positivists' criterion of metaphysical nonsense. He argues that it provides a very useful purgative to administer to science. The observational criterion of meaning

> is of great value, I think, as a purge, to rid our thinking of much that is meaningless ... An example given by Carnap illustrates this very well. From a book on metaphysics he selects a passage which runs somewhat as follows: 'outside being there is nothing. What is this nothing? It is that which ... ' And so on. Now here is a case, says Carnap, in which the imperfection of language has led to meaningless 'thought.' ((Dingle 1937a), p. 336)

Throughout the 1930s, Dingle attempted to administer his purgative to cosmology, focussing first and especially upon Milne, and only later upon Eddington.

Milne had announced his new theory in a short article appearing in an early June 1932 issue of *Nature*. (Milne 1932) A year later, Milne published a very long, very detailed exposition of his theory in *Zeitschrift für Astrophysik* (Milne 1933); immediately following his article were commentaries by Robertson (Robertson 1933) and Dingle. (Dingle 1933). Dingle's reaction was sharply negative, taking a stance against Milne's philosophy that he never gave up. The problem with Milne, Dingle claims, is that he repudiates the "fundamental principles of scientific method," namely, "Newton's principle of induction from phenomena." (Dingle 1933, 177–178). Relativity, at least as Dingle viewed it, fit squarely in the inductive and empiricist tradition stretching back to Newton. Milne violated both philosophical tenets: he was no inductivist, and, worse, he was no empiricist—indeed, he was an a priori metaphysician:

> The spirit of relativity is simply a re-affirmation of Newton's principle of induction from phenomena...Milne approaches the problems of physics in precisely the opposite way. He starts, not with phenomena, but with a hypothetical smoothed-out universe which must obey an arbitrary principle. (Dingle 1933, 178).

Rather than starting with observations over which inductive logic will range, Milne starts with an imagined[10] universe, and attempts "to deduce observed phenomena such as gravitation as we know it. But the possibility of doing so seems unlikely in the extreme." (Dingle 1933, 178). Yet, much as Dingle detested Milne's deductive logic, he found his a priori anti-empiricism even worse.

10.3.3 Those Metaphysical Aliens, Again

Of particular irritation to Dingle, exactly as it was to de Sitter, is Milne's cosmological principle. Indeed, Dingle agrees completely with de Sitter that Milne's principle is not an abstraction or extrapolation based in observation, but rather is an a priori assumption imposed upon the world model from the very start:

> the fundamental distinction between Milne's principle and the generally acknowledged principles of world structure, such as the principle of relativity and the laws of thermodynamics [is] that the former requires the events of nature to conform to it, whereas the latter are abstractions which are true (or false) whatever the events of nature are (Dingle 1933, 179).[11]

In exactly the same vein, Dingle, again just as de Sitter, did not in the least accept Milne's inventing alien observers on far away worlds:

> He introduces a 'cosmological principle,' which . . . means that the clock chosen by each observer in the universe must be such that all observers describe the universe in terms of its readings in precisely the same way. No such clock may be possible, of course; that depends on what the universe is like and what kinds of observers there are in it, and these things are beyond our knowledge. (Dingle 1937a, 247).

Alien observers, since they are beyond our knowledge—they are most certainly unobservable—are metaphysical constructs, according to Dingle; as such, they have no business in physics. Moreover, even if we could have knowledge of the existence of such observers, it still wouldn't be possible for all observers to observe the same universe with their bodily eyes, for Milne's construct was purely mental:

> the two observers will see the same universe not with the physical eye but only with a highly sophisticated mental eye which, so far as we can see, there is nothing to induce them to open (Dingle 1933, 173).

Milne's universe, according to Dingle, is so unsubstantial, so metaphysical, that his aliens couldn't be induced to even conceive it, let alone observe it.

Dingle's attack on the metaphysicians reached its high (or, perhaps, "low" is the better term!) point in 1937. In a deliberately provocative *Nature* article entitled "Modern Aristotelianism," Dingle took out against the Metaphysicians, the so-called 'modern Aristotelians,' most especially Milne and Eddington (Dingle 1937b).[12] Their crime? "The phenomenon may be described in broad terms as an idolatry of which 'The Universe' is the god ... This cosmolatry, as might be expected, came by metaphysics out of mathematics." (Dingle 1937b, 786). Metaphysics had beset cosmology like a vicious disease:

> Nor are we dealing with a mere skin disease which time itself will heal. Such ailments are familiar enough; every age has its delusions and every cause its traitors. But the danger here is radical. Our leaders themselves are bemused; so that treachery can pass unnoticed and even think itself fidelity. (Dingle 1937b, 786.

As we have seen before, Dingle, like de Sitter, believed steadfastly that the starting point for science was and must ever be, empirical observation. From thence, inductive logic—safe extrapolation—would provide the general principles and hypotheses needed for theoretical work. Dingle's statement of the issue is quite clear: "the first step in the study of Nature should be sense observation, no general principles being admitted which are not derived by induction therefrom." (Dingle 1937b, 786). Dingle saw it in essentially moral terms. The issue concerns what

> it was proper to do: Should we deduce particular conclusions from a priori general principles or derive general principles from observations? ... the question presented to us now is whether the foundation of science shall be observation or invention (Dingle 1937b, 785).

Obviously, the proper thing to do is not to "deduce particular conclusions from a priori general principles," and the foundation of science is not "invention." Milne and Eddington, just insofar as they both start from the "invention" of "a priori general principles," from whence to deduce "particular conclusions," are metaphysicians, they are modern Aristotelians, traitors to the Newtonian tradition, who deserve nothing but contempt from real scientists such as Dingle and de Sitter.

10.3.4 Undying fire

De Sitter died at about this time, and the European side of the anti-metaphysicalist campaign died with him. But even though basically alone, Dingle never did give up the fight. Although things got a bit quieter after the intense debate over "Modern Aristotelianism" finally settled down, Dingle himself did not settle down. He kept pushing his brand of basic empiricism as the only Real Science, attacking the hypothetico-deductivists whenever he could. His last gasp was the wild polemic he delivered against Milne's philosophical successor, Herman Bondi,[13] who began from 1948 putting forward another wild and crazy bit of rationalist metaphysics, namely, the steady state universe.[14] Heroic as always, Dingle left the lists as a Grand Old Curmudgeon.

10.4 Some (Very) Brief Concluding Remarks

De Sitter and Dingle behave in surprizingly similar ways in their attacks upon the metaphysicians. The similarities reach all the way from the identity of their targets to the language they use in their attacks. Such close resemblence calls out at least for an attempt at explanatory speculation. Herewith that attempt.

First and foremost must be mentioned their highly similar backgrounds and talents. Almost alone among their generation, the two men were thoroughly steeped in both the exquisite discipline of empirical astronomy, and the subtle arcana of relativistic theory. Moreover, and perhaps more importantly, when push came to shove, both scientists reverted to observation as the foundation of their science. While neither actually mistrusted theory, certainly neither gave it precedence either. At bottom, both men were empiricists, old-fashioned, classical empiricists, who trusted their observational skills above all else. Finally, both men relied upon their inductive ability to take them from their observations to their theoretical generalizations. Imagination played a role in their science, but only as subsidiary to empirical observations.

There is no doubt that their own suite of professional talent and experience put them together against the metaphysicians.

But a final influence must be remarked as well. Even cursory inspection of the language both men use to describe the issues in their disagreement with the metaphysicians reveals a common source: the waxing conceptual scheme parlayed by Carnap and others in the logical positivist movement. Dingle's reliance upon their ideas is explicit—he tells us about it. But de Sitter's choice of terms and perspectives of attack show similar reliance, even if he doesn't explictly tell us their provenance. A clear task for future scholarship would be the discovery and elucidation of de Sitter's connections with positivist thought.

In the end, it would seem that similar experience and talent provided de Sitter and Dingle the motivation for their campaign against the metaphysicians, this campaign provisioned and sustained by Carnap and his company of logical positivists. It was quite an alliance, and quite a campaign.

References

Bridgman, Percy W. (1928). *The Logic of Modern Physics*. Macmillan, New York.

Deprit, Andre (1995). Monsignor Georges Lemaitre. In *The Big Bang and Georges Lemaître*. A. Berger, ed. D. Reidel, Dordrecht.

de Sitter, Willem (1917). On Einstein's Theory of Gravitation, and its Astronomical Consequences. *Monthly Notices Roy. Astr. Soc.* **78**, 3–29.

— (1930). On the Distances and Radial Velocities of Extra-Galactic Nebul, and the Explanation of the Latter by the Relativity Theory of Inertia." Proceedings of the National Academy of Sciences 16, 474-488.

— (1931a). Contribution. In *The Evolution of the Universe*. Herbert, Dingle, ed. *Nature*, London.

— (1931b). The Expanding Universe. *Scientia* **49**, 1–10 .

— (1932). *Kosmos*. Harvard University Press, Cambridge.

— (1933a). Astronomical Aspect of the Theory of Relativity. In *University of California Publications in Mathematics*. Vol. II. D. N Haskell and M. W. J. H. McDonald Lehmer, eds. University of California Press, Berkeley.

— (1933b). On the Expanding Universe and the Time-scale. *Monthly Notices Roy. Astr. Soc.* **93**, 628–634.

— (1934). On the Foundations of the theory of Relativity, with Special reference to the Theory of the Expanding Universe. *Proceedings Royal Academy Amsterdam* **37**, 597–601 .

Dingle, Herbert (1931). *The Evolution of the Universe*. *Nature*, London.

— (1933). On E.A. Milne's theory of world structure and the expansion of the Universe. *Z. Astrophysik* **6**, 173–179.

— (1937a). *Through Science to Philosophy*. Clarendon, Oxford.

— (1937b). Modern Aristotelianism. *Nature*, London, **139**, 784–786.

— (1938). Science and the Unobservable. *Nature*, London, **141**, 21–28.

— (1953a). Science and Modern Cosmology. *Monthly Notices Roy. Astr. Soc.* **113**, 95–110.

— (1953b). The President's Address. *Monthly Notices Roy. Astr. Soc.* **113**, 393–407.

Eddington, Arthur S. (1930). On the Stability of Einstein's Universe. *Monthly Notices Roy. Astr. Soc.* **90**, 668–678.

— (1931). Contribution. In The Evolution of the Universe. Herbert, Dingle, ed. *Nature*, London.

— (1932). *The Expanding Universe*. Ann Arbor Paperbacks-U.Michigan Press, Ann Arbor.

Gale, George (1992). Rationalist Programmes in Early Modern Cosmology. *Astr. Qtly.* **8**, 4–13.

Gale, George and Urani, John (1999). Milne, Bondi and the 'Second Way' to Cosmology. In *The Expanding Worlds of General Relativity*. H. Goenner, ed. Birkhäuser, Boston.

Hins, C. H. (1935). In Memoriam Prof. Dr. W. De Sitter. *Hemel en Dampkring* **33**:1, 1–18.

Jones, Harold Spencer (1935). Willem De Sitter. *Monthly Notices Roy. Astr. Soc.* **95**, 343–347.

McCrea, William H. and McVittie, George C. (1930). On the Contraction of the Universe. *Monthly Notices Roy. Astr. Soc.* **91**: 128–133.

McVittie, George C. 1931. The Problem of n Bodies and the Expansion of the Universe. *Monthly Notices Roy. Astr. Soc.* **91**, 274–283.

— (1932). Condensations in an Expanding Universe. *Monthly Notices Roy. Astr. Soc.* **92**, 500–50.

— (1987). An Anglo-Scottish University Education, In *The Making of Physicists*. Rajkumari Williamson, ed. Hilger, Bristol.

Milne, Edward A. (1932). World Structure and the Expansion of the Universe. *Nature*, London 9–10.

— (1933). World-structure and the Expansion of the Universe. *Z. Astrophysik* **6**, 1–35.

Milne, Edward A. et al. (1937). On the Origin of Laws of Nature. *Nature*, London, **139**, 997–1007.

North, John D. (1994). *The Fontana History of Astronomy and Cosmology*. Fontana Press, London.

Robertson, Howard P. (1933). On E.A. Milne's Theory of World Structure. *Z. Astrophysik* **7**, 152–172.

Whitrow, Gerald J. (1980). Herbert Dingle, *Q. Jl. Roy. astr. Soc.* **21**, 333–338.

Notes

[1] Although de Sitter here uses the term "aesthetical," he more frequently says "philosophical." (Cf. e.g., (de Sitter 1932, 113).) Sometimes "metaphysical" is used instead, although this term more frequently is not used as neutrally as the present context requires. Typically, "metaphysical" serves de Sitter as a pejorative (de Sitter 1932, 5).

[2] For a more thorough discussion of Eddington's theory, cf. (Gale 1992).

[3] Two good obituaries are available: (Hins 1935), (Jones 1935).

[4] De Sitter received the 1931 Darwin Prize from the Royal Astronomical Society; his acceptance lecture was entitled "Jupiter's Galilean Satellites."

[5] It is obvious that de Sitter has adopted the positivist demarcation criterion. What is not so obvious is where he got it. In none of the published works reviewed herein have I found a reference to any of the usual suspects, Carnap, Bridgman, et al. It would be interesting to track down his sources. But, in any case, his reliance upon the demarcation criterion is clear and evident.

[6] The term functions as a technical one for de Sitter.

[7] As discussion during HGR6 pointed out, there is a wealth of material testifying to the close friendship of the two men, particularly their correspondence. It is to be strongly hoped that this valuable colleagueship will soon be studied.

[8] Cf. "the assertion that the universe is finite is a pure a priori assumption;" and "...only by making an a priori hypothesis, which is practically equivalent to the choice of a particular solution." (de Sitter 1933a, 183–184).

[9] The book is still in print 2002.

[10] Dingle was death on a priori imaginings. As he later declared "I hope I need not say that none of the considerations I have put before you should tend in the smallest

degree to diminish the importance that belongs to imagination in science, when the word is used in its true sense as referring to the ability to form vivid images of possible happenings, and not in the sense in which I have had sometimes to use it, as indicating the invention of arbitrary postulates ... By all means keep imagination free, but let it be directed, and let its products be examined and properly assessed before they are announced as discoveries of the order of nature. Even idle speculation may not be quite valueless if it is recognized for what it is." (Dingle 1953b, 404).

[11]Compare de Sitter: "the building of world structures on the supposed experiences of these fictitious observers is equivalent to the introduction in disguise of certain specific assumptions regarding the interpretation of our own observations." (de Sitter 1934, 598).

[12]Dingle had written to the editor of *Nature*, volunteering to write just such an article. His offer was instantly, and gratefully, taken (Dingle correspondence, Archives, Imperial College, London).

[13]For a full account of this episode, cf. (Gale and Urani 1999).

[14]By "last gasp" here, I intend to limit my case to the philosophical. As is well known, Dingle waged a fearsome battle against proponents of the special theory of relativity, whose principle spokesman was William McCrea. But this was a battle over content, not philosophy, at least as nearly as I can tell.

11

George Gamow and the 'Factual Approach' to Relativistic Cosmology

Helge Kragh

History of Science Department, University of Aarhus, DK-8000 Aarhus C, Denmark;
ivhhk@ivh.au.dk

11.1 Introduction: Styles of Cosmology

The somewhat slippery notion of "scientific style" has been widely discussed by historians, sociologists and philosophers, both when it comes to style associated with individuals, with schools and institutions, and—what is rather more controversial—with nations and cultures.[1] Although there is no agreement on the precise meaning of the term, in general scientific styles denote characteristic attitudes to and ways of doing science, indeed different conceptualizations of what science, or a particular science, is all about. In one meaning of the term, styles function like Kuhnean paradigms. They determine what legitimate or interesting science is, which may lead to foundational conflicts between different claims of scientific styles. The notion of style in science is admittedly loose and difficult to pin down, but it is not empty. In my view, the history of modern cosmology, especially the period from 1910 to 1970, offers good and convincing examples of different kinds of style. Thus, a large part of the protracted controversy between evolution and steady-state cosmologies can be seen as a conflict between two widely different styles of cosmological research (Kragh 1996b).

Whereas the cosmological controversy in the 1950s involved styles shared by two different groups of scientists, here I am more concerned with a particular cosmologist's style of science, which includes his conception and use of the general theory of relativity in cosmology. George Gamow was neither a contributor to nor a great specialist in the theory of relativity but he used Einstein's theory most effectively in his construction of a theory of the exploding universe, the first modern version of big-bang cosmology. In general it seems to me that scientists' use of theory has been somewhat neglected in the history of the physical sciences, where the creation of theory and its further development through testing and theoretical refinement have very much dominated. This is contrary to, e.g., the history of technology, where there is a long tradition for dealing not only with inventors' creative work but also with the ways in which the inventions were used and transformed in daily life situations. The ways in which scientists use a more or less well-established theory may differ considerably, and such uses often reflect different styles to no less a degree than what can be found in the creation of theories. In my view, they deserve more attention.

11.2 Gamow's Early Career

The Russian-American physicist George Antonovich Gamow has an important place in the annals of twentieth-century physical science, primarily because of his pioneering work in nuclear physics and its applications in astrophysics and cosmology (Reines 1972; Harper, Parke, and Anderson 1997; Harper 2001). After a year at the Novorussia University in Odessa, he enrolled as a physics student at what was still the University of St. Petersburg (or Petrograd) but was soon to be renamed the University of Leningrad. Young Gamow thrived in the intellectual atmosphere of the former capital of Tsarist Russia and became part of a group of physicists and physics students that included Lev Landau, Victor Ambarzumian, Matvei Bronstein, Dmitri Iwanenko, and Vladimir Fock. He later recalled: "The subject which fascinated me most from my early student days was Einstein's special, and especially general, theory of relativity, and I had quite a lot of somewhat uncoordinated knowledge in this field" (Gamow 1970, 41). His knowledge became more coordinated when he, in 1923–1924, attended lectures by Alexander Friedmann, who had recently shown that Einstein's cosmological field equations include non-static solutions. Unfortunately, we do not know precisely what subjects Friedmann taught and in which way, except that the title of his course was "Mathematical Foundations of the Theory of Relativity." It seems that Friedmann left a strong impression on the young student, who wanted him to be his supervisor, first for his diploma work and eventually also for his later dissertation. However, due to Friedmann's premature death in 1925, nothing came of the plan. Although Gamow was only 21 years old when Friedmann died, he continued to consider himself a pupil of Friedmann.[2]

During the 1920s, relativity received focal attention among the young theorists of the Leningrad school of physics, which included many of the future leaders of Soviet and international physics. The interest appeared in many ways and included the very first research paper that was published under Gamow's name. Together with Iwanenko, his fellow student and close friend, he submitted in the fall of 1926 a short paper to *Zeitschrift für Physik* on the five-dimensional formulation of wave mechanics (Gamow and Iwanenko 1926). Shortly after the introduction of wave mechanics, Oskar Klein had extended Schrödinger's theory by formulating it in the framework of five-dimensional relativity theory earlier proposed by Theodor Kaluza. During the years 1926–1928, the Kaluza–Klein approach attracted considerable attention, not least among the theoretical physicists in Leningrad (Kragh 1984). Independently of Klein, Fock obtained in the summer of 1926 a five-dimensional wave equation that corresponded to the one found by Klein a few months earlier. Gamow and Iwanenko adopted Fock's formulation and merely followed what at the time looked like becoming a trendy area of theoretical physics. They assumed microscopic space to be Euclidean and hence disregarded gravitational fields. Although their paper was of little significance, it does demonstrate Gamow's familiarity with the formalism of the general theory of relativity.

Gamow soon found his own, rewarding niche in the new quantum physics. Less than two years after his paper with Iwanenko, he published his important wave-mechanical explanation of alpha radioactivity, a pioneering paper in quantum nuclear physics that made him internationally known (Gamow 1928; Stuewer 1986). During the following decade, first in the Soviet Union and from 1933 onwards in the United States, he developed into one of the world's foremost experts of theoretical nuclear physics, a field in which general relativity was however irrelevant.[3] Still, during the many years that Gamow concentrated on the atomic nucleus, he did not entirely forget about his early interest in relativity. A glimpse of such interest may be inferred from a plan he proposed in 1931 for an Institute of Theoretical Physics under the Russian Academy of Sciences. He proposed the institute to be composed of four sections, one of which should deal with "theoretical astrophysics," an area that he divided into two parts, one being "the structure of the interior of stars" and the other "problems of cosmology" (Frenkel 1994, 784). Although nothing came of the plan, it is remarkable that there was at this early date a serious proposal to establish an institute with a section focusing on cosmology.

11.3 From Astrophysics to Cosmology

The situation of the general theory of relativity during the years 1925–55 has aptly been called a "low water mark," mostly because of the theory's failure in making connection to experiments (Eisenstaedt 1986, 1989). Cosmology was about the only area of observational science that seemed relevant to general relativity, and the role of observations in cosmology was still a matter of dispute. Whereas most physicists and astronomers at the time came to theoretical cosmology either through the general theory of relativity or rival theories of space and time, the case of Gamow was different. His route started in astrophysics, a field that in the 1930s became increasingly associated with the nuclear physics in which he was recognized as an authority (Nadyozhin 1995). Although I know of no direct documentary evidence, most likely he was also influenced by his friend Bronstein's interest in cosmology. For example, in 1933 Bronstein had examined Lemaître's theory of the expanding universe and suggested to modify it by including a time variation of the cosmological constant (Bronstein 1933). Incidentally, given the current interest in constants of nature changing with time (Uzan 2002), Bronstein's little-known work may attract new attention. It is possibly the first suggestion of a time-varying constant of nature in the history of modern physics.

A significant part of Gamow's work in the latter half of the 1930s dealt with astrophysical problems. They covered a wide range, from stellar energy over supernovas to galaxy formation, but soon Gamow focused on the formation of the chemical elements in stellar and, eventually, cosmological processes. He was well acquainted with Carl Friedrich von Weizsäcker's idea of element formation in a prestellar, highly compact state of the universe, a scenario that he had discussed with von Weizsäcker in the summer of 1938, before it appeared in print (Weizsäcker 1938; Kragh 1996b, 97–101). Although Gamow did not immediately take up the suggestion, it seems to have continued to occupy a corner of his mind. He referred to it in his popular book of 1940, *The Birth and Death of the Sun*, where he found that the idea of a superdense and supercompact early universe was supported by what he called "good physical reality" (Gamow 1940, 201). In an interesting but somewhat neglected paper of 1939, dealing with galaxy formation and written jointly with Edward Teller, his friend and colleague at George Washington University, Gamow dealt for the first time explicitly with the Friedmann–Lemaître equations. The two authors concluded that, "our theory of nebular formation requires that the velocity of expansion remain [sic] nearly constant while the distances between nebulae increase by a factor 600" (Gamow and Teller 1939, 656). In the last section, they considered the cosmological consequences which they discussed from the form of the fundamental (Friedmann–Lemaître) equation for the expanding universe as given by Richard Tolman in his important textbook of 1934 (Tolman 1934, 394–405), namely

$$\frac{de^{g/2}}{dt} = \left(\frac{8\pi G \rho}{3} e^g - \frac{c^2}{\mathfrak{R}^2} \right)^{1/2}. \tag{11.1}$$

$\mathfrak{R}$ is the radius of curvature, ρ the mean mass-energy density, and the factor $e^{g/2}$ expresses through $g = g(t)$ the time variation of the radius of the universe. In a slightly rewritten form, the equation can be written as

$$\frac{1}{R} \left(\frac{dR}{dt} \right) = \left(\frac{8\pi G \rho}{3} - \frac{kc^2}{R^2} \right)^{1/2}, \tag{11.2}$$

where R is the scale factor, k the curvature parameter, and the quantity on the left side denotes Hubble's constant for the expansion rate.[4] Substituting the then accepted present values for Hubble's constant ($H_0 = 1.8 \cdot 10^{-17}$ sec^{-1}) and the density of the universe ($\rho_0 = 10^{-30}$ gcm^{-3}), Gamow and Teller found that the density term in Equation (11.2) was much smaller than the curvature term; consequently the space curvature must be negative, or $k = -1$: "Thus in order to understand the formation of great nebulae and to satisfy the condition of continuity at the moment of their separation, it is necessary to accept the hypothesis that space is infinite and ever expanding" (Gamow and Teller 1939, 657). As they noted, this went against the conclusion that Edwin Hubble and Tolman had reached in 1935, namely, that the universe must be closed and uncomfortably small. However, Gamow and Teller argued that the discrepancy might be resolved if it was admitted that the absolute luminosities of very distant (hence young) galaxies were higher than those nearer by. This idea was to play an important role in cosmology in the 1950s, but in 1939 it was merely a suggestion that lacked independent support.

Gamow's work with Teller of 1939 was based on the expanding universe but did not presuppose any explosive event in the past, what ten years later would be coined the big bang. All they had to say about the state of the universe before the separation of the nebulae about 1.8 billion years ago was, "Before that time, space must have been uniformly populated by stars, or by gas molecules, if we suppose that the formation of stars took place after the separation of nebulae" (Gamow and Teller 1939, 655). Although still in the pre-big-bang tradition, the paper was in several respects to serve as a kind of blueprint for Gamow's later contributions to cosmology. Also noteworthy, if not exactly important, is the acknowledgment at the end of the paper, where "Mr. C. G. H. Tompkins" was thanked for having suggested the topic. The referee may have wondered who Mr. Tompkins was, but if so it became apparent later in the year when Cambridge University Press published the best-selling *Mr. Tompkins in Wonderland*.[5]

During the 1942 Washington Conference on Theoretical Physics, Gamow and the other participants discussed how heavier elements could be built up by nuclear processes, and they reached the important conclusion that "the elements originated in a process of explosive character, which took place at the 'beginning of time', and resulted in the present expansion of the universe" (Gamow and Fleming 1942, 580). Although the expansion of the early universe was an important ingredient in Gamow's first attempts to provide a cosmological explanation for the formation of the elements, until 1946 the Friedmann–Lemaître equations did not enter—nuclear physics and the general theory of relativity were still treated as if they belonged to separate worlds. The first evidence of what would eventually become the standard model of evolutionary cosmology appeared in a letter to Niels Bohr of October 24, 1945, in which Gamow wrote that he was presently engaged in "studying the problem of the origin of elements at the early stages of the expanding universe. It means bringing together the relativistic formulae for expansion and the rates of thermonuclear and fission reactions" (Kragh 1996b, 106).

11.4 General Relativity à la Gamow

Between 1946 and 1956, Gamow wrote nine papers on the cosmology of the early universe, either by himself or with his collaborators Ralph Alpher and Robert Herman. In this series of pioneering papers on physical big-bang cosmology (a term introduced in 1949 and which Gamow never used), he developed a research program that was very much his own.[6] After 1956, he largely stopped working on cosmology. In Gamow's program, the general theory of relativity, or rather the Friedmann–Lemaître equations, were used as an unproblematic tool, just like he, when working with nuclear physics, would use the Schrödinger equation as a tool. Gamow was a user of the relativistic theory of cosmology, not a contributor to it and certainly not a critic of it. He showed no interest in either its mathematical subtleties or philosophical problems, and he stuck to the cosmological field equations in their standard formulation. That is, he accepted the cosmological principle and therefore also the Robertson–Walker metric that led to the simple Friedmann–Lemaître equations in the form (11.2). Whether working in nuclear physics or in cosmology, he was a great believer in simplicity and always kept to concepts and mathematical techniques that were as simple and transparent as possible.

Although in some areas Gamow had an inclination toward speculation, when it came to Einstein's theory of general relativity he was orthodox. He never seriously considered rival theories, such as those proposed in the 1930s by Edward A. Milne, Paul Dirac, and Pascual Jordan, or the steady-state theory of Fred Hoyle, Thomas Gold, and Hermann Bondi that entered the cosmological scene in 1948. In 1937–38, Paul Dirac had suggested a new model of the expanding universe based on what he called the Large Number Hypothesis, from which he derived that the gravitational constant slowly decreases with the cosmic era as $G(t) \sim t^{-1}$. Although Gamow found

Dirac's $G(t)$ suggestion fascinating because of its foundation in the Large Number Hypothesis, he flatly denied that the idea could be physically correct. Gamow was greatly interested in dimensionless numbers formed by combination of natural constants, but his inclination toward empirical and testable physics was always stronger than his attraction by rationalist numerology. If a reading of his works, especially the more popular ones, may give a different impression, it is mainly a result of his rhetorical style, because Gamow enjoyed acting as a self-styled *agent provocateur*. Thus, in 1949 he referred to Dirac's Large Number Hypothesis as "philosophically ... a most satisfying one" only immediately thereafter to dismiss the varying-G hypothesis. Also late in life, when he corresponded with Dirac on the matter, he kept steadfastly to a constant G, in part because of empirical reasons but also because he realized that Dirac's hypothesis did not agree with the general theory of relativity in which he had absolute confidence. (See Kragh 1991 for details.)

In his series of papers on the explosive universe, Gamow used the Friedmann–Lemaître equations with almost no variation. The starting point was always the same as in his work with Teller of 1939, the equation adopted from Tolman's textbook. He invariably concluded that the present curvature of space must be negative and the universe thus be open and ever expanding. The greatest change in his conceptualization of the early universe occurred in 1948, when he, about the same time as Alpher, realized that the very early universe must be dominated by radiation energy rather than matter. The notion of a hot big bang appeared first in print in *Nature*, in a paper where Gamow argued that in the earliest phase of the universe the matter-energy and curvature terms would be negligible compared to the radiation energy ρ_r (Gamow 1948). Consequently he wrote the Friedmann–Lemaître equation as

$$\frac{1}{R}\left(\frac{dR}{dt}\right) = \left(\frac{8\pi G\rho_r}{3}\right)^{1/2}. \tag{11.3}$$

By ignoring the matter density term, he narrowly missed the opportunity to predict the cosmic background radiation. The opportunity was seized upon slightly later by Alpher and Herman, who were more careful in their work than the notoriously sloppy Gamow.

The only major variation in Gamow's handling of the Friedman–Lemaître equation appeared in the 1949 festschrift issue of *Reviews of Modern Physics* dedicated to Einstein's 70-year's birthday, an important source for the history of modern cosmology. Not only does it include Kurt Gödel's relativistic model of a universe allowing time travels, and Tolman's very last paper, published posthumously; it also includes review articles by Lemaître and Gamow that strikingly illustrate the different interests and styles of the two founding fathers of big-bang cosmology.

In his invited contribution to the festschrift, Gamow started by declaring that he intentionally disregarded non-Einsteinian cosmologies because his aim was "to see whether or not the problems of cosmology and cosmogony can be understood entirely on the basis of the 'old fashioned' general theory of relativity in its original form proposed by Einstein" (Gamow 1949b, 367). His attitude was much like that of Tolman, another orthodox relativist who had no doubts that Einstein's theory of gravitation was applicable in cosmology and who expressed himself rather more strongly on the matter than Gamow did. According to Tolman, to abandon the general theory of relativity as the secure basis of cosmology was an invitation to "speculations" that allowed "free rein to unbridled fancy." He was willing to consider non-homogeneous models or even the cosmological constant, but any attempt to solve cosmological problems just had to remain within the limits of standard general relativity:

> [I]t is reasonable to regard general relativity as a development which like others before it will sometimes find its place in some broader theoretical structure. Nevertheless, general relativity provides our present best theory of gravitation—and a very good one at that—and it is my opinion that this is the appropriate theory of gravitation to use in treating the motions of the nebulae (Tolman 1949, 377).

Like Tolman, Gamow was acutely aware of the age or time-scale problem, the paradoxical situation that the universe seemed to be younger than the earth and the stars (Kragh 1996b, 73–79, 271–275; Brush 2001). The problem was widely discussed at the time and it was one of the reasons behind the new alternative of a steady-state model of the universe. In 1949, Gamow suggested to take care of it by simply reintroducing the cosmological constant, a standard remedy that Lemaître had advocated since the early 1930s. That is, rather than basing the theory on Equation (11.2), he would base it on the extended equation

$$\frac{1}{R}\left(\frac{dR}{dt}\right) = \left(\frac{8\pi G\rho}{3} + \frac{\Lambda}{3} - \frac{kc^2}{R^2}\right)^{1/2} \tag{11.4}$$

where Λ is the cosmological constant. He noted that for $\Lambda = 0$ and Hubble's value of the expansion parameter, the ρ-term must be negligibly small, which leads to a linear dependence of R on the time and therefore to an age of the universe close to 1.8 billion years, which was smaller than the accepted age of the earth as based on radiometric dating methods. Gamow found that the problem could be solved by assigning to Λ a positive value so small (about $8.6 \cdot 10^{-34} s^{-2}$) that it did not affect local astronomy. Moreover, the curvature of space would still have to be negative and the universe thus be in an eternally open state. By 1950, Λ was generally an unwelcome parameter which was embraced only by Lemaître and a few other cosmologists. Gamow followed the mainstream in disregarding it, and so it may seem surprising that it appeared in his 1949 paper, and only there. He introduced it without further ado as a saving device, and he was apparently undisturbed by using it in what was clearly an ad hoc way. The cosmological constant might be an arbitrary quantity, but, on the other hand, it was at his disposal, so why not use it? Contrary to Lemaître, Gamow had no strong feelings about Λ, neither of an aesthetic nor a methodological kind. After 1952, when Walter Baade and others re-evaluated the Hubble constant, he quietly dropped Λ and never returned to it.

Gamow's inclination toward technical simplicity in his cosmological works is underlined by a comparison between his research papers in the period and his popular book of 1952, *The Creation of the Universe*. Remarkably, in this book he introduced the idea of a big-bang universe in a manner and at a technical level that did not differ drastically from his papers in *Physical Review*. He even included, albeit in an appendix, the Friedmann–Lemaître equation and discussed its classical analog in much the same way as he had done in his research papers. Contrary to most scientific authors, Gamow did not respect the strict border that is supposed to distinguish research papers from popular writings. As he admitted, he started writing popular works "probably because I love to see things in a clear and simple way, trying to simplify them for myself" (Gamow 1970, 155). His books were received with enthusiasm by the public, but they probably harmed his academic reputation among his colleagues in physics and astronomy. "Gamow committed an unforgivable sin," Wolfgang Yourgrau wrote, referring to his best-selling popular books. "Most scientists do not fancy the oversimplifying, popularizing of our science . . . it is tantamount to a cheapening of the sacred rituals of our profession . . . many of us considered him washed up, a has-been, an intemperate member of our holy order" (Yourgrau 1970, 38).

11.5 Broader Issues

Although physical cosmology can be traced back to the 1920s, in a modern sense, with nuclear and particle physics being of crucial importance in the study of the early universe, it was very much Gamow's invention. In his version, it included a robust, no-nonsense approach to the study of cosmology that was thoroughly permeated with instrumentalist ideas of science adopted from his work in nuclear physics. The approach differed in important respects from what had previously been the favored approach to the study of the universe. What may be called the Gamow style of cosmology rested on doctrines that were rarely mentioned but were tacitly accepted by Gamow and his few collaborators and sympathizers. Two such doctrines or themes may be singled out:

(1) *There is neither place nor need for philosophical and metaphysical questions; if such questions turn up, ignore or circumvent them.* Thus, Gamow and his coworkers Alpher and Herman simply started their calculations briefly after the magical moment $t = 0$, in a pre-existing mini-universe, and they did not concern themselves with the difficult question of the "beginning" at $t = 0$. Theirs was a creation cosmology, but not in the *creatio ex nihilo* sense, only in the sense of explaining matter and radiation as the creation from an earlier state. Gamow, who should probably be labeled either an atheist or an agnostic, did not associate at all the term "creation" with any theological connotation. It merely meant "making something shapely out of shapelessness," as he explained in 1952 (Gamow 1952, second printing, preface). Likewise, he refrained from speculating about a cosmic singularity and held with Einstein, Lemaître, and others that the gravitational field equations would break down very near $t = 0$.

On the other hand, Gamow could not entirely resist the temptation to speculate about the origin of the universe and what was possibly before it. In several of his works from about 1950, he pictured the history of the universe not as beginning at $t = 0$ but as an eternally existing, rebounding universe. That is, he imagined a hypothetical collapse that had preceded the present expansion. As he noted, although such a previous state cannot be known physically, it is allowed mathematically since it corresponds to a solution of the Friedmann–Lemaître equations. Gamow's speculation did not belong to the age-old tradition of cyclic universes, as revived by Friedmann who had first discussed such models from a relativistic point of view in his seminal paper of 1922. A cyclic or oscillatory universe would be inconsistent with Gamow's insistence on the present universe being open. He thought of a cosmic one-cycle process where the universe evolved from infinite rarefaction over a superdense state toward a new state of infinite rarefaction. In this way a causal explanation of the primordial universe could be imagined. However, Gamow was careful to point out that the idea was purely speculative and hence of no real scientific value. As he wrote in 1954: "From the physical point of view we must forget entirely about the precollapse period and try to explain all things on the basis of facts which are no older than five billion years—plus or minus five per cent" (Gamow 1954, 63).

(2) *Cosmology should, like any branch of physics, be based on accepted physical knowledge and the ordinary methods of science.* Thus, Gamow considered the very early universe merely to be an extremely hot and compact crucible, an exotic laboratory for nuclear-physical calculations. His approach was conservative in the sense that he saw no need to introduce new principles of physics. The universe should and in fact could be understood as any other physical system, that is, basically in terms of known particles and the laws of quantum mechanics, thermodynamics, and general relativity. In a remarkable address at a conference in Denver, he coined the name *factual cosmology* for the approach "to accept the physically established laws governing matter and radiation and look for cosmological models which are derived on the basis of these laws and are consistent with astronomical observations" (Kragh 1996b, 136). Gamow did not deny that cosmology posed difficult problems of a conceptual and methodological nature, but he found it unprofitable to dwell on these as long as progress could be obtained by the tested methods of physics. In his Denver address, he not only distanced himself from more rationalistic versions of cosmology—what he called *postulatory cosmology*—he also described his own position as an engineering attitude. The factual cosmologist studying the universe, he said, approached the problem in essentially the same way as "when an engineer wants to design an automobile, a jet plane, or a spaceship, he starts with the well-known physical and chemical properties of the materials he uses and looks for the arrangement of these materials which would satisfy his purposes." It was an attitude that differed drastically from the view of Bondi, William McCrea and most other steady-state cosmologists, who denied that cosmology was merely a branch of physics and to whom cosmology and engineering were worlds apart.

11.6 The Legacy of Gamow's Approach to Cosmology

About 1950, when Gamow developed his program of physical cosmology, there was a great deal of uncertainty and disagreement about the foundation of the field and its proper methodology. It was still at a time when it could be discussed, and in fact was discussed, whether or not cosmology was a science in the first place. Roughly speaking, one can place most of the cosmologists of the period from about 1940 to 1965 in a kind of methodological spectrum that ranges from, say, extreme pragmatism to extreme rationalism. Gamow and his collaborators Alpher and Herman clearly were at the extreme pragmatic end, whereas I would place Milne at the opposite, rationalist extreme, with most of the steady-state cosmologists belonging to the rationalist part as well (but with Hoyle more towards the center). I would also count Tolman, Lemaître, and George McVittie to the pragmatic or empiricist part, although they had in fact little in common with Gamow and his group, whose view of cosmology was still foreign to the majority of physicists and astronomers.

Cosmologists could favor a rationalist or an empiricist attitude in different ways, for example by subscribing more or less dogmatically to the standard theory of general relativity. To many cosmologists of the period, general relativity was the theoretical framework of cosmology and what mattered was to master the theory as applied to the universe as a whole. Such an attitude could lead to models anywhere on the methodological spectrum. Many cosmologists of a rationalist bend, including most of the steady-state theorists, valued general relativity highly, only they did not accept it as a valid theory of the universe. As illustrated by the case of Kurt Gödel, there was no guarantee that a theory safely anchored in general relativity did not lead to flights into rationalist fancies (Gödel 1949). Another case, of a different kind, is that of Oskar Klein, who had no doubt about the truth of general relativity but did not believe that the universe could be described consistently in terms of a cosmological model, relativistic or not (e.g., Klein 1968). In the figure below an attempt is made to show in an impressionistic manner the position of leading cosmologists in the period 1935-65 with regard to methodological preference and subscription to the general theory of relativity.

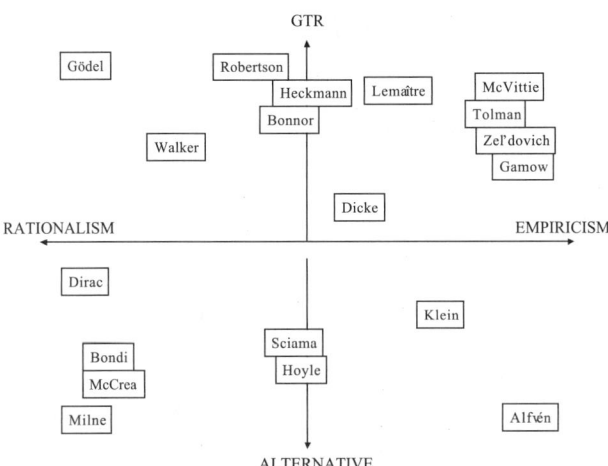

Gamow's style of cosmology may be illuminated by briefly comparing it with that of two of his great contemporaries, Georges Lemaître and George McVittie. As to

Lemaître, he failed to appreciate Gamow's research program, and that in spite of the obvious similarities between his own *atome primitif* and Gamow's compact inferno of nuclear particles. Lemaître was a relativity *afficionado* and he always considered Einstein's general theory of relativity the supreme paradigm of physical science, whereas particle and nuclear physics did not appeal to him. In these partly phenomenological branches of physics he could find nothing corresponding to the rigor and beauty of general relativity. Like Gamow, Lemaître believed that the universe was accessible to the human mind and that it could be described in simple terms. But his notion of simplicity was mathematical, derived from the theory of relativity and not from the nuclear physics which was at the heart of Gamow's view of cosmology.[7] Far from being a cosmo-engineer, he was a relativist natural philosopher. Characteristically, although the papers by Lemaître and Gamow in the Einstein festschrift of 1949 appeared side by side and carried very similar titles, and although they shared a common background in general relativity, they were quite different in nature. Lemaître did not refer to Gamow's theory of thermonuclear synthesis of the chemical elements but instead proposed that hydrogen and helium were produced by cosmic rays interacting with gaseous clouds (Lemaître 1949). Nor did Gamow refer to Lemaître.

McVittie's methodological position is interesting because he moved along the spectrum, starting within the rationalist tradition of Milne and Eddington and ending as a staunch advocate of what he called the empiricist or observational school. All along the route, he stuck to the field equations of general relativity as the sure basis of theoretical cosmology. In a paper of 1961, he distinguished between the empiricist school and the rationalists who, he charged, tended to substitute logic for observation. McVittie's classification was nearly the same as Gamow's distinction between factual and postulatory cosmology, and yet McVittie conceived of cosmology in quite a different light than Gamow did. According to McVittie, cosmology was essentially an interplay between the general theory of relativity and astronomical observations. He was an empirical cosmologist, but hardly a pragmatic in Gamow's sense, and just as little as Lemaître did he feel at home with Gamow's emphasis on the physical properties of the early universe. As he correctly pointed out, the general theory of relativity predicts no big bang, no nuclear explosion, it merely predicts —when supplemented with astronomical observations—that the expansion of the universe began from a state in which all matter was concentrated in a single point. Without mentioning Gamow's name, he referred scornfully to "imaginative writers" who had woven fanciful notions such as the big bang round the predictions of general relativity (McVittie 1961).

When the relativistic big-bang model was finally revived in the mid 1960s, in the wake of the discovery of the cosmic microwave background, the new generation of physical cosmologists followed or reinvented an approach that was markedly similar to the one that Gamow had introduced in the late 1940s. Not only were the early calculations of Dicke, James Peebles, Zel'dovich and others notably similar to those of Gamow, Alpher, Herman, and James Follin, they also expressed a style of cosmology that had much in common with that of Gamow's group. The similarity is particularly striking in the case of Yakov Zel'dovich, the eminent Russian chemist-turned-physicist-turned-cosmologist whose rhetoric as well as practise can be hard to distinguish from the way Gamow thought about and did cosmological research.

Just like Gamow, Zel'dovich was convinced that the universe can be examined by the means of general relativity and, for the early universe, nuclear and particle physics; he denied that there was any need to introduce in cosmology new laws or principles that violated established physics; and he used general relativity in much the same robust and simple way that Gamow preferred (Zel'dovich 1963). In short, although separated in space and time, Gamow and Zel'dovich shared the same style of cosmology.

References

Alpher, Ralph A. and Herman, Robert C. (1990). Early Work on 'Big-Bang' Cosmology and the Cosmic Background Radiation. In *Modern Cosmology in Retrospect*. Bruno Bertotti et al., eds., 129–158. Cambridge University Press, Cambridge.

Bronstein, Matvei. (1933). On the expanding universe. *Physikalische Zeitschrift der Sowjetunion* **3**, 73–82.

Brush, Stephen G. (2001). Is the Earth too Old? The Impact of Geochronology on Cosmology, 1929–1952. In *The Age of the Earth: From 4004 BC to AD 2002*. C. L. E. Lewis and S. J. Knell, eds., 157–175. Geological Society, London.

Chernin, Arthur D. (1994). Gamow in America: 1934–1968. (On the Ninetieth Anniversary of G. A. Gamow's Birth. *Soviet Physics Uspekhi* **37**, 791–801.

Eisenstaedt, Jean (1986). La relativité génerale l'étiage: 1925–1955. *Archive for the History of Exact Sciences* **35**, 15–85.

— (1989). The Low Water Mark in General Relativity, 1925-1955. In *Einstein and the History of General Relativity* (Einstein Studies, vol. 1). Don Howard and John Stachel, eds., 277–292. Birkhäuser Boston.

Frenkel, Victor Ya. (1994). George Gamow: World Line 1904–1933, *Soviet Physics Uspekhi* **37**, 767–789.

Friedmann, Alexander (2000). Die Welt als Raum und Zeit (Ostwald's Klassiker der Exakten Wissenschaften, Band 287). Harri Deutsch, Frankfurt am Main.

Gamow, George (1928). Zur Quantentheorie des Atomkernes. *Zeitschrift für Physik* **51**, 203–220.

— (1939). *Mr. Tompkins in Wonderland*. Cambridge University Press, London.

— (1940). *The Birth and Death of the Sun*. Viking Press, New York.

— (1948). The evolution of the universe. *Nature* **162**, 680–682.

— (1949a). Any physics tomorrow? *Physics Today* **2**, 16–21.

— (1949b). On relativistic cosmogony. *Reviews of Modern Physics* **21**, 367–373.

— (1952). *The Creation of the Universe*. Viking Press, New York.

— (1954). Modern Cosmology. *Scientific American* **190**, 55–63.

— (1970). *My World Line: An Informal Autobiography*. Viking Press, New York.

Gamow, George and Fleming, J. A. (1942). Report on the Eighth Annual Washington Conference of Theoretical Physics, April 23-25, 1942. *Science* **95**, 579–581.

Gamow, George and Iwanenko, Dmitri (1926). Zur Wellentheorie der Materie. *Zeitschrift für Physik* **39**, 865–868.

Gamow, George and Teller, Edward (1939). On the Origin of Great Nebulae. *Physical Review* **55**, 654–657.

Gödel, Kurt (1949). An Example of a New Type of Cosmological Solutions of Einstein's Field Equations of Gravitation. *Reviews of Modern Physics* **21**, 447–450.

Gorelik, Gennadi E. and Frenkel, Victor Ya. (1994). Matvei Petrovich Bronstein and Soviet Theoretical Physics in the Thirties. Birkhäuser, Basel.

Hacking, Ian (1992). 'Style' for Historians and Philosophers. *Studies in the History and Philosophy of Science* **23**, 1–20.

Harper, Eamon (2001). George Gamow: Scientific Amateur and Polymath. *Physics in Perspective* **3**, 335–372.

Harper, Eamon, Parke, W.C., and Anderson, G.D., eds., (1997). The George Gamow Symposium (Astronomical Society of the Pacific Conference Series, vol. 129). ASP, San Fransico.

Klein, Oskar (1968). On the Foundations of General Relativity Theory and the Cosmological Problem. *Arkiv för Fysik* **39**, 157–170.

Kragh, Helge (1984). Equation with the Many Fathers. The Klein–Gordon Equation in 1926. *American Journal of Physics* **52**, 1024–1033.

— (1991). Cosmonumerology and Empiricism: The Dirac–Gamow Dialogue. *Astronomy Quarterly* **8**, 109–126.

— (1996a). Gamow's Game: The Road to the Hot Big Bang. *Centaurus* **38**, 335–361.

— (1996b). *Cosmology and Controversy: The Historical Development of Two Theories of the Universe*. Princeton University Press, Princeton.

— (2003). Expansion and Origination: Georges Lemaître and the Big Bang Universe. *Sedes Scientiae: L'émergence de la recherche à l'Université*. Patricia Radelet-de la Grave and Brigitte van Tiggelen, eds.

Lemaître, Georges (1949). Cosmological Applications of Relativity. *Reviews of Modern Physics* **21**, 357–366.

McVittie, George (1961). Rationalism versus Empiricism in Cosmology. *Science* **133**, 1231–1236.

Nadyozhin, D. K. (1995). Gamow and the Physics and Evolution of Stars. *Space Science Reviews* **74**, 455–461.

Okun', L. B. (1991). The Fundamental Constants of Physics. *Soviet Physics Uspekhi* **34**, 818–826.

Reines, Frederick, ed. (1972). *Cosmology, Fusion and Other Matters: George Gamow Memorial Volume*. Colorado Associated University Press, Boulder.

Stuewer, Roger H. (1986). Gamow's Theory of Alpha-Decay. In *The Kaleidoscope of Science*. E. Ullmann-Margalit, ed., 147–186. Reidel, Dordrecht.

Tolman, Richard C. (1934). *Relativity, Thermodynamics, and Cosmology*. Oxford University Press, Oxford.

— (1949). The Age of the Universe. *Reviews of Modern Physics* **21**, 374–378.

Uzan, Jean-Philippe (2002). The Fundamental Constants and their Variation: Observational Status and Theoretical Motivations. xxx.lanl.gov/abs/hep-ph/0205340.

Vallarta, Manuel S. (1925). Sommerfeld's Theory of Fine structure from the Standpoint of General Relativity. *Journal of Mathematics and Physics* **4**, 65–83.

Vicedo, Marga (1995). Scientific Styles, Toward some Common Ground in the History, Philosophy, and Sociology of Science. *Perspectives on Science* **3**, 231–253.

Weizsäcker, Friedrich von (1938). Über Elementumwandlungen im Innern der Sterne, II. *Physikalische Zeitschrift* **39**, 633–646.

Yourgrau, Wolfgang (1970). The Cosmos of George Gamow. *New Scientist* **48**, 38–39.

Zel'dovich, Yakov B. (1963). Problems of Present-Day Physics and Astronomy. *Soviet Physics Uspekhi* **6**, 128–137, 193–197.

Notes

[1] See, e.g., (Hacking 1992) and the overview in (Vicedo 1995), which includes references to the literature.

[2] (Gamow 1970, 41–44); (Frenkel 1994, 770–771). We may assume that Gamow was acquainted with Friedman's semipopular book on "The World as Space and Time" that was published in 1923. (See the German translation in (Friedmann 2000), which includes a valuable introductory essay by Georg Singer.)

[3] During the 1920s, there were several attempts to apply general relativity to atomic and nuclear physics, for example by determining the metric form of the field of the nucleus. For an example, see (Vallarta 1925). When nuclear physics took off about 1930, these attempts largely stopped.

[4] Gamow and Teller did not refer to Hubble's name in relation to either the linear law of expansion or the constant. With one or two exceptions, such epynomous terminology only began about 1950 and became common at the end of the decade. In Gamow 1949, he wrote of "Hubble's constant" but not of "Hubble's law," a name that he first used in his book of 1952. Most other cosmologists followed a similar usage.

[5] Gamow 1939. Of course, Mr. Tomkins' initials referred to the fundamental constants of nature (c, G, h) and their interrelationship. This was a topic that greatly interested Gamow and on which he had written a paper back in 1928 together with Iwanenko and Landau. See (Gorelik and Frenkel 1994, 91–92) and (Okun 1991). For Gamow's continual interest in the subject, see (Gamow 1949a).

[6] I have traced Gamow's road to the hot big-bang model in (Kragh 1996a) and (Kragh 1996b, 101–141), which include references to the literature. See also (Alpher and Herman 1990 and Chernin 1994.).

[7] On the different styles of Lemaître and Gamow, see (Kragh 1996b, 58–59) and (Kragh 2003). Lemaître's lack of interest in the new approach to cosmology that Gamow pioneered is indirectly documented by the letters kept in his archive in Louvain-la-Neuve, Belgium. They include no correspondence with the new generation of nuclear astrophysicists and cosmologists, such as Gamov, Alpher, Herman, Follin, Hoyle, Edwin Sapleter, Chusiro Hayashi, and Robert Dicke.

George McVittie, The Uncompromising Empiricist

José M. Sánchez-Ron

Departamento de Física Téorica, Universidad Autónoma de Madrid, Cantoblanco, 28049 Madrid, Spain; josem.sanchez@uam.es

For obvious reasons, the history of relativity has been centered, with not many exceptions, on the contributions of the great names; even more, on those who besides their greatness were among the first in exploring the Einsteinian special and general theories: Einstein, of course, as well as, for example, Lorentz, Poincaré, Eddington, Hilbert, Schwarschild, de Sitter, Friedmann, Weyl or Lemaıtre.[1] However, we know that history is a complex (not a foreign) world inhabited by many and different sorts of "elements:" persons, ideas, presuppositions, problems or expectations. Indeed, the variety and abundance of such elements increases the greater the protagonist of the history in question, like, in the present case, Einstein's general theory of relativity.

The present paper is dedicated to a scientist who was neither among the pioneers on the elaboration and early development of the general theory of relativity (he could not be so just by age: he was born in 1904), nor among the most well known, although certainly he was a competent professional in the field: George Cunliffe McVittie (1904–1988). We can call him a relativist of the "second generation," a scientist whose approach to relativity theory was somewhat different in important aspects from those of his teachers. Of him (and others) John North (1994, 529) wrote: "Richard Chase Tolman (1881–1948) is a good example of a new type of cosmologist. A graduate of California Institute of Technology, Pasadena. The author of the first American textbook on (special) relativity, Tolman took a strong interest in the work of Hubble—who in cosmological terms was one of his nearest neighbours—and he wrote a brilliant and influential study of ways in which thermodynamics could be introduced into relativist cosmology. Others with similarly broad physical interest were Eddington, McVittie and William McCrea."And a few pages after, he went on (North 1994, 532–533): "The 1930s saw a movement in cosmology of great value to scientific practice generally. It was not for nothing that Eddington's general writings aroused great interest among the philosophers, especially about the nature of theoretical entities. Of course some of the problems raised came directly from fundamental physics and theories of relativity, but the question of whether or not the spectral red shifts were true Doppler shifts indicative of velocities was seen to depend on what was meant by distance, and as soon as this question was pondered, the entire network of interrelations between observational data and the concepts of cosmological theory was seen to be highly problematic. In no other branch of science was so much care given to the analysis of the concepts employed in it, and here the names of Eddington, E. T. Whittaker, R. C. Tolman, E. A. Milne and G. C. McVittie are among those deserving mention."

More specifically, and remarking one of the outstanding characteristics (which I will comment later on) of McVittie's scientific approach, Stamatia Mavridès (1973, 7) called him "l'empiriste irréductible" (the uncompromising empiricist").[2]

It is precisely because of all this that McVittie is worth our attention; no doubt, his case can help to understand better the history of general relativity. Moreover, it happens that his professional life and contributions also offers some light in other questions, like the situation of relativity, cosmology and astrophysics both in Britain and in the United States, disciplines to which he contributed and countries in which he lived.

Early education and the Edinburgh years

George Cunliffe McVittie was born in Smyrna (Turkey), on June 5 1904, where his father, of Scottish ancestors although born himself in Blackpool, Lancashire, had built a trading company. The family was on holiday in England when Kemal Attaturk drove the Greeks out of Asia Minor, which they had occupied, destroying Smyrna in the middle of September 1922. In view of the situation, the McVittie's family settled in England, and after an interruption of one year which he spent helping his father, George resumed his studies, entering Edinburgh University.

While still in Turkey, McVittie became interested in relativity. He was what we could call one of the "sons of the 1919 eclipse expedition," in the sense that his interests in relativity first arose soon after 1919 by reading articles published in the wake of the enormous social popularity of Einstein's relativity theories immediately after the results of the eclipse expedition were announced. "My father," he wrote in an "Autobiographical sketch" (McVittie ca. 1975) prepared at the request of the Royal Society of Edinburgh, "was the Honorary Secretary of the British Chamber of Commerce in Smyrna and in the spring of 1922 I was employed as Secretary of the Chamber of Commerce. During this period I learnt typing and shorthand but I continued reading mathematics on my own. My father had obtained for me through his book-suppliers a book on Einstein's theory of relativity supposedly at a semi-popular level. Though it excited my curiosity, its contents seemed to me to be not only unintelligible but remarkably close to being nonsensical!."[3]

In Edinburgh, McVittie had as teachers scientists like Edmund Whittaker, Professor of Mathematics there since 1912, Charles G. Darwin, Professor of Natural Philosophy, and Edward T. Copson, Lecturer in Mathematics. Of the three, the only one who did not work on general relativity was Darwin: special relativity was an important element of several of his papers, but only dealing with different aspects of quantum physics, such as the wave equations of the electron. Although he cannot be considered a "relativist," Copson was sufficiently interested in the field to publish in 1929 a paper on electrostatics in a gravitational field (Copson 1928). Nothing compared, however, with the many contributions of Whittaker to Einstein's November 1915 theory, on topics like Hilbert's world function, electrical phenomena in gravitational fields or unified theories. Besides his investigations, Whittaker lectured on Einstein's theory of gravitation to senior year undergraduate and postgraduates.[4] In the late 1920s he lectured on the unification of gravitation and electromagnetism. Indeed, in the academic year 1927–1928, already a graduate student, McVittie followed one of those courses on unified theories,[5] an education that served him well when he went to Cambridge to work under Eddington.

In his autobiographical notes, McVittie (ca. 1975) recalled something of his years as a student in Edinburgh, characterizing some of his teachers in the following manner: "In due course I graduated with First Honours in Mathematics and Natural Philosophy in 1927. During this time, the men teaching who influenced me most were (Sir) Edmund T. Whittaker (1873–1956), the professor of pure mathematics, (Sir) Charles G. Darwin (1887–1962), the Tait Professor of Natural Philosophy, and N. Kemp Smith (1872–1958), Professor of Logic and Metaphysics. Whittaker had a highly polished lecturing style and persuaded his audience that every topic was easily comprehensible. A subsequent reading of one's notes showed that this was not so, at least, not until much further work was done. Darwin's lecturing style was untidy but his asides on the nature of applied mathematics — and of applied mathematicians — and his obvious enthusiasm for the subject intrigued me. I often came away from one of his lectures having understood very little but determined to find out what my chaotic notes meant and what it was that aroused such interest in this man. My introduction to relativity theory came through a course that Whittaker gave in 1926/27. To Kemp Smith's discourses on Locke, Hume, Berkeley and Kant I perhaps owe the germ of my attitude to mathematical physics which, many years later, Stamatia Mavridès (1973) was to describe as that of an 'empiriste irréductible' (uncompromising empiricist)."

To the historian David DeVorkin, McVittie (1978, 12) told something revealing about his attitude towards mathematics, a point that it is worthy of attention, inasmuch as a characteristic of general relativity for a long time was that it was "appropriated" by mathematicians. "I wasn't much good at pure mathematics," he recalled to DeVorkin. "I remember we used to call it 'Epsilonology' because the lecturer, and E. T. Copson, used that word too, so it must have come from Oxford, I should think, or it must have been current there. But Epsilonology consisted of doing what I now realize was the proper proof of theorems, with all the logical rigor that was needed. That kind of thing always made me impatient. I felt, and still have felt this all my life. I'm quite sure that it is a very good thing, that Bertrand Russell . . . , in the Principia Mathematica, wrote two volumes of a thousand pages each and at the end of the second volume, I believe he finally concluded that one plus one was equal to two. Well, I decided that it was a very good thing for Bertrand Russell to have taken all that trouble, but for goodness sake, I wasn't going to try to understand how he did it!"

After graduating in 1927, McVittie was awarded the Charles Maclaren Mathematical Scholarship (£200 for three years) and the Nicol Foundation (£50 for one year). The second award involved doing some teaching in the Physics Department and therefore he spent the year 1927/28 as a research student at Edinburgh, attending, as already mentioned, Whittaker's postgraduate lectures. Which unified theory version Whittaker considered in his 1927–1928 course was something McVittie could not recall when, many years later, he wrote his autobiographical notes or was interviewed by DeVorkin.

Anyhow, Whittaker's lectures gave him "the idea of what unified field theory was, so it was very easy to go over to the Einstein version of 1928, and to Levi-Civita's" (McVittie 1978: 15).

In his 1928 version of unified theory, Einstein (1928) introduced a new geometry, characterized by the property of distant parallelism (or "parallelism at a distance," as others, like McVittie (1929: 1033), called it), expressed in terms of "n-beings," i.e., orthogonal tetrads, while Levi-Civita (1929a,b) modified Einstein's theory by discarding the concept of distant parallelism with respect to four orthogonal vectors of reference, using the concept of congruence and introducing the concept of a "world lattice," which was equivalent to a field of tetrads.[6]

It is no surprise that Whittaker taught such a course; as William McCrea (1990, 53) wrote:

> Whittaker and his pupils in Edinburgh were probably the only workers in Britain who had been interested in the attempts by Weyl and by Eddington to extend general relativity to accommodate electromagnetism into a unified system. Einstein had not much liked these particular attempts—which indeed had not got anywhere much ... Almost certainly it was Whittaker who gave McVittie the idea of trying to test this new theory by comparing its consequences with an exact particular example of standard Maxwell–Einstein theory.

Whittaker would have given McVittie the idea of what in due course would be his first publications, but it will be not in Edinburgh but in Cambridge and under Arthur Eddington that he would carry out the idea. That McVittie went to Cambridge was not due to the fact that Eddington had been the first scientist to study unified field theories in Britain, but a consequence of the status – a sort of "scientific imperialism" – that Cambridge had then in British science, a status that implied that even those of Cambridge University former students who had successfully settled in another university wanted their best students to go to their old alma mater to further their scientific careers. "Whittaker," recalled McVittie (1978, 15–16), "was all for sending people he regarded as his bright students to do something at Cambridge, either to take the Tripos, or to do research. For instance, W. V. D. Hodge, the geometer, had left for Cambridge a year or two before I went there. Robin Schlapp went as a research student. He never did very much the rest of his life except be a first class teacher, and run the department of theoretical physics. And there were a number of others... It was taken for granted that if you showed promise as a mathematician outside Cambridge, you would go, in some capacity or other, to Cambridge to finish off, so to speak." This did not apply, McVittie added, to Oxford, "because Oxford regarded themselves as just as good as Cambridge. Copson for instance, did not go from Oxford to Cambridge."

In Cambridge with Eddington

Before going to Cambridge, to do his Ph.D., McVittie had written to Eddington, who answered on February 1, 1928:[7]

Dear Mr. McVittie, I think I have some acquaintance with the work you have been doing, and I shall be glad to act as your supervisor here.

However, when in Cambridge, McVittie (who entered Christ's College) found Eddington remote and unapproachable. Here is what he recalled of him (McVittie 1978, 18–19): "I was Eddington's research student, it's true. This meant that perhaps twice a term, I would cycle out on my bicycle to the observatory on Madingley Road, and be shown by the maid into Eddington's study... Eddington would look up from his desk, and I always had the feeling that he was thinking, 'Now, who is this young man and why does he come to see me?' But he always was pleasant. We would chat about something."

In Edinburgh, McVittie had enjoyed Whittaker's and Darwin's manners, to the extent that he was not prepared "for Eddington's remoteness and unapproachability" (McVittie ca. 1975). "It is true," McVittie added on that occasion (his autobiographical sketch), that "by 1928 he was entering those mystical realms of thought that were eventually to produce *Fundamental Theory*.[8] He was preoccupied with these matters to the extent that at one point he set me to work on the cosmological problem, forgetting that G. Lemaître, who had worked with him a year or two earlier, had already solved it."

The history of how Eddington did not remember Lemaître's contribution has been told many times, for example, by Godart (1992), Eisenstaedt (1993) and, specially, by Kerszberg (1989, 335-337), who did not forget to mention McVittie's role, a role that McVittie (ca. 1975) himself recalled in his autobiographical sketch when he stated: "In a letter to W. de Sitter posted in Cambridge on 19 March 1930, Eddington writes misspelling my name –: 'A research student McVitie and I had been worrying at the problem and made considerable progress; so it was a blow to us to find it done much more complete by Lemaître'.[9] I well remember the day when Lemaître's letter arrived and Eddington rather shamefacedly showed it to me."

Although not with a thesis dedicated to cosmological models, McVittie was able to write a dissertation on unified field theories (a subject in which he was, as we have seen, well prepared), which was accepted for the Cambridge Ph.D. degree in 1930. Out of his thesis arose McVittie's first papers, in particular two he published in 1929: "On Einstein's unified field theory" and "On Levi-Civita's modification of Einstein's unified field theory" (McVittie 1929a, b). In the first, he investigated whether an exact solution (which corresponded to an electrostatic field uniform in direction and nearly constant) of Einstein's 1915 general theory of gravitation was also an exact solution of his 1929 unified field theory, finding out that it was not and that the new theory was not equivalent to the old beyond the first order of approximation. In the second paper he made the same comparison with the theory developed by Levi-Civita, a modification, as it was mentioned, of Einstein's unified theory.

One may think that on the whole the years he spent at Cambridge were not at all favorable to McVittie: Eddington's remoteness, unapproachability and absent mind, and the need to resort, at the end, to Whittaker's ideas and interests, could be taken as good arguments in this sense. However, the Cambridge period was also, as he himself wrote in his autobiographical sketch, "the only time in my life when I received any formal instruction on astronomy. It consisted of Eddington's course in Stellar Structure. Otherwise," he added, "I was self-taught. The books by Cecilia Payne (later Payne-Gaposchkin), Russell, Dugan and Stewart's textbook, the writings of Harlow Shapley and Edwin Hubble were my main sources of information. I also profited greatly by listening to the discussions at the meetings of the Royal Astronomical Society from 1931 onwards." [10]

However, it was not until McVittie left Cambridge that he began to change his scientific outlook: [11]

> Then there was more unified theory, and more and more theoretical solutions, either of Einstein's original equations, or modified ones, and I began to say to myself: 'There is no way out of this multitude. There is no reason for preferring one rather than another, the way these chaps are going about it'.
>
> And it then occurred to me, slowly, that there is surely a way of getting some order into this confusion, and that is to look at the observational data, and pick out things by that criterion, and not by what seems reasonable or mathematically elegant, or combines relativity and electromagnetism, or as Eddington wanted to do, combines relativity and quantum mechanics. Let's try and pin it down by observations.

This last word, 'observations,' is important, because it has been behind many of McVittie's works in cosmology, although not immediately after leaving Cambridge. Indeed, as I already indicated part of McVittie's importance as a relativist lies precisely in this dimension of his work, which allowed that he be called, as we have seen, an "uncompromising empiricist." We shall, however, return to these points later on.

Further career in Britain

Coming back to McVittie's career after leaving Cambridge, we have that immediately after getting his Ph.D. degree he went to Leeds University, as Assistant Lecturer in Mathematics, a position that he held until 1933. There he did some works dealing with Lemaître's model of the universe. Together with William McCrea (whom he met in June 1930 in Edinburgh, where he was a lecturer in Whittaker's department) McVittie pointed out that Lemaître's theory did not distinguish between contracting and expanding solutions (McVittie and McCrea 1931). Following Eddington's idea of instability, they investigated the effect of a single condensation and showed that it would produce continual contraction, i.e., cause the universe to collapse. Here is how McCrea (1990, 56-57) recalled the origin of this joint work: "we had all recently learned of Hubble's discovery of the expansion of the actual Universe. Somehow we came to know that Eddington had inferred that the Universe had started as an Einstein static universe which had been disturbed and was now expanding for ever more. So when I got home from Scotland I began to wonder whether the formation of condensations in an initially uniform Einstein universe would serve to initiate expansion. I got some results and sent them to McVittie for his opinions which show that he was the one to consult on the subject... McVittie had evidently been looking at the same problem. He replied very quickly showing great interest, but saying that he preferred his own approach but also suggesting that we should write a joint paper.... But we inferred that the formation of condensations would initiate contraction! Or rather, the work seemed to show that the formation of a condensation in a uniform Einstein model would make it begin to contract. McVittie himself then published a paper that seemed to show that the formation of more than a single condensation would initiate expansion."

Indeed, a few months later McVittie (1932) was led to an empirically more satisfactory result by considering an initial state with a large number of randomly located condensations. His analysis showed that the Einstein world would turn into an indefinitely expanding Lemaître world and that the formation of condensations thus might be the cause of the initial expansion.[12]

The expanding universe would be in the future one of McVittie's favorite themes, although he considered it connected with several questions, such as spherically symmetric solutions of Einstein's equations, observational data or Milne's kinematical relativity. Precisely because of such plurality of subjects, and the way in which McVittie combined them, the analysis of his works is not easy at all. Contrary to those scientists who make of a specific subject their life-program, McVittie pursued quite a large number of problems within the field of general relativity, cosmology and astrophysics. Indeed, although there are permanent traits in his scientific personality (above all, his insistence in relating theoretical entities with observable magnitudes), one of the characteristics of his scientific career was that he was always alert and receptive to new developments. Indeed, one is tempted to say that his critical nature needed new ideas and theories to develop himself, to flourish as a scientist. His was a sort of "dialectic scientific personality." Precisely because of this it is so interesting to study McVittie's works; that is, because such works provide a sort of mirror, a critical mirror, in which to observe what happened to relativity throughout his life.

The examples in this sense are so many that their study would take too much effort and pages. However in what follows, and while I continue reviewing his life and career, a few of them will be considered. Let us, then, proceed with his career at the point we had left it.

After Leeds, McVittie went to the University of Edinburgh, where he was temporary lecturer in Mathematics during the academic year 1933–34. There followed the University of Liverpool (lecturer in Applied Mathematics, 1934-36), and King's College London (reader in Mathematics, 1936–48), where he carryied out duties mainly connected with undergraduate teaching and with little administrative obligations.

McVittie and Milne's kinematical relativity

I said that among the topics McVittie became interested in was Milne's kinematical relativity, which implied the abandonment of some of the fundamental assumptions of general relativity, in particular the principle of covariance. As has been rather extensively documented, Milne's theory aroused much attention since its appearance, especially during the second half of the 1930s and early 1940s,[13] and McVittie was no exception, although he tried to impress his own points of views, through a reinterpretation of the theory, a fact that finally would provoke controversies with Milne and his followers.[14]

"During the 1930s," McVittie (ca. 1975) wrote in his autobiographical notes, " E. A. Milne's kinematical relativity was vigorously advanced as an alternative to general relativity. I was able to show that the theory of teleparallelism could be used to generalize kinematical relativity." This he did in papers he published in 1935 (McVittie 1935a, b), as well as in his little book *Cosmological Theory* (McVittie 1937), a work which deserves a few words before proceeding.

"Cosmological theory," wrote McVittie (1937, 5) in the Preface, "is that branch of physics which deals with the structure of matter in its most bulky and massive state, the whole physical universe being regarded as a single system whose broad features are to be investigated. The subject is necessarily highly mathematical, but, in this introductory account, attention has been concentrated on those developments most easily comparable with observation to the exclusion of others of a purely mathematical interest." That is, he wanted a public not limited to scientists.

Differently from other accounts of cosmology, McVittie began his book with a chapter on the extra-galactic nebulae, in which the first question to tackle was that of stellar magnitudes and distances: "The first problem connected with the extra-galactic nebulae," he wrote (McVittie 1937, 2), "is the determination of their distance." As there was fundamentally only one method available, viz. the identification in the nebulae of stars of known brightness, or 'luminosity,' and a comparison of the apparent with the true luminosities of the stars, McVittie explained what was involved in such a method, obtaining the formula

$$\text{Log}_{10}D = 0, 2(m - M) + 1$$

where m is the apparent magnitude of the star, M the absolute, and D its distance in parsecs. Soon after he deduced another important formula, this one for the distribution of nebulae in space

$$\log_{10}N = 0, 51m - 2, 758$$

in which N is the number of nebulae over the whole sky of apparent magnitude $\leq m$.

Only once these questions were settled, did McVittie go into more traditional topics, like tensor calculus and the principles of general relativity (chapters 2 and 3), which enabled him to discuss the subject, *his subject*, of the expanding universe (chapter 4), both at a theoretical level as well as in its observational dimension (luminosity-distance relations, the $N - \delta$-relation, δ expressing the Doppler shift). We remark that he took the opportunity to insist that "the term 'distance' in an expanding universe is ambiguous so long as the method of measurement is not specified. Distance in terms of measurements with rigid scales is not the same thing as luminosity-distance deduced from apparent magnitudes."[15] "This dependence of the meaning of distance on the process of measurement," he added, "we shall find emphasized on the kinematical theory of the universe" to which he dedicated the final chapter of his book, but not to Milne's own version, but a generalization of Milne's theory to any Riemannian space that he had constructed.

Why, can we ask ourselves, was McVittie attracted by Milne's approach? The answer was put forward quite clearly in that chapter of *Cosmological Theory* (McVittie 1937, 70):

In the previous chapters the general theory of relativity provided us with a scheme of ideas which had already achieved success as a theory of gravitation. We found that this scheme, which accounted for the small-scale gravitational motions observed in nature, was equally capable of dealing with the structure of the whole universe. But a moment's reflection will convince the reader that the most striking phenomenon exhibited by the universe, the recession of the spiral nebulae, has very little resemblance to gravitational phenomena as exhibited in the motions of planetary systems, double stars, &c. It is therefore legitimate to inquire whether a theory of the universe can be constructed without an *a priori* appeal to a theory of gravitation. The problem which was set by E. A. Milne, and of which he gave one solution, was that of first building up a theory of the whole universe and then, if possible, of deducing from it the necessity of small-scale gravitational motion.

Faithful to his own empiricist principles, McVittie tried to relate his version of Milne's theory to observations, producing an expression for the relation between the number of nebulae, N, with Doppler shift, δ, to which he had dedicated, as we have seen, attention previously. However, when comparing such a relation and the corresponding one in general relativity, he concluded (McVittie 1937, 95) that "we have here a *too rapid* increase of nebulae with Doppler shift as compared with observation. But it is very satisfactory that the theory does predict an apparent outward increase in the number of nebulae of the type which the observed counts suggest. It will require much greater certainty with regard to the observed nebular-count before [the relation between N and δ that he had developed] can be definitely rejected as contrary to observation."

It was different with the problem of the meaning of gravitation in kinematical relativity, of which McVittie (1937, 100), after reviewing the ideas that had been put forward until them, said that "it still awaits a completely satisfactory solution." Nevertheless, he also concluded that "even if a gravitational theory comparable with that of the general relativity is never attained, yet kinematical theory will have served the important purpose of showing that the recession phenomenon is essentially distinct from the gravitational phenomena of the universe."

However, finally kinematical relativity did not survive. It is interesting in this sense to quote what Otto Heckmann had to say in a review of cosmological theories he prepared for a volume edited, precisely, by McVittie in 1962. "Milne," wrote Heckmann (1962, 436-437), "discovered as early as 1931 that if one adopts a definite continuous group of transformations and demands a world model to be invariant with respect to this group, then the laws governing the motion of the so-called fundamental observers and of free particles are fixed to a large extent. Milne chose the Lorentz group as a starting point. He found new ways of deriving it from a certain set of axioms, it is true. But he considered his main achievement to be a world-model which was constructed 'more arithmetico' as he, himself, said and which he claimed to coincide with nature. The basic group was the rigid frame into which everything had to fit, and exactly this rigidity made it impossible to get away from a completely smooth distribution of matter. No clustering, no individual forms, could be described in the theory without inner contradictions to the basic invariance. Milne gave the proof of this impossibility in 1935; but he, nevertheless, later offered a theory of spiral structure based on the time-dependence of this constant of gravitation. He offered two time-scales and ascribed each of them to definite microscopic and macroscopic phenomena. But he did not see that his two time-scales were possible in relativistic cosmology also, and that their separate connection with two definite phenomena was a dream. A very artistic and elaborate mathematical building was erected, but its relation to nature was not here understood."

Nevertheless, and looking from a distance, in his autobiographical sketch, McVittie (ca. 1975) conceded that "E. A. Milne exercised a considerable influence on me, particularly through his emphasis on what is nowadays called 'radar distances.' This drew my attention to the problems associated with the notion of distance in general relativity." [16]

McVittie and the notion of distance

We arrive here at one of the notions to which McVittie would in the future pay more attention: the notion of distance (we have found already some evidence of such interest). Although in one way or another such preoccupation was present early in his investigations, especially when he tried to relate observations with theoretical expressions, an aspect of his interests which we have found already, it was especially during the 1950s and after that he insisted on the importance of being extremely precise when talking about distance. Making use once more of his autobiographical sketch, we have (McVittie ca. 1975): "If distance in a model universe could not be defined in an absolute fashion, yet the concept could not simply be ignored. By the middle 1950s I had proposed that the notion of distance in cosmology could be made precise by defining it with reference to the method to be used for measuring it."

When referring to "the middle of the 1950s," McVittie added a note with a reference to chapter 8 of his influential book of 1956 (second edition of 1965), *General Relativity and Cosmology* (McVittie 1956, 1965a) There he stated quite clearly that the "problem of distance is a complicated one in cosmology largely because astronomers in their ordinary work are accustomed to using classical theories in which a Newtonian absolute distance is pre-supposed" (McVittie 1965a, 160), proceeding then to define the notions of "Mathematical distances," "Distance by apparent size," "Luminosity-distance," and which definition of distance must be used so that the term "velocity of recession" of a source be not ambiguous. His book would not be the only occasion in which McVittie studied from a technical point of view the notion of distance in astrophysics and cosmology, but it would take us too far to enter in this point.[17] Instead, it is convenient to recall that he did not limit himself to technical discussions, but that he was also a frequent general expositor of the necessity of being careful with the notion of distance in relativity, astrophysics and cosmology, as in the book he published in 1961, *Fact and Theory in Cosmology*, dedicated to "weld together the astronomical observations relevant to cosmology with cosmological theory without entering into detailed mathematical proofs" (McVittie 1961, 9), and in which he concentrated on general relativity, the steady-state theory and, to some extent, kinematical relativity. Indeed, in very few other problems could he better emphasize the necessity that scientists be good empiricists. It is worth quoting in this sense what he wrote in an article adapted from a paper that he presented on 26 December 1957 at the meeting of the American Association for the Advancement of Science held at Indiana (McVittie 1958: 501):

> What does an astronomer mean when he says the sun and the earth are, on the average, 92,900,000 miles apart? And what bearing can the two theories of relativity have on the matter? It is questions of this kind that I shall try to answer in this article, but I must warn you that I have been described by my scientific colleagues as an uncompromising empiricist. I daresay that this is true, but it is also a little strange, for all my training and all my research work have lain in theoretical astronomy and not at all in the extremely difficult and fundamental task of making astronomical observations. Perhaps, however, the theoretician does have an advantage over his colleagues who are engaged in observational or experimental work. He can stand slightly to one side and ask himself: What exactly are these men doing, what kind of significance can be attached to the results of their efforts, and in what way are their data really conditioned by theories? To speak in generalities would, I think, be profitless, and it is for this reason that I propose to concentrate on the question of distance in the solar system and to leave the equally intricate and fascinating problem of distance in the universe at large.

The cosmological constant: Einstein and McVittie

There is another general relativity and cosmologic topic of some importance in which McVittie made clear his point of view rather early in his career: the cosmological constant. Indeed, as we shall see immediately, he had an interchange with Einstein which deserves to be remembered.

The starting point of this story was a paper McVittie published in 1933 under the title "The mass-particle in an expanding universe."[18] There, we read (McVittie 1933):

> It has been suggested by certain investigators [Einstein (1931), Einstein and de Sitter (1932)] that the constant, λ, which appears in these formulae is merely a mathematical device and that its value is indifferent from the physical point of view. They propose putting $\lambda = 0$. This cannot be done, however, without introducing difficulties with regard to the expansion.

And here he argued that as the values of the density of pressure calculated in the observer's system at an instant could not be negative, because of the specific characteristics of the equations he had obtained; and, theorefore, the observer "must conclude that the expansion is proceeding subject to a retardation. Exactly the same result follows if $\lambda < 0$. Hence, in either case, he must conclude, firstly, that at some time in the past the expansion started instantaneously with a finite velocity; secondly, that there is a 'retarding force' slowing up the expansion which, obviously, cannot be the initial cause that started the latter. No attempt has been made to account for these properties of the expansion, nor is it easy to imagine how they could arise. We are driven to the conclusion that our observer would necessarily take λ to be a positive constant.

Einstein was not happy at all with such a conclusion, and in a letter to McVittie written in May 16, 1936, and after mentioning that "your critical research about the situation of the cosmological problems seems to me very interesting," he cautioned McVittie on a few points.[19] The first one that: "In your investigations you are always careful to introduce the cosmological constant Λ [λ as written by McVittie] in the gravitational equations," after which he included as a footnote in his letter the expressive comment: "mea culpa." Such procedure, Einstein went on, "was considered necessary because it was believed that the quasi-stationary character of the space-time metric should be preserved." "However," he went on, "from a formal point of view, the introduction of the Λ term is something absolutely unnatural and odious *[Vom formalen Standpunkte aus ist aber die Einführung des Λ Gliedes eine durchaus unnatürliche und hässliche Sache]*, and also seems physically unjustified after the discovery of the expansion motion of matter. In view of this, it would be most desirable that from the very beginning you do not introduce the Λ term in your researches, that is, that you made it zero. It seems to me that in this way you could obtain results somewhat more secure."

There was at least another occasion on which he corresponded with Einstein: among his papers there is a letter from Einstein, dated June 7, 1939, in which the creator of relativity, thanking McVittie for a previous letter (which I have not located), reiterated his opposition to introducing the cosmological constant in cosmology. However, Einstein was not able to convince McVittie, who would remain faithful to the cosmological constant all his life, or at least most of it. An example in this sense is what he stated (McVittie 1962a, 446) in the summary which closed the volume that he himself edited in 1962, *Problems of Extra-Galactic Research*, and that was the proceedings of a Symposium of the International Astronomical Union that was held at the Santa Barbara campus of the University of California from August 10 to 12, 1961:

> I deprecate the identification of the cosmology of general relativity with the special cases in which the constant Λ is zero and the pressure is zero also. It is only if we make this quite arbitrary preliminary selection that we can agree with Baum that the red-shift apparent-magnitude relation leads to a single conclusion regarding the model universe, or that we can accept most of Minkowski's numerical results. The restriction to these models out of an infinity of possibilities is equivalent to asserting that the problem of the nature of the universe has already been solved, except for the relatively minor detail of selecting one out of a few very similar alternatives. This procedure appears to be in complete contradiction to the assertions of the observers that there are hardly any reliable data from which the nature of the universe can be deduced — to their satisfaction at least!
>
> I agree with Heckmann that any mathematical sound derivation of Einstein's gravitational equations shows that Λ is present. The theory cannot de-

termine the value of Λ, but we can see how it could be found from observation."

And at this point he wrote the equations that sustained his arguments.

We see that McVittie, as uncompromising an empiricist as always, wanted any argument in favor of or against the cosmological constant carefully appraised in connection with observations. His approach was very different from that employed by Hermann Bondi, who opened the discussion following McVittie's intervention with the following, rather disdainful, words (McVittie, ed. 1962, 448): "I feel that a discussion of the cosmological constant is unnecessary. While there clearly are arguments in favor of the term, they are not accepted by every student of relativity. Indeed, Einstein himself was very much opposed to the Λ term."

It is not impossible that Bondi's sharp manifestation would be influenced by McVittie's critical stand as regards a theory that had been cherished by Bondi, the steady state theory.

The influence that Hermann Bondi, Thomas Gold and Fred Hoyle's 1948 steady-state theory exerted on the general relativity and cosmology scenario during the late 1940s and most of the 1950s has been the subject of several studies.[20] Considering that when the theory was formulated, McVittie was an active member of the British general relativity and cosmology communities, and that he already had shown his interest in comparing different cosmological theories with observation data, the case of his relationship with the new theory seems attractive. And indeed, it is, as we will see.

McVittie and the steady-state theory

According to his own recollections, McVittie (1978) learnt about the steady-state theory directly from Bondi: "I was at King's College [remember that he was reader there until 1948] and Bondi came to see me, before he and T. Gold went to a meeting of the Royal Astronomical Society which was held in Edinburgh, as far as I remember.[21] He came and told me about this theory, and I said, 'Well, yes, Hermann, do it that way, but this is much more restrictive than general relativity.' I didn't show any great enthusiasm. However, I was taken aback when, after the meeting of the Astronomical Society where this was first expounded publicly by Bondi and Gold, E. T. Whittaker wrote to me and said that he'd heard the most interesting account, from two youngish men called Bondi and Gold, about a new theory of the universe and so on. So I said to myself, well, dear me, have I missed something?"

I have been able to locate Whittaker's letter to McVittie.[22] It is dated "Nov. 2, 1948." Here is what it says:

> We had a good meeting of the R.A.S. in Edinburgh. Cosmology was prominent, as besides a paper by E. A. Milne (showing how his theory of special nebulae could be extended so as to account for star-streaming) there was a most suggestive paper, rich in original ideas, by two Austrian Jews who are now Fellows of Trinity [Cambridge], Bondi and Gold. Although I enjoyed and appreciated both Milne and Bondi–Gold, they didn't convince me. I still

think that the foundations of the true cosmology were well and truly laid by Eddington and that his 'cosmical number' N is the one discovery so far that is both likely to be permanent.

So, Whittaker was not as enthusiastic as McVittie recalled 30 years later. Anyhow, when the steady-state theory appeared in print, McVittie (1978) "looked at it, and the more I looked at it, the less I liked it. For one thing, it was very restrictive, compared to general relativity, and for another, it contained this most mysterious creation process."

Indeed, the new cosmological theory, which won lots of adepts in Britain (especially there) during the 1950s — until Martin Ryle's radioastronomical counts began to turn the tide — contributed to McVittie's emigration to the United States. "In fact," he recalled (McVittie 1978) decades later, "one of the reasons why I went to the United States, I think, was to get away from the atmosphere of the steady-state theory [in Britain]. There was such a hullabaloo about the new revelation! Everybody."

In the years to come, McVittie would remain a firm antagonist to the steady-state theory. Leaving aside his publications, there is ample evidence of such antagonism among his papers. A few examples deserve to be mentioned, although they refer to the period when he was already settled in America, a period I will consider later on.

The first example concerns Martin Ryle, perhaps the greatest opponent to the steady-state theory in Britain; it was, we must remember, Ryle's radio star counts, with its "logN-logI curves,"[23] on the basis of which he saw "no way in which the observations can be explained in terms of a Steady-State theory" (Ryle 1955, 137), that began to undermine Bondi–Gold–Hoyle's approach. On 6th January, 1957, McVittie wrote to Ryle:[24]

> I am sending you by Air Mail parcel post a copy of a paper which I have written on the theory of the distribution in space of extra-galactic radio-sources and which may interest you. I have been working on this on and off since the I.A.U. [International Astronomical Union] meeting and was stimulated to finish the work by the receipt of the Cambridge and Sydney catalogues. I have sent the top copy of the MS to [Joseph L.] Pawsey.
>
> Do not assume that, because I have used the Australian data for illustrative purposes, I thereby commit myself to accepting them in preference to the Cambridge ones! On this controversial question, I do not have the necessary technical knowledge to take sides.[25] My reason for selecting Mills and Slee's counts rather than yours is that the former are more easily dealt with by means of what I have called 'first order logN models" than are the latter. You will see from p. 19 of the MS that, to get the Ryle and Scheuer slope for the $\log N$-$\log S$ curve, would need so small a negative value of b_1 that second-order models, at least, would have to be used. I feel that the labour involved in using them would not be justified until the Sydney–Cambridge controversy is settled.

And he finished his letter with the following sarcastic words: "Perhaps I must apologize for using general relativity model universe in preference to the Bondi and Gold creation of matter theory, which, as I see it from this distance, is now regarded as almost the final word on cosmology in Britain. Apart from the question of creation, the Bondi and Gold theory so restricts $R(t)$ and k that it is always running up against contradictions with observations."

The second example is another letter, this time one McVittie wrote to Joseph Pawsey, the scientific leader of the Sydney radio astronomers, on April 16,1958:[26]

Dear Joe,

I have had some correspondence with Fred Haddock, after he wrote me saying that I was not to [be] a speaker on the programme of the Radio-Astronomy Symposium. I gather that there is to be one formal speaker only on cosmological matters and that he is going to be Hoyle. It seems to me that this is tantamount to giving the approval of the Symposium to the 'creation of matter' point of view in cosmology and only permitting others' views to be expressed in any discussion-period there may be and provided opponents of the 'creation of matter' theory can get a word in. I believe that a similar policy was adopted at the Manchester Symposium.

Let me say at once that I am not suggesting a modification of the Committee's decision, which would be humiliating for all concerned. I would like your views on a personal question that the decision brings up. Perhaps rashly, I took your remark whilst we were motoring to Yerkes last October that you wanted me to read a paper at the Symposium, at its face value and have given a good deal of thought to the problem you spoke about last year, namely, how can one deal with a scatter in the intrinsic flux-density of extra-galactic radio-sources? I have got to the stage of modifying the theory I put forward in the Austr. J. of Physics paper,[27] to allow for an arbitrary mixture of standard sources. Essentially this is to be done by a step-function distribution of flux-densities, to replace the single standard flux-density used previously. The work still needs a good deal of polishing up, which I had intended to do in May and June. But Fred's letter contains the remark about the Symposium that 'a small amount of time should be allocated to cosmology because of the very uncertain and tentative nature of the observational radio data upon which to base cosmological conclusions.' I suppose that this is a hint that the kind of investigations I have carried out — or in the course of doing — which are intended to show how radio data could be useful, are not of interest to radio-astronomers, but that 'creation of matter' speculations are. If this is the correct interpretation of the view of radio-astronomers, I would drop the work I have been doing like a hot brick and turn to something else. But I wanted to check with you first that this was so.

The Radio-Astronomy Symposium of which McVittie was complaining was held in Paris in August (1958), and we have a copy of a report to the Office of Naval Research, that financed his attendance to it, as well as to the Xth General Assembly of the International Astronomical Union, held the same month in Moscow, which deserves to be quoted, as it provides an interesting personal, but informed, perspective on events that were happening at the time in the field of gravitation, astrophysics and cosmology.[28]

After complaining about the accommodation facilities provided by the Paris Symposium, McVittie summarized his opinion about the scientific content of the meeting:

The Symposium revealed that the leadership in radio-astronomy is still held by the groups at Leiden (Holland), Cambridge and Manchester (England) and Sydney (Australia). In my own fields of interest — galactic and extragalactic studies — the most remarkable piece of work carried out in the U.S. was by D. W. Dewhirst during a visit to the Mount Wilson and Palomar Observatories from Cambridge, England. His identifications of peculiar galaxies with faint radio-sources is a considerable step forward in this difficult subject. A feature of the groups from Leiden, Cambridge, Manchester and Sydney was the existence of teams of young workers in each place. The proximity of a University to each of these centers of radio-astronomy is significant. The University can provide the training in physics, mathematics, astronomy and electrical engineering needed by young men who can then become radio-astronomers.

I read a paper entitled 'Remarks on Cosmology' at the Symposium, which will be published in its Proceedings.

The development of radio-astronomy in France is startling. The equipment at Nancy is lavish and the men are keen and able. The same may be said of optical astronomy at the Observatoire de Haute Provence, where I spent three days. There the new reflector of approximately 72 inches had just come into use and excellent trial photographs had been obtained. The optical image-amplifier due to Lallemand and Duchesne appears to be a workable and very valuable device. At the Observatoire de Haute Provence Laffineur has constructed a radio interferometer using two parabolic cylinder antennae. It has a base-line of 1 km. The instrument was crudely constructed of chicken-wire on wooden posts.

Of the Moscow meeting, McVittie, who acted as Secretary of one of the Commissions (No. 28, dedicated to "Extragalactic Nebulae"), selected to say that "the most remarkable work reported on at the Commission 28 meetings was by a Russian, A. L. Zelmanov, who has been working on non-homogeneous models of the universe."[29]

More informal and direct is what McVittie told Allan Sandage in a letter of 23 September 1958, which will serve as my final example, and in which McVittie's antagonism to the "steady-state boys" is evident:[30]

"I had a notice the other day of the Neighbors meeting at Perkins on Oct. 4 together with a covering note from Sletteback urging me to come because 'Allan Sandage especially mentioned you in his letter.' I like to think that this really happened – even if it did not – because I have been feeling recently that what I might have to say about cosmology is not of interest to the younger generation! Perhaps it is because I am still feeling a bit low as a result of a parting gift from our Russian friends in Moscow which was a mild bout of pneumonia from which I recovered in England. In any case the summary of your proposed talk interested me greatly and I should much have liked to be able to come on Oct. 4. But I am supposed to take things a bit easily for a time and I have the long journey to Green Bank to face on Oct. 14–16.

In the summary of your lecture there is a reference to a 'theory of gravitation' which could be used, in part, to find the 'mass of the universe'. I was intrigued to know what this theory could be: is it a new one? Also you seem to imply that the mass of the universe is finite and therefore that the curvature is positive. In the general relativity interpretation I long ago concluded that today's data, whether on red-shift or by counts of faint radio-sources, cannot determine the sign of the curvature directly as it is a third-order effect. Of course, one can help oneself out with Einstein's equations but then the balance of evidence seems to be in favor of negative curvature.

The status of the deceleration also puzzles me. The steady-state boys (Burbidge, Gold, Hoyle) at the Radio-Astron. Paris symposium noisily and rather rudely insisted that the deceleration no longer existed. Mayall and I were both puzzled by this hullabaloo. Also in Moscow, Baum gave us a talk on this large red-shifts and drew the conclusion that the deceleration occurred. So what were the steady-state boys shouting about, and on whose authority?

Apparently the new party line is that faint radio-sources are not extragalactic, at least so Gold told us making little effort to conceal his irritation. This after arguments by Mills, Ryle and Dewhirst to show that they either all were or that at least some of them were (Dewhirst). I think I am right in saying that Dewhirst's work impressed greatly most of the participants in the symposium. So again I am puzzled by what may be cooking!

General relativity and cosmology

In 1948, McVittie was appointed professor of Mathematics and Head of the Mathematics Department at Queen Mary College, a position that offered him a little more scope as regards advanced teaching work than what he had enjoyed before.[31] There he had Clive W. Kilmister as his first Ph.D. student, who in due course, years afterwards, would become professor at the Mathematics Department at King's College London, the same place which accommodated other noted relativists, like Hermann Bondi and Felix Pirani.

It was during his years at Queen, almost at the end of them, that began the process that would lead to the publication of one of his most well-known and influential works: the book *General Relativity and Cosmology*. The origins of this work can be dated on March 29, 1950, when M. A. Ellison, from the Edinburgh Royal Observatory wrote McVittie telling him that "Lowell and I are co-operating as Editors, with Messrs. Chapman & Hall in the publication of a series of monographs, under the general title 'International Astrophysics Series.' The first of these books, on 'Aurora,' will be out in a few months' time and we have promises of another half-dozen subjects," adding that they "would much like to include a volume dealing with the present outlook in Cosmology, and we feel that you are the obvious person to write it." [32]

McVittie's reply (dated April 2, 1950) contains interesting paragraphs: "Very many thanks for your letter of the 28th March and for your flattering remark that I was the person to write the Monograph in your series on cosmology. As it happens I am going to Harvard for the last three months of this year to lecture on relativity and its applications to cosmology and I have no doubt that an expanded version of these lectures would make a book of the kind you suggest. But there are two complications." One of such "complications" was not very important: his possible obligations with Methuen, for which he had written *Cosmological Theory* in 1937: in his contract he observed that it was stated that were he to write another book on the subject, then he must offer it first to Methuen for publication. The second is more interesting for my purposes:

> The second complication is a more serious one. My feeling is that it would be a little unwise to write a book on cosmology just *before* the observations of the 200 becomes available. As you perhaps know, I view cosmology in the same way as I look upon any other branch of mathematical physics, viz. it is a theory intended to interpret observations. The Milne+Hoyle–Bondi school look upon cosmology as an exercise in speculation and mathematical ingenuity, such observed data as there are being dealt with on the principle 'since the observations are inaccurate, it is sufficient if they do not contradict my theory too glaringly.' I have a hunch that a lot of cosmology is going to look pretty silly as soon as the 200 comes into production and I do not want to be one of those whose faces are going to turn red.

Here McVittie must be referring to the 200-inch Hale reflector telescope at the Palomar Observatory of the California Institute of Technology. That the telescope be named "Hale" was in honor of George Ellery Hale, who proposed its construction in 1928. Then, Hale pointed out the need for a 200-inch telescope, that will surpass the instruments at Lick, Yerkes, Hooker, and Carnegie, whose "possibilities," in his opinion, "have passed out."[33] It would take, however, two decades for its construction: the dedication ceremony took place in 1948. Therefore, in 1950, when McVittie was writing to Ellison, not much definite results had come from the new powerful instrument.

Continuing with McVittie's letter, we have that he suggested that "if the Methuen difficulty can be overcome," as it finally was, he should be given the opportunity of publishing the new book in 1952 or 1953, "Meanwhile I should be getting all the theoretical side ready, so that as the 200" observations appear, I could feed them in to the theoretical formula and draw the necessary conclusions. In this way I feel sure that the work should be a worthwhile one when completed, instead of still another arid exercise in speculation."

Finally McVitties's plans had to be somewhat modified because, as he wrote in a letter to Ellison (April 7, 1951), during his stay in America, "apart from Stebbins & Whitford's work, nothing new seems yet to have come from the observational side and H. P. Robertson, when I saw him in Washington in the autumn, seemed doubtful whether anything would for some years." He was, "therefore more than ever doubtful about writing a book exclusively devoted to cosmology. One could, I suppose, catalogue the ad hoc theories (kinematical relativity, creation of matter, F. Jordan's numerology, etc.) and compare them with general relativity on a 'speculation' basis, using the inconclusive comparison with observation. But this seems to me to be a gloomy prospect and one which I should personally regard as a waste of time."

Instead, McVittie proposed "to write a book with some such title as 'Astronomical applications of General Relativity' to deal with Einstein's theory alone. The aim of the book would be to show how local gravitation (Sun, etc.) and cosmology can be dealt with by means of a single theory."

When it finally was published, in 1956 (the second edition came out in 1965; it was also translated into Russian in 1961), the book dealt in some detail with "Observational Cosmology," but only in one of the chapters, the last one (chapter 8), mostly from the standpoint of general relativity and assuming above all that the universe can be assimilated to a perfect fluid. As it was natural of him, taking into account his previous interests, there was ample discussion about questions related to red-shift, apparent magnitudes or count of galaxies and radio-galaxies. It was, in any case, a text widely used during many years, a text different from the majority of general relativity books, which did not pay the same attention as McVittie's to the relationship of theory with observation.

Settled in the United States

In spite of the advance that McVittie's new situation as professor at Queen Mary College meant, "there was not much opportunity for anything beyond routine administrative work," McVittie (ca. 1975) wrote in his autobiography; "times were difficult and the Principal of the College, Dr. (later Lord) B. Ifor Evans, and I were temperamentally unsuited to produce a fruitful collaboration." However, the College sent him as its delegate to the International Congress of Mathematicians that took place at Harvard, August 30-September 6, 1950. This event would become decisive in his life.

Indeed, on February 22, 1950, Harold Shapley, then head of the Harvard Observatory, wrote to McVittie on the following terms:[34]

Dear Professor McVittie:
We have heard that you plan to attend the Mathematical Congress in Harvard at the beginning of September of this year, and also that you might be interested in prolonging your stay in America.

We need someone to give a graduate course during the first semester of the next academic year, since Dr. Bok has gone to South Africa to work at our southern station. At a meeting of the Observatory Council yesterday, which includes the members of the Department of Astronomy in Harvard University, it was voted to ask if you could not stay on through the first semester, which begins (so far as academic duties are concerned) about the 25th of September and continues to the middle of January. We had in mind that in addition to giving one advanced course suitable to our graduates students you would also take part in the general activities of the Observatory, including the consultation with individual students about their thesis problems. By general activities of the Observatory I mean of course only the colloquia and special conferences that occur on the average perhaps once a week.

As to the nature of the course, Shapley pointed out that it could be decided later but that "possibly it might concern the general problems of cosmogony, which have not been lectured on here at the Observatory in recent years." Finally, he mentioned that Whipple, chairman of the Department of Astronomy, and Cecilia Gaposchkin "join me in the hope that you will be able to come to the Harvard Observatory."

As it turns out, McVittie did not accept Shapley's offer and decided to stay in England. However, he did not stay long there: in 1951 he accepted an offer from the University of Illinois, in Urbana. As a matter of fact, and in an indirect way, Shapley had also something to do with the move.[35] He explained his participation to McVittie in a letter dated September 17, 1951, when the negotiations were still going on:[36]

> I shall enclose an excerpt from one of my letters to Dean Henning Larsen of the University of Illinois. He had written to me, asking for general sugges-tions, primarily on the issue of closing down the department of astronomy and letting someone in mathematics do astronomy on a part-time basis. You will see from the enclosure that in a polite way I tried to point out that they were on the wrong track. In some other letters and especially in personal conversa-tions I emphasized the point that what they should have is not a suspension of astronomical interest but a tremendous enlargement. I have tried to argue with them that there should be at least three men in a department even if there is no expensive telescopic equipment. Their big computing machine, their distin-guished department of physics and chemistry and engineering, and the strong competition of the first-class universities of the Middle West, and their other assets make it seem advisable to take astronomy and astrophysics seriously.
>
> Dean Larsen tells me that my letter was reproduced and sent around the University quite a bit, for it awakened sympathetic interest and considerable understanding. It seems to me, therefore, that your going to Illinois is the first step in something that may go further. I should like to have you look into the possibility that Dr. Herget and his asteroidal computing enterprises might be transferred to Illinois. He gets very poor support from the extremely poor (financially) university in Cincinnati. But if not the highly competent Herget, possibly someone else, like Ivan King, who has a deep interest and experience in the application of computing machinery to astronomical and astrophysical problems.

Besides these possibilities, there were other "coordinating opportunities," Shapley went on, "if the Illinois program and budget can stand expansion. The borders of geochemistry, geophysics, meteorology, microwave theory and exploration — all are of astronomical interest, and should be of interest to such a great institution as the University of Illinois."

It was, indeed, a great institution, as Shapley himself had said before to McVittie, when he first heard of the possibility of his move to Illinois (July 11, 1951):[37] "Urbana is indeed a tremendous educational plant – the largest budget in America but in some respects not very lush. It probably has more members in the National Academy than any other state university (except California), and is superb in two or three fields."

In America McVittie would flourish. Not that he would become one of the uncon-tested leaders of the general relativity, cosmology or theoretical astrophysics there, but no doubt he was active and on the general better considered than in Britain, where, according to many indications, he seems to have had, as we have seen, a certain number of influential "enemies."[38] Although "I [am] not indissolubly tied to the U.S.," he wrote to Whittaker on July 1, 1953, commenting about the possibilities of being the successor of Max Born in Edinburgh, "I have found here far greater opportunities for research than in London."[39] In 1952, he was elected to the American Astronomical Society, serving as its Secretary during the period 1961–1969, a position that entailed travel to many cities of the continental United States and Canada, with long-range excursions to Fairbanks, Alaska and Hawaii.[40] He had also a long association with Commission 28 (Galaxies) of the International Astronomical Union, in which he was successively Secretary (1958–1964), Vice-President (1964–1967) and President (1967–1970).

McVittie and radio astronomy at Urbana University

Among the opportunities McVittie found in America there was an easier access to observations. We have seen repeatedly that his scientific and philosophical outlook was such that he was not satisfied with a purely theoretical approach to gravitation and cosmology. However, in Britain observational possibilities remained distant. Not so in Illinois. Here is how he referred in his autobiography (McVittie ca. 1975) to what happened there:

> At the instigation of Professor Edward C. Jordan, shortly to become the head of the Electrical Engineering department at Illinois, I was sent by the University in 1954 to conferences in Washington on radio astronomy. Eventually the plan materialized in the National Radio Astronomy Observatory, at Green Bank, W. Virginia. However, at the Washington conference there was a group to which I belonged that emphasized the necessity of smaller radio astronomy projects at Universities as well. In these, young radio astronomers would be trained. In pursuing this idea, Jordan and I appointed in 1956, George W. Swenson, Jr., an electronics expert specializing in antenna design, to half-time professorships in each of our departments. At that time extragalactic radio sources would provide a good criterion for selecting the appropriate model of the universe. I urged Swenson to plan an instrument suitable for survey work and able to detect faint, and therefore presumably very distant, radio sources.

The project was founded by the Office of Naval Research. Construction of a 600×400 ft. parabolic cylinder dish, fixed to the ground, which operated at 610.5 MHz, began in September 1959. Late in 1962, the instrument came into operation, at a location which they named Vermilion River Observatory, 30 miles east of the Urbana campus. However, different problems, including the cutting back of support by the Office of Naval Research and the Federal Government, and Swenson's lack of interest in the reduction of the survey data from the instrument, led to the fact that the sky survey planned remained uncompleted when the instrument was retired by June 1970. But this is, however, another history. What I want to emphasize is how different was for George McVittie the American setting from the British. The frontiers between the theoreticians and the experimentalists were not as strict as, in several aspects, in

Britain, and in general in Europe, a fact that has been pointed out by Michael Eckert (1996) with respect to solid state physics in Germany and the United States. Eckert argued that such lack of — or difficulty in — communications and interchanges explain the ultimate failure of the Sommerfeld school on becoming the world leader in the new field, then emerging, of solid state physics. I guess that something similar can be said about important areas of gravitation physics, astrophysics and cosmology.

A proof, albeit indirect, of such differences is what McVittie wrote to Allan Sandage on December 21, 1966: [41]

> Dear Allan,
>
> I hope to see you at the AAS meeting next week but meanwhile I wanted to raise a point with regard to a preprint of an article entitled 'Radio Astronomy and Cosmology' by P. A. G. Scheuer for Vol. IX of 'Stars and Stellar Systems" of which you are the editor. I wonder if you, as editor, think it quite fair that the work of my pupil and myself and the work done at the VRO [Vermilion River Observatory] should be totally ignored? The papers I have more particularly in mind are
>
> > G. C. McVittie, Austr. J. of Physics, 10, 331, 1957.
> >
> > G. C. McVittie and R. C. Roeder, Ap. J. 138, 899, 1963.
> >
> > G. C. McVittie and L. Schusterman, A. J., 137, 1966.
>
> Yet I observe that Scheuer gives numerous references to the work of W. Davidson who always seems to me to re-write my papers in an intricate notation and publish them in M. N. [*Monthly Notices*] Of course, I know that I am not persona grata in Britain and particularly in Cambridge where our work at VRO on surveys is regarded as an unwarranted intrusion on their private preserve! But I wondered if you would feel quite happy about this attitude appearing in an American publication. I might add that the second of the above papers drew favorable comments from Minkowski at Padua in 1964.

McVittie and the Royaumont General Relativity and Gravitation Congress (GR2)

The documents conserved among McVittie's papers include several reports in which the British relativist informed of his travels and meetings to which he attended. To finish the present paper I will consider one of those reports, the one in which McVittie reviewed the Colloquium dedicated to Relativistic Theories of Gravitation held at Royaumont, near Paris, from June 18 till August 8, 1959.

That meeting is especially worth our attention because it was one of the first General Relativity and Gravitation Conferences, which came to be known as GR. Following the nomenclature introduced by André Mercier (1992), the Royaumont Conference would be GR2 (Royaumont 1962), after GR0, the "Jubilee of Relativity Theory" meeting held in Bern, July 11–16, 1955 (Mercier and Kervaire, eds. 1956), and GR1, the Conference on the Role of Gravitation in Physics, held at the University of North Carolina, Chapel Hill, January 18–23, 1957 (DeWitt, ed. 1957). It was, therefore, one of the meetings which contributed most to the institutionalization of general relativity. And it was also the first of the GR Congresses which McVittie attended.

As to the report mentioned, we have that McVittie dated it on August 26, 1959, and entitled it: "Report on Travel sponsored by Office of Naval Research (Electronics Branch, Contract ONR 1834 (22)), June 18–August 8, 1959." Here is what it said:[42]

> (1) Colloquium on 'Les Theories Relativistes de la Gravitation,' Royaumont, near Paris, France, June 21–27, 1959.
>
> This week-long symposium was attended by some 100 persons drawn from 14 countries. The U.S contingent was the largest, 40 in number, followed by France with 21. Three came from the USSR, four from Poland and three from E. Germany.
>
> Much time was devoted to the concept of energy in general relativity. Since no tensor definition of energy is to be found in this theory, attention was concentrated on non-tensorial definitions that seemed plausible. Of these the most interesting was that of P. A. M. Dirac (Cambridge, England) who suggested that energy was an integral of the equations of motion which had the correct physical dimensions and was useful in studying the equations.
>
> Gravitational waves were much discussed. Exact solutions of Einstein's equations certainly exist in which gravitational effects can be regarded as propagated with the speed of light. Little interest was evidenced in the astronomical consequences of such a finite speed of propagation for gravitation or in its physical detection by experiments. There was however one paper by J. Weber (Univ. of Maryland, USA) on the possibility of detecting gravitational waves experimentally. One proposed method was to observe the relative motion of masses which interact with a gravitational wave. The second was to employ the strains set up in a solid by such a wave. It was not clear that the effects to be observed would be distinguishable from the 'noise' inherent in the proposed experimental methods.
>
> The quantization of general relativity received some attention though no significant progress appears to have been made since the first of these conferences in 1950. With the death of Einstein, the search for a unified field theory of gravitation and electromagnetism has apparently faded into the background.
>
> The group of French mathematicians whose leader is A. Lichnerowicz (Institut H. Poincaré, Paris) produced much interesting pure mathematics on the propagation of discontinuities in the space-times of general relativity and on the properties of various tensors that can be defined therein.
>
> Cosmology was treated in two papers only: one of these was the lecture by the present writer (copy attached) which dealt mainly with the distribution in space of faint radio sources. The other was by D. N. Sciama (London, England) on the present observational situation in cosmology. These two papers, together with another one having astronomical implications by A. H. Taub (U. of Illinois, USA) on small motions of spherically symmetric distributions

of matter, were discussed at an extra session. The writer acted as chairman of this session.[43]

The conference was well attended indeed: 119 participants, McVittie among them, and with the noted mathematician and contributor to general relativity André Lichnerowicz presiding.

As McVittie stated, there were a significant number of contributions devoted to the concept of energy in relativity: those, for instance, of Christiaan Møller, Felix Pirani, Paul Dirac and Stanley Deser. Thus Møller's contribution (which opened the volume of proceedings) was entitled "The energy-momentum complex in general relativity and related problems." Møller's opening words serve well to illustrate the nature of the problem:[44] "Within the framework of Einstein's theory of gravitation it is possible to define a large number of algebraic functions of the field variables which satisfy 'conservation laws,' and the problem arises how to determine which of these functions represent quantities with a physical meaning."

Gravitational radiation was, as McVittie also mentioned, another of the topics dealt with in several interventions: like those of André Lichnerowicz, of his student the Spanish physicist who settled in Paris, Luis Bel, of "Bel–Robinson tensor" fame, of V. Fock, and, on the experimental side of Joseph Weber, whose talk was "On the possibility of detection and generation of gravitational waves."

We know that during decades, general relativity was dominated by mathematics. It was still so, generally, when GR2 was held, and this is something that can be appreciated simply reading its proceedings. Precisely because that, surely McVittie ought not to have been very satisfied with the Royaumont congress: too many mathematical papers were read there. Papers like: Jürgen Ehlers, "Transformations of static exterior solutions of Einstein's gravitational field equations into different solutions by means of conformal mappings;" Yves Thiry, "Sur les théories pentadimensionnelles;" Jean-Marie Souriau, "Relativité multidimensionnelle non stationnaire;" Olivier Costa de Beauregard, "Quelques remarques d'analyse dimensionnelle pouvant intéresser les futures théories unitaires;" Cécile DeWitt, "Grandeurs relatives a plusieurs points. Tenseurs generalizes;" David Finkelstein and Charles Misner, "Futher results in topological relativity;" or Roger Penrose, "General relativity in spinor form."

Apparently, he should have been happier with Peter Bergmann and Arthur Komar's contribution: "Observables and commutation relations." The appearance of the word "observable" must have sounded well to McVittie's ears, even though the paper was in fact dedicated to a highly theoretical subject: the quantization of general relativity. What, however, did Bergmann and Komar mean by "observables"? In their words (Bergmann and Komar 1962, 313): "In principle, invariant quantities represent intrinsic properties of a physical situation, properties that are independent of the (equivalent) modes of description. If our theory deals with physically meaningful quantities, invariants should possess expectation values; and to the extent that in our theory observable quantities can be predicted, invariants should be predictable from sufficiently complete data given on one space-like hypersurface. This, then, is the motivation for talking of observables and for proposing to formulate the whole physical theory as far as possible in terms of observable theory." And then they added: "In general relativity many observables are constants of the motion outright. From any that are not we can construct constants of the motion, by expressing the value of an observable at the coordinate time x_0^0 in terms of observables at another (variable) coordinate time x^0." Later on, they dedicated themselves to the hard task of constructing observables.

However, it seems that, not surprisingly, McVittie was not completely satisfied with their approach, and one can perceive a point of irony in his intervention (Royaumont 1962, 323) at the discussion that followed the presentation of the paper (made by Bergmann): "I find absolutely convenient to consider the observables as essential in general relativity. An example is provided by the determination of the distance by means of the apparent value of a luminous extended source, in which case an operational prescription of the measure can be done. Could Mr. Bergmann give us an example of observable: a) which be the natural result of the association of general relativity and the quantum theories; b) and for which it could be possible to indicate an operational measure process?" According to the proceedings, his question remained unanswered.

As to his own contribution to the congress (entitled "Cosmology and the interpretation of astronomical data" [McVittie 1962b]), in it he discussed on one side three cosmological problems: the applicability of uniform cosmological models to the observed universe, the status of the problem of the expansion of the universe, and the distribution in space of extragalactic Class III radio sources. At the same time, he considered the applicability to those problems of different gravitational theories: general relativity, steady state, and kinematical relativity. "The formulae through which a comparison of theory and observation is made," he stated (McVittie 1962 b: 253), "are indeed for the most part so nearly identical in form that all three theories can be treated together."

In his conclusions, McVittie (1962b, 265) considered several questions, among them the use of the cosmological constant, that he still defended ("attention should be concentrated on those [uniform cosmological models] having hyperbolic space and a negative cosmical constant"), but it was especially against the steady-state cosmology that he addressed his critics:

> I think it is illusory to claim, as is done in the steady-state theory and in kinematical relativity, that one single highly specialized model universe can be chosen to represent the observed universe
>
> Supporters of the steady-state theory suggest that new kinds of observations are needed [F. Hoyle, 'Paris Symp. on Radio Astronomy,' p. 529. University Press, Stanford (1959)] in order to 'test' their theory as against general relativity. The observers [M. Ryle, *Proc. Roy. Soc. A*, 248, 289 (1958); J. L. Pawsey, Trans. I.A.U. (Moscow 1958), 10 (in press). Report of Commission 40] also appear to believe that the question is still open. In fact, it has been known since 1956 that the steady-state theory predicts the wrong sign for the acceleration parameter, We have also pointed out that the predicted average density of matter is rather too high. And lastly, one result of the present paper has been to show that the steady-state theory fails to reproduce the empirical law of distribution of Class II radio sources.

And then he concluded: "In view of these considerations, it is not clear how further 'tests' could validate the steady-state theory. Its model universe simply does not agree with observation whereas, as we have seen, certain general relativity model universes do."

Besides reading a paper, McVittie participated in a number of discussions, always faithful to his critical personality, that brought him a not small number of enemies all through his life. Thus, during the discussion after Joshua Goldberg's intervention ("Conservation laws and equations of motion" [Goldberg 1962]), which was of course mainly mathematical, McVittie asked a question which was perfectly fitted to his own philosophy of science (Royaumont 1962, 43): "Would you like to tell us which is the reason of this work? Is it conceived as an exercise of analysis or perhaps it can throw some light on specific physical problems?"

Goldberg's reply deserves to be quoted, at least in part (Royaumont 1962, 43): "The aim of one theory of motion in general relativity is not to obtain more efficiently, or with more precision, the perihelion precession, but to explore till what point the general relativity theory contains those predictions. There are a certain number of fundamental questions in the [General Relativity] theory of motion which are not yet completely clarified. "[45]

We know what he really thought of the Royaumont congress through a letter he wrote to his British colleague Gerald J. Whitrow, on August 18, 1959:[46] "The Paris relativity conference was indeed boring in parts: endless re-hashing of 'energy' considerations, much talk about gravitational waves, hankering after unification of quantum mechanics and general relativity, etc. But the French group under Lichnerowicz produced some very elegant pure mathematics." And in a new note against the steady-state theory: "If such a secular effect [a greater rate for the formation of rich clusters than the one predicted by the steady-state theory] exists, it would provide yet another observational argument against the steady-state theory. Not that such evidence will have any effect on the 'cosmological' climate in England! The steady-state boys, Bondi, Hoyle, Pirani, etc., are far too good publicists to let such points (or those listed at the end of my Paris paper) throw them off their stride."

Such was that man who, fittingly, was called "the uncompromising empiricist."

References

Bergia, Silvio (1993). Attemps at Unified Field Theories (1919–1955). Alleged Failure and Intrinsic Validation/Refutation Criteria. In Earman, Janssen, and Norton, eds., 274–307.

Bergmann, Peter G. and Komar, Arthur B. (1962). Observables and commutation relations. In (Royaumont 1962), 309–323.

Bertotti, B., Balbinot, R., Bergia, S. and Messina, A., eds. (1990). *Modern Cosmology in Retrospect*. Cambridge University Press, Cambridge, U.K.

Bondi, Hermann (1952). *Cosmology*. Cambridge University Press, Cambridge, U.K.

Copson, Edward T. (1928). On electrostatics in a gravitational field, *Proceedings of the Royal Society of London A* **118**, 184–194.

DeVorkin, David (2000). *Henry Norris Russell. Dean of American Astronomers.* Princeton University Press, Princeton, N.J.

DeVorkin, David, ed. (1999). *The American Astronomical Society's First Century.* American Institute of Physics. Washington D.C.

DeWitt, Cécile M., ed. (1957). *Conference on the Role of Gravitation in Physics.* WADC Technical Report 57-216, ASTIA Document No. AD 118189. Ohio.

Earman, J., Janssen, M. and Norton, John D., eds. (1993). *The Attraction of Gravitation.* Birkhäuser Boston.

Eckert, Michael (1996). Theoretical Physicists at War: Sommerfeld Students in Germany and as Emigrants. In (Forman and Sánchez-Ron, eds. 1996), 69–86.

Eddington, Arthur S. (1914). *Stellar Movements and the Structure of the Universe.* London.

— (1953). *Fundamental Theory.* Cambridge University Press, Cambridge, U.K.

Einstein, Albert (1928). Riemann-Geometrie mit Aufrechterhaltung des Begriffes des Fern-parallelismus. *Preussische Akademie der Wissenschaften. Phys.-mathe. Klasse. Sitzungsberichte*, 217–221.

— (1931), Zum kosmologischen Problem der allgemeinen Relativitätstheorie. *Preus - sische Akademie der Wissenschaften. Phys.-mathe. Klasse. Sitzungsberichte*, 235– 237.

Einstein, Albert and de Sitter, Wilhelm (1932), On the Relation Between the Expansion and the Mean Density of the Universe. *Proceedings of the National Academy of Sciences* **18**, 213–214.

Eisenstaedt, Jean (1993). Lemaître and the Schwarzschild Solution. In (Earman, Janssen and Norton, eds. 1993), 353–389.

Eisenstaedt, Jean and Kox, Anne J., eds. (1992). *Studies in the History of General Relativity.* Einstein Studies, Vol. 3. Birkhäuser Boston.

Forman, Paul and Sánchez-Ron, José M., eds. (1996). *National Military Establishments and the Advance of Science and Technology.* Kluwer Academic Publishers, Dordrecht.

Gale, George and Urani, John Milne, Bondi and the 'Second Way' to cosmology. In (Goenner, Renn and Sauer, eds. 1999), 343–375.

Godart, Odon (1992), Contributions of Lemaître to General Relativity (1922-1934). In (Eisenstaedt and Kox, eds. 1992), 437–452.

Goenner, Hubert, Renn, Jürgen and Sauer, Tilman, eds. (1999). *The Expanding Worlds of General Relativity.* Einstein Studies, Vol. 7. Birkhäuser Boston.

Goldberg, Joshua N. (1962), Conservation Laws and Equations of Motion. In (Royaumont 1962), 31–43.

Guide (1994). *Guide to the Archival Collections in the Niels Bohr Library at the American Institute of Physics, International Catalog of Sources for History of Physics and Allied Sciences*, Report No. 7. College Park, MD.

Heckmann, Otto (1962). General Review of Cosmological Theories. In (McVittie, ed. 1962), 429–439.

Hetherington, Norris S. (1993). *Encyclopedia of Cosmology.* Garland, New York.

Hide, Raymond (1990). Brief Comments on George McVittie's Meteorological Papers. *Vistas in Astronomy* **22**, 63–64.

Kerszberg, Pierre (1989). *The Invented Universe. The Einstein–De Sitter Controversy (1916-17) and the Rise of Relativistic Cosmology.* Oxford University Press, Oxford, U.K.

Knighting, E. (1990). War Work, *Vistas in Astronomy* **33**, 59–62.

Kragh, Helge (1993). Steady State Theory. In (Hetherington, ed., 1993), 629–636.

— (1996). *Cosmology and Controversy.* Princeton University Press, Princeton N.J.

— (1999). Steady-state Cosmology and General Relativity: Reconciliation or Conflict? In (Goenner, Renn and Sauer, eds. 1999), 377–402.

Kransinski, Andzej (1990). Early Inhomogeneous Cosmological Models in Einstein's Theory. In (Bertotti, Balbinot, Bergia and Messina, eds. 1990), 115–127.

Levi-Civita, Tullio (1929a). Vereinfachte Herstellung der Einstenschen einheitlichen Feldgleichungen. *Preussische Akademie der Wissenschaften. Phys.-math. Klasse. Sitzungsberichte*, 137–153.

— (1929b). A Proposed Modification of Einstein's Field Theory. *Nature* **123**, 678–679.

Lilley, A. E. (1957). Radio Astronomical Measurements of Interest to Cosmology. In (DeWitt, ed. 1957), 55–59.

Mavridès, Stamatia (1973). *L'Univers relativiste.* Editions Albin Michel, Paris.

McCrea, W. H. (1990). George Cunliffe McVittie (1904-88) OBE, FRSE, Pupil of Whittaker and Eddington: Pioneer of Modern Cosmology. *Vistas in Astronomy* **33**: 43–58.

McVittie, G. C. and McCrea, W. H. (1931). On the Contraction of the Universe. *Monthly Notices of the Royal Astronomical Society* **91**: 128–133.

McVittie, G. C. (1929a). On Einstein's unified field theory. *Proceedings of the Royal Society of London* A **124**, 366–374.

— (1929b). On Levi-Civita's Modification of Einstein's Unified Field Theory. *Philosophical Magazine* **8**, 1033–1044.

— (1932). Condensations on an Expanding Universe. *Monthly Notices of the Royal Astronomical Society* **92**, 500–518.

— (1933). The Mass-Particle in an Expanding Universe. *Monthly Notices of the Royal Astronomical Society* **93**, 325–339.

— (1935a). Absolute Parallelism and Milne's Kinematical Relativity. *Monthly Notices of the Royal Astronomical Society* **95**, 270–279.

— (1935b). Absolute Parallelism and Metric in the Expanding Universe Theory. *Proceedings of the Royal Society of London* A **151**, 357–370.

— (1937). *Cosmological Theory.* Methuen, London.

— (1940). Kinematical Relativity. *The Observatory* **63**, 273.

— (1941). Kinematical Relativity—a Discussion (with E. A. Milne and A. G. Walker). The Observatory 64, 11.

— (1956). *General Relativity and Cosmology.* Chapman and Hall, London.

— (1957). Counts of Extra-Galactic Radio Sources and Uniform Model Universes. *Australian Journal of Physics* **10**, 331–350.

— (1958). Distance and Relativity. *Science* **127**, 501–505.

— (1959). Distance and Time in Cosmology: the Observational Data. *Handbuch der Physik.* **53**, 445 (Springer-Verlag, Berlin).

— (1961). *Fact and Theory in Cosmology*. Eyre and Spottswood, London.

— (1962a). Galaxies as Members of the Universe: Summary. In (McVittie, ed. 1962), 441–450.

— (1962b). Cosmology and the Interpretation of Astronomical Data. In (Royaumont 1962), 253–274.

— (1965a). *General Relativity and Cosmology*. Second ed. Chapman and Hall, London.

— (1965b). Some Consequences of Large Redshifts. *Astrophysical Journal* **142**, 1637.

— (1974), Distance and Large Redshifts. *Quarterly Journal of the Royal Astronomical Society* **15**, 246.

— (ca. 1975). Autobiographical Sketch, prepared at the request of the Royal Society of Edinburgh, copy deposited at the Niels Bohr Library, American Institute of Physics.

— (1978). *Transcript of an Interview [made by David DeVorkin] taken on a Tape Recorder*. American Institute of Physics, Center for History of Physics, New York.

— (1987). An Anglo-Scottish University Education. In (Williamson, ed. 1987), 66–70.

McVittie, George C., ed. (1962). *Problems of Extra-Galactic Research*. MacMillan, New York.

Mercier, André (1992). General Relativity at the Turning point of its Renewal. In (Eisenstaedt and Kox, eds. 1992), 109–121.

Mercier, André and Kervaire, Michel, eds. (1956). *Fünfzig Jahre Relativitätstheorie, Helvetica Physica Acta*, Supplement IV. Basel.

Møller, C. (1962). The energy-momentum complex in general relativity and related problems. In (Royaumont 1962), 15–29.

North, John (1990). *The Measure of the Universe*. New York.

— (1994). *The Fontana History of Astronomy and Cosmology*. London.

Payne-Gaposchkin, Cecilia (1925). *Stellar Atmospheres*. Cambridge University Press, Cambridge, U.K.

Robertson, H. P. (1928). On Relativistic Cosmology. *Philosophical Magazine* **5**, 835–848.

— (1929), On the Foundations of Relativistic Cosmology. *Proceedings National Academy of Sciences* **15**, 822–829.

Royaumont (1962). *Les théories relativistes de la gravitation*. CNRS, Paris.

Russell, Henry N., Dugan, Raymond S. and Stewart, John Q. (1926–1927). *Astronomy*. 2 vols. Ginn & Co., Boston.

Ryle, Martin (1955). Radio Stars and Their Cosmological Significance. *The Observatory* **75**, 137–147.

Sadler, D. H. (1987). The Decade 1940–50. In (Tayler, ed. 1987), 98–147.

Sánchez-Ron, José M. (1990). Steady State Cosmology, the Arrow of Time and Hoyle and Narlikar's Theories. In (Bertotti, Balbinot, Bergia and Messina, eds. 1990), 233–243.

— (1992). The Reception of General Relativity Among British Physicists and Mathematicians (1915–1930). In (Eisenstaedt and Kox 1992), 57–88.

Stachel, John (2002). *Einstein from 'B' to 'Z'*. Birkhäuser Boston.

Tayler, R. J. (1987). *History of the Royal Astronomical Society*. Vol. 2, 1920–1980. Oxford.

Tropp, E. A., Frenkel, V. Ya. and Chernin, A. D. (1993). *Alexander A. Friedmann: the Man who Made the Universe Expand*. Cambridge University Press, Cambridge, U.K.

Vizgin, Vladimir P. (1994). *Unified Field Theories*. Birkhäuser, Basel.

Williamson, Rajkumari, ed. (1987). *The Making of Physicists*. A. Helger, Bristol.

Wright, Helen, Warnow, Joan N. and Weiner, Charles, eds. (1972). *The Legacy of George Ellery Hale*. Cambridge, MA.

Notes

[1] See, for instance, Stachel 2002, Kerszberg 1989, Tropp, Frenkel and Chernin 1993.

[2] "Il nous apparaît," wrote Mavridès in the sentence in which the previous characterization appears, "que la Cosmologie n'a de sel que si on la confronte aux resultants d'observation. Cette discipline n'a de sens que fondée sur le faits. 'L'émpiriste irréductible' qui'est McVittie—ainsi que le désignent ses collèges scientifiques—a mis l'accent, à diverses reprises et notamment dans son livre: Fact and Theory in Cosmology, sur cette indispensable dualité de la Cosmologie."

[3] A copy of these autobiographical notes is deposited at the Niels Bohr Library of the American Institute of Physics: see Guide (1994: 132–133). I am grateful to Alexei Kojevnikov for giving me access to this document. In an interview that David DeVorkin made with McVittie (1978: 4) in March 1978 he was a bit more precise, recalling that he had read "an article on relativity in the periodical *Engineering*, a journal chiefly for engineers [that] for some reason, my father used to import."

[4] I have commented on Whittaker and general relativity in (Sánchez-Ron 1992, 74–75).

[5] As stated in McVittie 1978, 1987.

[6] Einstein's paper and use of teleparallelism is studied in (Bergia 1993, 292–294) and (Vizgin 1994, 234–258), who also discusses Levi-Civita's contribution.

[7] This letter, as well as other materials to which I will refer later on in the present paper, is deposited among the "George C. McVittie Papers," University Archives, University of Illinois at Urbana-Champaign. In what follows I will refer to these materials as "McVittie papers, Urbana."

[8] McVittie was referring here, of course, to Eddington (1953), that, as it is well-known was published after Eddington's death (which happened in 1944). Whittaker was the editor of the book, adding in such capacity a preface and a few notes. E. T. Copson, then at University College, Dundee, in the University of St. Andrews, and George Temple, read the proof-sheets.

[9] This letter is also quoted by Kerszberg (1989, 336), in a slightly more complete form: "A research student McVittie and I had been worrying at the problem and made

considerable progress; so it was a blow to us to find it done much more completely by him (a blow softened, as far as I can concerned, by the fact that Lemaître was a student of mine)." What Eddington was working on with McVittie, Kerszberg added, "was the question of whether Einstein's cylindrical world is stable, using two papers by Robertson (1928, 1929) as a basis.

[10]Most probably, the books McVittie was thinking about are: (Eddington 1914), (Payne-Gaposchkin 1925), and (Russell, Dugan and Stewart 1926–1927). About this last, and influential textbook, see (DeVorkin 2000, 224-229).

[11](McVittie 1978, 23).

[12]Both papers, (McVittie and McCrea 1931) and (McVittie 1932), have been commented on by Kragh (1996).

[13]A good review of Milne's theory is included in (North 1990, chapter 8.) See also (Gale and Urani 1999), which contains a discussion of some of the controversies kinematical relativity aroused, and the influence that kinematical relativity exerted on Bondi's philosophy in the steady state theory.

[14]See in this regard, (McVittie 1940, 1941).

[15](McVittie 1937, 68–69).

[16]Related to these points, is what Hermann Bondi (1952, 69) wrote in the first edition of his influential book on cosmology:

> General relativity bases itself on the concept of the rigid ruler which enters into its fundamental assumption of the metric. As will be seen later this concept leads in cosmology to the mathematically well-defined but physically somewhat nebulous picture of the 'absolute distance' between 'simultaneous events'. This measurement of intergalactic distances with rigid rulers is much further removed from physical practice than the definition of distance adopted by Milne as fundamental for kinematic relativity, which is, at least in principle, capable of being carried out. Milne proposed that an observer, in order to measure the distance of a second observer, should send out a light pulse, and that the distant observer should respond by sending out a similar pulse as soon as he receives the first one (or, alternatively, by reflecting it)."

[17]Thus, in 1959, he described the methods employed by astronomers to determine distances in the galactic and extragalactic domains; in 1965 the luminosity-distance relations for objects of large redshift in various models of the universe, and in 1974 he tabulated the luminosity-distance, the distance by apparent size and the so-called U-distance of an object of given redshift (up to $z = 6$) for seven models of the universe (McVittie 1959, 1965b, 1974).

[18]McVittie's model has been summarily studied by Andrzej Krasinski (1990, 118–119).

[19]McVittie papers, Urbana.

[20]See, for instance, (Kragh 1993, 1996, 1999) and (Sánchez-Ron 1990).

[21] Such meeting took place on October 29–30, 1948. It was, according to an official history of the Royal Astronomical Society (Sadler 1987, 119), "a resounding success." This reference contains a photograph of the group of scientists who attended the meeting.

[22] McVittie papers, Urbana.

[23] N is the number of radio stars per unit solid angle having intensity greater than I.

[24] McVittie papers, Urbana, carbon copy of the letter.

[25] Ryle's results were received with caution. Speaking at an important conference, held at the University of North Carolina, Chapel Hill, in January 1957, A. E. Lilley (1957, 55–56) could say: "[Ryle and Scheuer's observations] will suggest departures from an isotropic and uniform universe and the results, if valid, are not consistent with a steady-state universe. However, the interpretation of his $[\log N - \log I]$ curve has been discussed by Bolton, who has suggested that when one has observational errors which increase with decreasing intensity, even an isotropic distribution can produce a curve of the form [given by Ryle and Scheuer]."

[26] McVittie papers, Urbana, carbon copy of the letter.

[27] He must refer to (McVittie 1957), where he made use of both the Cambridge and the Sydney catalogues of radio sources.

[28] G. C. McVittie, "Report on attendance at Radio-Astronomy Symposium, Paris, and Xth General Assembly International Astronomical Union, Moscow, August 1958." McVittie papers, Urbana.

[29] It must be recalled that he was unable to attend the meetings held during the last three days, because he contracted a mild case of pneumonia.

[30] McVittie papers, Urbana.

[31] The fact that McVittie was attached to a Mathematics Department must be understood in the light that during many years it was frequent that relativists were members of such departments (a notable example is the Department of Mathematics of King's College London, in which worked during the 1960s and 1970s men like Herman Bondi and Felix A. E. Pirani).

[32] As all the documents I made use of in the present paper, this letter is deposited at the McVittie papers, Urbana.

[33] Quoted in (Wright, Warnow and Weiner, eds. 1972, 98), which also includes the history of the origins of this telescope and the role Hale played in it.

[34] McVittie papers, Urbana.

[35] In the meantime, McVittie and Shapley were in contact. Through some of the letters that have survived we can imagine that Shapley's opinion about McVittie must have been increasingly positive, as he received testimony that the British relativist was interested and competent not only on general relativity but in astrophysics as well. Thus, on May 31, 1951 Shapley wrote McVittie (McVittie papers, Urbana): "I was

glad to have your letter and note your interest in some of my galaxy work," after which he entered into technical details related to the problem of how "to get an approximately true picture of the distribution of galaxies over an area of the sky and complete to a given apparent magnitude," a problem on which he was working with Hubble ("Both Hubble and I have found," he pointed out), although as we know he did not publish anything with Hubble.

[36] McVittie papers, Urbana.

[37] McVittie papers, Urbana.

[38] These comments do not mean that he had not received any recognition while in Britain. Thus, in 1931 he was elected a Fellow of the Royal Astronomical Society, on whose Council he served during 1942–1946. He was elected also Fellow of the Royal Society of Edinburgh (1943), and of the Royal Meteorological Society (1948), as recognition of the meteorological work he had done during War World II; see in this regard (Knighting 1990) and (Hide 1990).

[39] McVittie papers. Urbana.

[40] Some of McVittie's activities as Secretary of the American Astronomical Society are mentioned in (DeVorkin, ed. 1999).

[41] McVittie papers, Urbana.

[42] McVittie papers, Urbana

[43] Next, McVittie summarized visits he had made to the Max-Planck Institut für Physik und Astrophysik, Munich, June 29–July 2, where he delivered a lecture, and to the Jodrell Bank Experimental Station, near Manchester, July 13–16. About the last he wrote: "I was shown over the facilities at this Observatory, particularly the 250-foot dish, and found the whole installation even more impressive than I anticipated. Some 200 to 250 extragalactic radio-sources have been measured for angular diameter with the 250-ft. dish. The highlight of the visit was a long discussion with R. C. Jennison on his investigations of the radio-source Cygnus A. He has shown that the source consists of two sources of almost equal intensity, with the pair of colliding galaxies between them"

[44] (Møller 1962, 15).

[45] There were other participants whose interventions were dedicated to conservation laws in general relativity; for instance, John L. Synge and Andrzej Trautman.

[46] McVittie papers, Urbana.

False Vacuum: Early Universe Cosmology and the Development of Inflation

Chris Smeenk

University of California, Los Angeles, Department of Philosophy, 321 Dodd Hall, Los Angeles, CA 90095, U.S.A.; smeenk@humnet.ucla.edu

Inflationary cosmology has been widely hailed as the most important new idea in cosmology since Gamow's pioneering work on nucleosynthesis, or perhaps even since the heady early days of relativistic cosmology in the 1920s. Popular accounts typically attribute the invention of inflation to Alan Guth, whose seminal paper (Guth 1981) created a great deal of excitement and launched a research program. These accounts typically present Guth and a small band of American particle physicists as venturing into untouched territory. More careful accounts (such as Guth's memoir, Guth 1997) acknowledge that inflation's central idea, namely that the early universe passed through a brief phase of exponential expansion, did not originate with Guth. Reading this earlier research merely as an awkward anticipation of inflation seriously distorts the motivations for these earlier proposals, and also neglects the wide variety of motivations for such speculative research. Below I will describe several proposals that the early universe passed through a de Sitter phase, highlighting the different tools and methodologies used in the study of the early universe.

The early universe was the focus of active research for over a decade before Guth and other American particle physicists arrived on the scene in the late 1970s. The discovery of the background radiation in 1965 brought cosmology to the front page of the *New York Times* and to the attention of a number of physicists. In his influential popular book *The First Three Minutes*, Steven Weinberg characterized the effect of the discovery as follows:

> [Prior to discovery of the background radiation]...it was extraordinarily difficult for physicists to take seriously *any* theory of the early universe. ... The most important thing accomplished by the ultimate discovery of the 3°K radiation background in 1965 was to force us all to take seriously the idea that there *was* an early universe. (Weinberg 1977, 131–132)

Taking the early universe seriously led to efforts to extend the well understood "standard model" of cosmology developed in the 1960s, accepted by a majority of mainstream cosmologists and presented in textbooks such as Peebles (1971); Weinberg (1972), to ever earlier times. According to the standard model, the large scale structure of the universe and its evolution over time are aptly described by the simple

Friedmann–Lemaître–Robertson–Walker (FLRW) models. Extrapolating these models backwards leads to a hot, primeval "fireball," the furnace that produced both the background radiation and characteristic abundances of the light elements. Finally, the theory included the general idea that large scale structure, such as galaxies and clusters of galaxies, formed via gravitational clumping. But the standard model was not without its blemishes. In particular, it was well known that extrapolating the FLRW models led to arbitrarily high energies and a singularity as $t \to 0$.

The paper proceeds as follows. The first section below focuses on efforts by a number of Soviet cosmologists to eliminate the initial singularity. Their abhorrence of the singularity was strong enough to motivate a speculative modification of the FLRW models, namely patching on a de Sitter solution in place of the initial singularity. Gliner and Sakharov arrived at the idea by considering "vacuum-like" states of matter, whereas Starobinsky found that de Sitter space is a solution to Einstein's field equations (EFE) modified to incorporate quantum corrections. These proposals highlight two problems facing any modification of the early universe's evolution: what drives a change in the expansion rate near the singularity, and how does an early de Sitter phase lead into the standard big bang model? Section 2 turns to the influx of ideas into early universe cosmology from particle physics, focusing in particular on symmetry breaking. A group of physicists in Brussels proposed that the "creation event" could be understood as a symmetry breaking phase transition that sparked the formation of a de Sitter-like bubble, which eventually slowed to FLRW expansion. The more mainstream application of symmetry breaking to cosmology focused on the consequences of symmetry breaking phase transitions. Early results indicated a stark conflict with cosmological theory and observation. Despite this inauspicious beginning, within a few years early universe phase transitions appeared to be a panacea for the perceived ills of standard cosmology rather than a source of wildly inaccurate predictions.

13.1 Eliminating the Singularity

Cosmologists have speculated about the nature of the enigmatic "initial state" since the early days of relativistic cosmology. Research by Richard Tolman, Georges Lemaître and others in the 1930s established the existence of an initial singularity in the FLRW models, but this was typically taken to represent a limitation of the models rather than a feature of the early universe. Debates about exactly how to define a "singularity" continue to the present, but in early work singularities were usually identified by divergences in physical quantities (such as the gravitational field or curvature invariants).[1] Tolman argued that the presence of a singular state reflects a breakdown of the various idealizations of the FLRW models (Tolman 1934, 438 ff.). But by the mid-1960s cosmologists could not easily dismiss singularities as a consequence of unphysical idealizations. New mathematical techniques developed primarily by Roger Penrose, Stephen Hawking, and Robert Geroch made it possible to prove the celebrated "singularity theorems." These theorems established that singularities, signalled by the presence of incomplete geodesics,[2] are a generic feature of solutions to the field equations of general relativity that: satisfy global causality constraints (ruling out pathologies such as closed time-like curves), contain matter fields satisfying one of the energy conditions, and possess a point or a surface such that light cones start converging towards the past. The precise characterization of these assumptions differed for various singularity theorems proved throughout the 1960s, but in general these assumptions seemed physically well motivated (see, e.g., Hawking and Ellis 1968). Thus these powerful theorems dashed the hope that a singularity could be avoided in "more realistic" cosmological models.

The prominent Princeton relativist John Wheeler described the prediction of singularities as the "greatest crisis in physics of all time" (Misner et al. 1973, 1196–1198). Confronted with this crisis many of Wheeler's contemporaries took evasive maneuvers. A number of prominent Soviet physicists (including Lev Landau, Evgeny Lifshitz, Isaak Khalatnikov, and several collaborators) analyzed the (allegedly) general form of cosmological solutions to Einstein's field equations (EFE) in the neighborhood of the singularity, with the hope of showing that the singular solutions depend upon a specialized choice of initial conditions.[3] Although this group (eventually) accepted the results of the singularity theorems, there were other ways of evading an initial singularity. Approaching the initial singularity (or singularities produced in gravitational collapse) leads to arbitrarily high energies, and theorists expected the as yet undiscovered theory of quantum gravity to come into play as energies approached the Planck scale, undercutting the applicability of the theorems.[4] But there was another obvious escape route: deny one of the assumptions. Another line of research made denial of the energy conditions more appealing: the "vacuum" in modern field theory turned out to be anything but a simple "empty" state, and in particular a vacuum state violated the energy conditions. Several Soviet cosmologists, who apparently abhorred the singularity more than the vacuum, proposed that an early vacuum-like state would lead to a de Sitter bubble rather than a singularity.

13.1.1 Λ in the USSR

Two Soviet physicists independently suggested that densities reached near the big bang would lead to an effective equation of state similar to a relativistic vacuum: Andrei Sakharov, the famed father of the Soviet H-bomb and dissident, considered the possibility briefly in a study of galaxy formation (Sakharov 1966), and a young physicist at the Ioffe Physico-Technical Institute in Leningrad, Erast Gliner, noted that a vacuum-like state would counter gravitational collapse (Gliner 1966). Four further papers over the next decade developed cosmological models on this shaky foundation (Gliner 1970; Sakharov 1970; Gliner and Dymnikova 1975; Gurevich 1975), in the process elaborating on several of the advantages and difficulties of an early de Sitter phase.

Gliner's paper took as its starting point an idea that has been rediscovered repeatedly: a non-zero cosmological constant Λ may represent the gravitational effect of vacuum energy.[5] Einstein modified the original field equations of general relativity by including a Λ term to vouchsafe cherished Machian intuitions (Einstein 1917), but later thought it marred general relativity's beauty. Even for those who didn't share Einstein's aesthetic sensibility, observational constraints provided ample evidence that Λ

must be *very close* to zero. Yet, as Gliner (1966) and others noted, Λ could be treated as a component of the stress-energy tensor, $T_{ab} = -\rho_V g_{ab}$ (where "V" denotes vacuum); a T_{ab} with this form is the only stress energy tensor compatible with the requirement that the vacuum state is locally Poincaré invariant.[6] The stress-energy tensor for a perfect fluid is given by

$$T_{ab} = (\rho + p)u_a u_b + p g_{ab}, \tag{13.1}$$

where u^a represents the normed velocity of the perfect fluid, ρ is the energy density and p is pressure. The vacuum corresponds to an ideal fluid with energy density $\rho_V \left(= \frac{\Lambda c^2}{8\pi G} \right)$ and pressure given by $p_V = -\rho_V$; this violates the strong energy condition, often characterized as a prerequisite for any "physically reasonable" classical field.[7] Yakov Zel'dovich, whom Gliner thanked for critical comments, soon published more sophisticated studies of the cosmological constant and its connection with vacuum energy density in particle physics (Zel'dovich 1967, 1968). The main thrust of Gliner's paper was to establish that a vacuum stress-energy tensor should not be immediately ruled out as "unphysical," whereas Zel'dovich (1968) proposed a direct link between Λ and the zero-point energy of quantum fields.

The novelty of Gliner's paper lies in the conjecture that high density matter somehow makes a transition into a vacuum-like state. Gliner motivated this idea with a stability argument (cf. Gliner 1970), starting from the observation that matter obeying an ordinary equation of state is unstable under gravitational collapse. For normal matter and radiation, the energy density ρ increases without bound during gravitational collapse and as one approaches the initial singularity in the FLRW models.[8] However, Gliner recognized that the energy density remains constant in a cosmological model with a vacuum as the only source. The solution of the field equations in this case is de Sitter space, characterized by exponential expansion $a(t) \propto e^{\chi t}$, where $(\chi)^2 = (8\pi/3)\rho_V$ and the scale factor $a(t)$ represents the changing distance between fundamental observers. During this rapid expansion the vacuum energy density remains constant, but the energy density of other types of matter is rapidly diluted. Thus extended expansion should eventually lead to vacuum domination as the energy density of normal matter becomes negligible in comparison to vacuum energy density.[9] It is not clear whether Gliner recognized this point. But he did argue that if matter undergoes a transition to a vacuum state during gravitational collapse, the result of the collapse would be a de Sitter "bubble" rather than a singularity. This proposal avoids the conclusion of the Hawking–Penrose theorems by violating the assumption that matter obeys the strong energy condition. In effect, Gliner prefered a hypothetical new state of matter violating the strong energy condition to a singularity, although he provides only extremely weak plausibility arguments suggesting that "vacuum matter" is compatible with contemporary particle physics.[10]

By contrast with Gliner's outright stipulation, Sakharov (1966) hoped to derive general constraints on the equation of state at high densities by calculating the initial peturbations produced at high densities and then comparing the evolution of these perturbations to astronomical observations. Sakharov argued that at very high densities (on the order of 2.4×10^{98} baryons per cm^3!) gravitational interactions would need to be taken into account in the equation of state. Although he admitted that theory was too shaky to calculate the equation of state in such situations, he classified four different types of qualitative behavior of the energy density as a function of baryon number (Sakharov 1966, 74–76). This list of four included an equation of state with $p = -\rho$, and Sakharov noted that feeding this into FLRW dynamics yields exponential expansion. But the constraints Sakharov derived from the evolution of initial perturbations appeared to rule this out as a viable equation of state. In a 1970 preprint (Sakharov 1970), Sakharov again considered an equation of state $\rho = -p$, this time as one of the seven variants of his speculative "multi-sheet" cosmological model.[11] This stipulation was not bolstered with new arguments (Sakharov cited Gliner), but as we will see shortly Sakharov discovered an important consequence of an early vacuum state.

Three later papers developed Gliner's suggestion and hinted at fruitful connections with other problems in cosmology. Gliner and his collaborator, Irina Dymnikova, then a student at the Ioffe Institute, proposed a cosmological model based on the decay of an initial vacuum state into an FLRW model, and one of Gliner's senior colleagues at the Institute, L. E. Gurevich, pursued a similar idea. According to the Gliner and Dynmikova's model, an initial fluctuation in the vacuum leads to a closed, expanding universe. The size of the initial fluctuation is fixed by the assumption that $\dot{a} = 0$ at the start of expansion. The vacuum cannot immediately decay into radiation. This would require joining the initial fluctuation to a radiation-dominated FLRW model, but as a consequence of the assumption this model would collapse rather than expand—the closed FLRW universe satisfies $\dot{a} = 0$ only at *maximum* expansion.[12] Gliner and Dymnikova (1975) stipulated that the effective equation of state undergoes a gradual transition from a vacuum state to that of normal matter.[13] The scale factor and the mass of the universe both grow by an incredible factor during this transitional phase, as Gliner and Dymnikova (1975) noted; however, there is no discussion of whether this is a desirable feature of the model.

This proposal replaces the singularity with a carefully chosen equation of state, but Gliner and Dymnikova (1975) give no physical motivation guiding these choices. Instead, details of the transition are set by matching observational constraints. As a result of this phenomenological approach, Gliner and Dymnikova (1975) failed to recognize one of the characteristic features of a de Sitter-like phase. In particular, the following equation relates parameters of the transition (the initial and final energy densities, ρ_0 and ρ_1, and the "rate" set by the constant α) to present values of the matter and radiation density (ρ_p, ρ_{rp}):[14]

$$\sqrt{\frac{\rho_1}{\rho_{rp}}} \exp\left(\frac{2(\rho_0 - \rho_1)}{3\gamma\rho_1(1 - \alpha)}\right) = \frac{\rho_0}{\rho_p}\left(1 - \frac{3H^2}{8\pi G\rho_p}\right)^{-1}. \tag{13.2}$$

This equation indicates how the length of the transitional phase effects the resulting FLRW model: for a "long" transitional phase, ρ_1 is small, and the left-hand side of the equation is exponentially large. This forces the term in parentheses on the right-hand side to be exponentially small, so that H^2 approaches $\frac{8\pi G\rho_p}{3}$, the Hubble constant for a flat FLRW model. Four years later, Guth would label his discovery of this feature a "Spectacular Realization," but Gliner and Dymnikova (1975) took no notice of it.

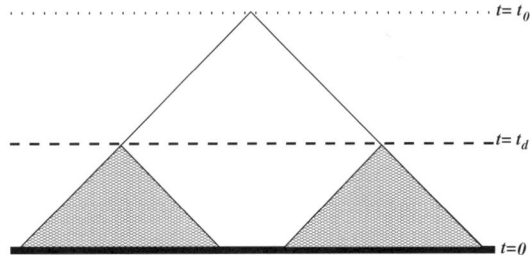

Fig. 13.1. This conformal diagram illustrates the horizon problem in the FLRW models. The singularity at $t = 0$ is stretched to a line. The lack of overlap in the past light cones at points on the surface $t = t_d$ (both within the horizon of an observer at $t = t_0$) indicates that no causal signal could reach both points from a common source.

Gurevich and Sakharov both had a clearer vision of the possible cosmological implications of Gliner's idea than Gliner himself. Gurevich (1975) noted that an initial vacuum dominated phase would provide the "cause of cosmological expansion." Gurevich clearly preferred an explanation of expansion that did not depend on the details of an initial "shock" or "explosion," echoing a concern first voiced in the 1930s by the likes of Sir Arthur Eddington and Willem de Sitter.[15] Gurevich aimed to replace various features of the initial conditions — including the initial value of the curvature, the "seed fluctuations" needed to form galaxies, and the amount of entropy per baryon — with an account of the formation and merger of vacuum-dominated bubbles in the early universe. The replacement was at this stage (as Gurevich admitted) only a "qualitative picture of phenomena" (Gurevich 1975, 69), but the goal itself was clearly articulated.

Gurevich failed to recognize, however, the implications of a vacuum-dominated phase for a problem he emphasized as a major issue in cosmology: Misner's horizon problem (Misner 1969). Horizons in relativistic cosmology mark off the region of space-time from which light signals can reach a given observer. The "particle horizon" measures the maximum distance from which light signals could be received by an observer at t_0 as the time of emission of the signal approaches the initial singularity:[16]

$$d_{ph} = \lim_{t \to 0} a(t_0) \int_t^{t_0} \frac{dt}{a(t)}. \tag{13.3}$$

This integral converges for $a(t) \propto t^n$ with $n < 1$ (satisfied for matter- or radiation-dominated FLRW models), leading to a finite horizon distance. A quick calculation shows that regions emitting the background radiation at nearly the same temperature lie *outside* each other's particle horizons. The horizon problem refers to the difficulty in accounting for this observed uniformity given the common assumption that the universe began in a "chaotic" initial state (see Figure 13.1). Misner (1969) suggested that more realistic models of the approach to the singularity would include "mixmaster oscillations," effectively altering the horizons to allow spacetime enough for causal interactions, but by 1975 a number of Gurevich's comrades (along with British cosmologists and Misner himself) had put the idea to rest (see, e.g., Criss et al. 1975, for a *post mortem*). But mixmaster oscillations were unnecessary to solve the horizon problem; as Sakharov recognized, an odd equation of state would suffice:[17]

> If the equation of state is $\rho \approx S^{2/3}$ [where S is baryon number density; this is equivalent to $p = -\frac{\rho}{3}$], then $a \approx t$ and the Lagrangian radius of the horizon is

$$\int_{t_0}^{t_1} \frac{dt}{a} \to \infty \quad \text{as} \quad t_0 \to 0, \tag{13.4}$$

> i.e., the horizon problem is resolved without recourse to anisotropic models.

To my knowledge this is the earliest "solution" of the horizon problem along these lines. (It is a solution only in the sense that altering the horizon structure makes causal interactions possible, but it does not specify an interaction that actually smooths out chaotic initial conditions.) Sakharov's colleagues at the Institute of Applied Mathematics in Moscow, notably including Igor Novikov and Zel'dovich, were probably aware of this result. But it appeared buried in the Appendix of a preprint that was only widely available following the publication of the *Collected Works* in 1982.

13.1.2 Starobinsky's model

During a research year in Cambridge in 1978–79, Zel'dovich's protegé Alexei Starobinsky developed an account of the early universe based on including quantum corrections to the stress-energy tensor in EFE. Starobinsky clearly shared Gliner and Dymnikova's willingness to replace the initial singularity with an early de Sitter phase. But there the similarity with Gliner and Dymnikova's work ends. Unlike their sterile phenomenological approach, Starobinsky's model drew on a rich source of ideas: recent results in semi-classical quantum gravity.

Throughout the 1970s Starobinsky was one of the main players in Zel'dovich's active team of astrophysicists at the Institute of Applied Mathematics, focusing primarily on semi-classical quantum gravity. Starobinsky brought considerable mathematical sophistication to bear on Zel'dovich's insightful ideas, including the study of particle production in strong gravitational fields and the radiation emitted by spinning black holes (a precursor of the Hawking effect). The relationship between the energy conditions and quantum effects was a recurring theme in this research. In response to an alleged "no go theorem" due to Hawking, Zel'dovich and Pitaevsky (1971) showed that during particle creation the effective T_{ab} violates the dominant energy condition.[18] Energy conditions might be violated as a consequence of effects like particle creation, but Starobinsky was unwilling to introduce new fields solely to violate the energy conditions. Shortly before developing his own model, Starobinsky criticized Parker and Fulling's (1973) proposal that a coherent scalar field would violate the strong energy condition and lead to a "bounce" rather than a singularity, pointedly concluding that "there is no reason to believe that at ultrahigh temperatures the main contribution to the energy density of matter will come from a coherent scalar field" (Starobinsky 1978, 84).[19]

Starobinsky's (1979, 1980) model accomplished the same result without introducing fundamental scalar fields. By incorporating quantum effects Starobinsky found a class of cosmological solutions that begin with a de Sitter phase, evolve through an oscillatory phase, and eventually make a transition into an FLRW expanding model. In the semi-classical approach, the classical stress-energy tensor is replaced with its quantum counterpart, the renormalized stress-energy tensor $\langle T_{ab} \rangle$, but the metric is not upgraded. Calculating $\langle T_{ab} \rangle$ for quantum fields is a tricky business due to divergences, but several different methods were developed to handle this calculation in the 1970s. Starobinsky's starting point was the one-loop correction to $\langle T_{ab} \rangle$ for massless, conformally invariant, non-interacting fields. Classically the trace for such fields vanishes, but due to regularization of divergences $\langle T_{ab} \rangle$ includes the so-called "trace anomaly."[20] Taking this anomaly into account, Starobinsky derived an analog of the Friedman equations and found a set of solutions to these equations.[21] This establishes the existence (but not uniqueness) of a solution that begins in an unstable de Sitter state before decaying into an oscillatory solution. Using earlier results regarding gravitational pair production, Starobinsky argued that the oscillatory behavior of the scale factor produces massive scalar particles ("scalarons"). Finally, the matter and energy densities needed for the onset of the standard big bang cosmology were supposedly produced via the subsequent decay of these scalarons.

In the course of describing this model, Starobinsky mentioned an observational constraint that simplifies the calculations considerably (Starobinsky 1980, 101):

> If we want our solution to match the parameters of the real Universe, then [the de Sitter stage] should be long enough: $Ht_0 >> 1$, where t_0 is the moment of transition to a Friedmann stage. This enables us to neglect spatial curvature terms ... when investigating the transition region.

The published version of a paper delivered in 1981 at the Moscow Seminar on Quantum Gravity (Starobinsky 1984) repeated a portion of this earlier paper with a page of new material added.[22] This added material explains that an extended de Sitter phase drives the universe very close to a "flat" FLRW model, with negligible spatial curvature. But Starobinsky did not present this aspect of the model as an important advantage: he commented that an extended de Sitter phase is necessary simply to insure compatibility with observations, and he did not further comment on whether an extended de Sitter phase is a *natural* or *desirable* feature of his model. Starobinsky's approach requires *choosing* the de Sitter solution, with no aim of showing that it is a "natural" state; as Starobinsky put it (Starobinsky 1980, 100), "This scenario of the beginning of the Universe is the extreme opposite of Misner's initial 'chaos'." In particular, his model takes the *maximally symmetric* solution of the semi-classical EFE as the starting point of cosmological evolution, rather than an *arbitrary* initial state as Misner had suggested.[23] In this assumption he was not alone: several other papers from the Moscow conference similarly postulate that the universe began in a de Sitter state (see, e.g., Grib et al. 1984; Lapchinksy et al. 1984).

Starobinsky's model led to two innovative ideas that held out some hope of observationally testing speculations about the early universe. The first of these was Starobinsky's prediction that an early de Sitter phase would leave an observational signature in the form of gravitational waves. Starobinsky (1979) calculated the spectrum of long-wavelength gravitational waves, and argued that in the frequency range of $10^{-3} - 10^{-5}$ Hz an early de Sitter phase would produce gravitational waves with an amplitude not far beyond the limits of contemporary technology. Zel'dovich was thrilled at the prospect (Zel'dovich 1981, 228): "For this it would be worth living 20 or 30 years more!" Mukhanov and Chibisov (1981) introduced a second idea that would carry over to later early universe models: they argued that zero-point fluctuations in an initial vacuum state would be amplified during the expansion phase, leading to density perturbations with appropriate properties to seed galaxy formation. Both of these ideas would prove crucial in later attempts to identify a unique observational footprint of an early de Sitter-like phase.

Starobinsky's proposal created a stir in the Russian cosmological community: it was widely discussed at the Moscow Seminar on Quantum Gravity 1981, and Zel'dovich — undoubtedly the dominant figure in Soviet cosmology, both in terms of his astounding physical insight and his institutional role as the hard-driving leader of the Moscow astrophysicists — clearly regarded the idea as a major advance. Zel'dovich (1981) reviewed the situation with his typical clarity. One of the appealing features of Starobinsky's model, according to Zel'dovich, was that it provided an answer to embarassing questions for the big bang model, "What is the beginning? What was there before the expansion began [...]?" In Starobinsky's model the "initial state" was replaced by a de Sitter solution, continued to $t \rightarrow -\infty$. But Zel'dovich noted two other important advantages of Starobinsky's model. First, it would solve the horizon problem (Zel'dovich 1981, 229):[24]

> An important detail of the new conception is the circumstance that the de Sitter law of expansion solves the problem of causality in its stride. Any two points or particles (at present widely separated) were, in the distant de Sitter past, at a very small, exponentially small distance. They could be causally connected in the past, and this makes it possible, at least in principle, to explain the homogeneity of the Universe on large scales.

Second, perturbations produced in the transition to an FLRW model might produce gravitational waves as well as the density perturbations needed to seed galaxy formation. But Zel'dovich also emphasized the speculative nature of this proposal, concluding optimistically that "there is no danger of unemployment for theoreticians occupied with astronomical problems" (Zel'dovich 1981, 229).

13.1.3 Common Problems

These proposals illustrate common problems faced by speculative theories of the early universe's evolution. First, what is the physical source of an early vacuum-like state? Second, how could an early de Sitter-like phase make a transition into FLRW expansion, during which the vacuum is converted to the incredibly high matter and radiation densities required by the hot big bang model? Gliner's outright stipulations leave little room to refine or enrich the proposal by incorporating believable physics. The contrast with Starobinsky's model is stark: in 1980, Starobinsky's model appeared to be on the verge of being developed systematically into a detailed model of the early universe based on speculative but actively studied aspects of semi-classical quantum gravity. As we will see in the next section, cosmologists would instead develop a detailed model of an early de Sitter phase based on a rich new idea from particle physics: symmetry breaking phase transitions.

13.2 Symmetries and Phase Transitions

This section focuses primarily on the study of early universe phase transitions, but this line of research was just one of many threads tying together cosmology and particle physics. In the 1970s the particle physics community began to study several different aspects of the "poor man's accelerator," as Zel'dovich called the early universe. Following the consolidation of the Standard Model of particle physics in the mid 1970s, nearly every bit of data from accelerator experiments had fallen in line. The drive to understand physics beyond the Standard Model led to exorbitantly high energies: the relevant energy scales for Georgi and Glashow's $SU(5)$ GUT proposed in 1974 was 10^{15} GeV, far beyond what would ever be accessible to earth-bound accelerators. Any sense that cosmology was too data-starved to compete with the precise science of accelerator physics was dispelled by a trio of young researchers well-versed in cosmology and particle physics. In 1977 Gary Steigman, David Schramm, and Jim Gunn argued that the number of lepton types had to be less than 5 for particle physics to be consistent with accounts of nucleosynthesis (Steigman et al. 1977). Unlike earlier cases of interaction between particle physics and cosmology, the three answered a fundamental problem in particle physics on the basis of cosmological constraints. In a time of decreasing support for ever-larger accelerators, the price tag of the poor man's accelerator must have been appealing; and Steigman, Schramm, and Gunn showed that even this bargain accelerator could be used to address fundamental issues.

   The first intensive study of GUTs applied to the early universe focused on "baryogenesis." For a given GUT, one can directly calculate an observable feature of the early universe — the baryon-to-photon ratio usually denoted η — and in 1978 Motohiko Yoshimura argued that an SU(5) GUT predicted a value of η compatible with observations. Yoshimura (1978) kicked off a cottage industry focused on developing an account of baryogenesis similar in its quantitative detail to the account of light element nucleosynthesis. The account of baryogenesis has been widely hailed as one of the "greatest triumphs" of particle cosmology (Kolb and Turner 1990, 158).[25] Below I will focus on another aspect of GUTs in cosmology, the study of symmetry breaking and restoration in the early universe.

13.2.1 Symmetries: broken and restored

The understanding of symmetries in quantum field theory (QFT) changed dramatically in the 1960s due to the realization that field theories may exhibit spontaneous symmetry breaking (SSB). A typical one-line characterization of SSB is that "the laws of nature may possess symmetries which are not manifest to us because the vacuum state is not invariant under them" (Coleman 1985, 116). Symmetry breaking in this loose sense is all too familiar in physics: solutions to a set of differential equations almost never share the full symmetries of the equations. The novel features of symmetry breaking in QFT arise as a result of a mismatch between symmetries of the Lagrangian and symmetries which can be implemented as unitary transformations on the Hilbert space of states $\mathcal{H}$. Roughly, systems for which a particular symmetry of the Lagrangian *cannot* be unitarily implemented on $\mathcal{H}$ exhibit SSB. This failure has several consequences: observables acquire non-invariant vacuum expectation values, and there is no longer a unique vacuum state. Physicists first studied symmetry breaking in detail in condensed matter systems displaying these features, but Yoichiro Nambu and others applied these ideas to problems in field theory starting in the early 1960s (see Brown and Cao 1991, Pickering 1984 for historical studies).

The introduction of SSB led to a revival of interest in gauge theories of the weak and strong interactions. Yang–Mills style gauge theories seemed to require massless gauge bosons (like the photon), in stark conflict with the short range of the weak and strong interactions. Adding mass terms for the gauge bosons directly to the Lagrangian would break its gauge invariance and, according to the conventional wisdom, render the theory unrenormalizable.[26] SSB garnered a great deal of attention in the early 1960s, but a general theorem due to Jeffrey Goldstone seemed to doom symmetry breaking in particle physics barely after its inception: SSB implies the existence of spin-zero massless bosons (Goldstone 1961; Goldstone et al. 1962).[27] Experiments ruled out such "Goldstone bosons," and there seemed to be no way to modify the particle interpretation of the theory to "hide" the Goldstone bosons along the lines of the Gupta–Bleuler formalism in QED.[28] Goldstone et al. (1962) concluded by reviewing the dim prospects for SSB; Weinberg added an epigraph from *King Lear* — "Nothing will come of nothing: speak again" — to indicate his dismay, which was (fortunately?) removed by the editors of *The Physical Review* (Weinberg 1980, 516). But there was a loophole: Goldstone's theorem does not apply to either discrete or local gauge symmetries.[29]

Philip W. Anderson was the first to suggest that breaking a gauge symmetry might cure the difficulties with Yang–Mills theory (by giving the gauge bosons mass) without producing Goldstone bosons. Anderson noted that this case may resemble condensed matter systems exhibiting SSB, in that the Goldstone bosons "become tangled up with Yang–Mills gauge bosons, and, thus, do not in any true sense really have zero mass" (Anderson 1963, 422; cf. Anderson 1958). He speculated that this "tangling" between Goldstone and gauge bosons could be exploited to introduce a massive gauge boson, but he supported these provocative remarks with neither a field theoretic model nor an explicit discussion of the gauge theory loophole in Goldstone's theorem. Within a year of Anderson's suggestive paper, Brout, Englert, Guralnik, Kibble and Higgs all presented field theoretic models in which gauge bosons acquire mass by "tangling" with Goldstone bosons (Englert and Brout 1964; Guralnik et al. 1964; Higgs 1964).

In the clear model presented by Peter Higgs, the massless Goldstone modes disappear from the physical particle spectrum, but in their ghostly gauge-dependent presence the vector bosons acquire mass.[30] Higgs began by coupling the simple scalar field of the Goldstone model with the electromagnetic interaction. Take a model incorporating a two component complex scalar field, such that $\phi = \frac{1}{\sqrt{2}}(\phi_1 - i\phi_2)$ with an effective potential

$$V(\phi) = \frac{1}{2}\lambda^2|\phi|^4 - \frac{1}{2}\mu^2|\phi|^2. \tag{13.5}$$

The effective potential includes all the terms in the Lagrangian other than the kinetic terms, and it represents the potential energy density of the quantum fields.[31] At first glance the second term appears to have the wrong sign; with the usual $+$ sign, $V(\phi)$ has a unique global minimum at $\phi = 0$. The "incorrect" sign leads to degeneracy of the vacuum state; with a $-$ sign, $V(\phi)$ has minima at $\phi_0 = \frac{\mu}{\lambda}$. Including the electromagnetic interaction leads to the following Lagrangian:

$$\mathcal{L} = (D_\mu\phi)^\dagger(D^\mu\phi) - V(\phi) - \frac{1}{4}F_{\mu\nu}F^{\mu\nu}, \tag{13.6}$$

where $F_{\mu\nu} = \partial_\mu A_\nu - \partial_\nu A_\mu$, and D is the covariant derivative operator defined as $D_\mu = \partial_\mu + ieA_\mu$. Rewriting the effective potential $V(\phi)$ by expanding the field ϕ around the "true vacuum" ϕ_0 shows that the ϕ_1 field acquires a mass term whereas ϕ_2 is the massless "Goldstone boson." Higgs realized that a clever choice of gauge can be used to "kill" the latter component, which then appears *not* as a massless boson but instead as the longitudinal polarization state of a massive vector boson. The Lagrangian is invariant under the following gauge transformations:

$$\phi(x) \to e^{-i\theta(x)}\phi(x), \tag{13.7}$$

$$A_\mu \to A_\mu + \frac{1}{m}\partial_\mu\theta(x), \tag{13.8}$$

where m is a constant. The "Higgs mechanism" involves choosing a value of $\theta(x)$ to cancel the imaginary part of ϕ. This choice of $\theta(x)$ also effects the vector potential, leading to the following Lagrangian:

$$\mathcal{L} = (\partial_\mu\phi)(\partial^\mu\phi) + m^2\phi^2 A_\mu A^\mu - V(\phi) - \frac{1}{4}F_{\mu\nu}F^{\mu\nu}. \tag{13.9}$$

The vector field A_μ has acquired a mass term (the second term), as has the "Higgs boson" (although it is buried in the expression for $V(\phi)$), and the dreaded "Goldstone boson" has disappeared from the Lagrangian.

The Higgs mechanism could be used to fix and combine two appealing ideas, ridding both Yang–Mills style gauge theories and SSB of unwanted massless particles. Several theorists hoped that the trail blazed by Higgs et al. would lead to a gauge theory of the strong and weak interactions.[32] Three years after Higgs' paper, Weinberg incorporated the Higgs mechanism in a unified theory of the electromagnetic and weak interactions (Weinberg 1967), and a similar theory was introduced independently by Abdus Salam. These theories faced a roadblock, however: although several theorists suspected that such theories are renormalizable, they were not able to produce convincing arguments to that effect (Weinberg 1980, 518). Without a proof of renormalizability or direct experimental support the Salam–Weinberg idea drew little attention.[33]

Although theories with *unbroken* gauge symmetries were known to be renormalizable term-by-term in perturbation theory, it was not clear whether SSB would spoil renormalizability. Progress in the understanding of renormalization (due in large part to the Nobel Prize winning efforts of the Dutch physicists Gerard 't Hooft and Martinus Veltman) revealed that the renormalizability of a theory is actually *unaffected* by the occurrence of SSB. In his 1973 Erice lectures, Sidney Coleman advertised this as the main selling point of SSB (Coleman 1985, 139).

Testing the Higgs mechanism required a venture into uncharted territory. Although accelerator experiments carried out throughout the 1970s probed various aspects of the electroweak theory (see, e.g., Pickering 1984), they did little to constrain or elucidate the Higgs mechanism itself. Physicists continue to complain three decades later that the Higgs mechanism remains "essentially untested" (Veltman 2000, 348). Although the Higgs mechanism was the simplest way to reconcile a fundamentally symmetric Lagrangian with phenomenology, physicists actively explored alternatives such as "dynamical" symmetry breaking.[34] Indeed, treating the fundamentally symmetric Lagrangian as a formal artifact rather than imbuing it with physical significance was a live option. However, several physicists independently recognized that treating the Higgs mechanism as a description of a physical transition that occurred in the early universe, rather than as a bit of formal legerdemain, has profound consequences for cosmology. Weinberg emphasized at the outset that this line of research "may provide some sort of answer to the question" of "whether a spontaneously broken gauge symmetry should be regarded as a true symmetry" (Weinberg 1974b, 274).

In the condensed matter systems that originally inspired the concept of symmetry breaking, a variety of conditions (such as high temperature or large currents) lead to restoration of the broken symmetry. Based on a heuristic analogy with superconductivity and superfluidity, David Kirzhnits and his student Andrei Linde, both at the Lebedev Physical Institute in Moscow, argued that the vacuum expectation value ϕ_0 in a field theory with SSB varies with temperature according to $\phi_0^2(T) = \phi_0^2(T = 0) - c\lambda T^2$, where c and λ are non-zero constants (Kirzhnits 1972; Kirzhnits and Linde 1972). Symmetry restoration occurs above the critical temperature T_c, defined by $\phi_0^2(T_c) = 0$ (for $T > T_c$, $\phi_0(T)$ becomes imaginary). In the Weinberg model $\phi_0(0) \approx G^{1/2}$ (G is the weak interaction coupling constant), and (assuming that $c\lambda \approx 1$) Kirzhnits and Linde estimated that symmetry restoration occurs above $T_c \approx G^{-1/2} \approx 10^3 \; GeV$. They concluded that the early universe underwent a transition from an initially symmetric state to the current broken symmetry state at the critical temperature, which corresponds to approximately 10^{-12} seconds after the big bang in the standard hot big bang model.

Within two years Kirzhnits and Linde and a group of Cambridge (Massachusetts) theorists had developed more rigorous methods based on finite-temperature field theory to replace this heuristic argument.[35] Finite-temperature field theory includes interactions between quantum fields and a background thermal heat bath at a temperature T.[36] These more detailed calculations showed that, roughly speaking, symmetry restoration occurs as a consequence of the temperature dependence of quantum corrections to the effective potential. The full effective potential includes a zero-temperature term along with a temperature-dependent term, $\bar{V}(\phi, T)$. Symmetry breaking occurs

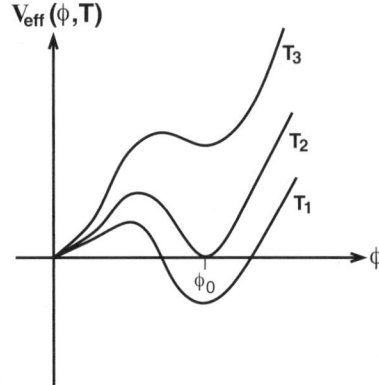

Fig. 13.2. This figure illustrates the temperature dependence of the effective potential of the Higgs field $V_{eff}(\phi, T)$ in the Weinberg–Salam model. T_2 is the critical temperature (approximately 10^{14} GeV), and $T_3 > T_2 > T_1$.

in a theory with $V(\phi) = \frac{1}{2}\lambda^2|\phi|^4 + \frac{1}{2}\mu^2|\phi|^2$, for example, if $\bar{V}(\phi, T)$ includes a mass correction that changes the sign of the second term above a critical temperature. Whether symmetry restoration occurs depends upon the nature of $\bar{V}(\phi, T)$ and the zero temperature effective potential in a particular model.[37] In the Weinberg–Salam model (with suitable choices for coupling constants), the global minimum at $\phi = 0$ for temperatures above the critical temperatures develops into a local minimum with the true global minimum displaced to ϕ_0 (see Figure 13.2 for an example). Determining the nature and consequences of such phase transitions drew an increasing number of particle physicists into the study of early universe cosmology throughout the 1970s, as we will see in Section 2.3. But before continuing with the discussion of this line of research, I will briefly turn to more speculative uses of SSB in cosmology.

13.2.2 Conformal Symmetry Breaking

By the late 1970s symmetry breaking was an essential piece in the field theorists' technical repertoire, and its successful use in electroweak unification and the development of the Standard Model encouraged more speculative variations on the theme. The "Brussels Consortium" (as I will call Robert Brout, François Englert, and their various collaborators) described the origin of the universe as SSB of conformal symmetry, but this imaginative line of research led to an increasingly rococo mess rather than a well constrained model. At roughly the same time, Anthony Zee developed an account of gravitational symmetry breaking motivated by the desire to formulate a "unified" gravitational theory with no dimensional constants other than the mass term of a fundamental scalar field.

   Like their countryman Lemaître decades earlier, the Brussels Consortium focused on a quantum description of the "creation" event itself. Brout et al. (1978) aimed to replace "the 'big bang' hypothesis of creation—more a confession of desperation and bewilderment than the outcome of logical argumentation" with an account of the "spontaneous creation of all matter and radiation in the universe. [...] The big bang is replaced by the fireball, a rational object subject to theoretical analysis" (Brout et al. 1978, 78). As with Tyron's (1973) earlier proposal, this account of spontaneous creation did not violate conservation of energy. Their theoretical analysis builds on an alleged "deep analogy" between relativistic cosmology and conformally invariant QFT, which in practice involves two fundamental assumptions.[38] First, the Consortium assumes that the universe must be described by a conformally flat cosmological model, which implies that the metric for any cosmological model is related to Minkowksi space-time by $g_{ab} = \phi^2(x^i)\eta_{ab}$, where η_{ab} is the Minkowski metric.[39] The conformal factor $\phi(x^i)$ is treated as a massless scalar field conformally coupled to gravitation. Second, a fluctuation of $\phi(x^i)$, which breaks the conformal symmetry of the pristine initial state (constant $\phi(x^i)$ in a background Minkowski space-time), bears the blame for the creation of the universe.

The devil is in providing the details regarding the outcome of the "rational fireball" triggered by such a modest spark. The Consortium's original script runs as follows: the fluctuation initially produces a de Sitter-like bubble, with the expansion driven by an effective equation of state with negative pressure. This equation of state is due to particle creation via a "cooperative process": the initial fluctuation in $\phi(x^i)$ perturbs the gravitational field; variations in the gravitational field produce massive scalar particles; the particles create fluctuations in the gravitational field; and so on. Eventually the cooperation ends, and the primeval particles decay into matter and radiation as the universe slows from its de Sitter phase into FLRW expansion. Although the details of these processes are meant to follow from the fundamental assumptions, a number of auxiliary conditions are needed to insure that the story culminates with something like our observed universe. The evolution of the Consortium's program belies the malleability of the underlying physics: Brout et al. replace the earlier idea regarding "cooperative processes" with the suggestion that particle production is a result of a "phase transition in which the 'edge of the universe' is the boundary wall between two phases" (Brout et al. 1980, 110).

Despite these difficulties, the Consortium often attributed a great deal of importance to their "solution" of the "causality problem." The basis for this solution was buried in an Appendix of Brout et al. (1978), but mentioned more prominently in later papers, including the title of Brout et al. (1979) — "The Causal Universe." Brout et al. (1978) note that in their model the integral in equation (13.3) diverges. There are no horizons. But there is also no pressing horizon *problem* in Misner's sense: conformal symmetry is stipulated at the outset, so there is simply no need to *explain* the early universe's uniformity via causal interactions. However, the absence of horizons is still taken to solve the "causality problem," in the sense that the universe and all its contents can ultimately be traced back to a simple single cause, the initial fluctuation of $\phi(x^i)$. Whatever the appeal of this solution, the Consortium ultimately failed to develop a believable model that realized their programmatic aims. However, the Princeton theorist J. Richard Gott III developed a variation of the Consortium's idea that would eventually lead to the development of "open inflation" models (Gott 1982).

Anthony Zee also solved the horizon problem with a variation on the theme of SSB. Zee (1979, 1980) proposed that incorporating symmetry breaking into gravitational theory (by coupling gravitation to a scalar field) leads to replacing the gravitational constant G with $(\epsilon \phi_v^2)^{-1}$, where ϵ is a coupling constant and ϕ_v is the vacuum expectation value of the scalar field.[40] If the potential (and the minima) of this field varies with temperature, then the gravitational "constant" varies as well. Zee (1980) argues that $\phi^2 \approx T^2$ at high temperatures, so that $G \propto 1/T^2$. This alters the FLRW dynamics so that $a(t) \propto t$; and it will come as no suprise that the integral in equation (13.3) diverges as a result. According to Guth's recollections (Guth 1997, 180–81), a lunchtime discussion of Zee's paper in the SLAC cafeteria led him to consider the implications of his own ideas for horizons.

13.2.3 Phase Transitions

The study of early universe phase transitions held out the promise of deriving stringent observational constraints from the cosmological setting for aspects of particle physics far beyond the reach of accelerators. Throughout the 1970s physicists studied three different types of consequences of symmetry breaking phase transitions: (1) effects due to the different nature of the fundamental forces prior to the phase transition, (2) defect formation during the phase transition, (3) effects of the phase transition on cosmological evolution. As we will see below, initial results ran the gamut from disastrous conflict with observational constraints to a failure to find any detectable imprint.

The first type of effect drew relatively little attention. Kirzhnits (1972); Kirzhnits and Linde (1972) briefly mentioned the possible consequences of long-range repulsive forces in the early universe. Prior to the electroweak phase transition any "weak charge" imbalance would result in long-range repulsive forces, and according to Kirzhnits and Linde such forces would render both a closed, positive curvature model and an isotropic, homogeneous model "impossible" (Kirzhnits and Linde 1972, 474).[41] By way of contrast, a group of CERN theorists suggested that interactions at the GUT scale would help to *smooth* the early universe. Ellis et al. (1980) consider the possibility that a "grand unified viscosity" would effectively insure isotropization prior to a symmetry breaking phase transition; they conclude that although these interactions damp some modes of an initial perturbation spectrum, they will not smooth a general anisotropic cosmological model.

The study of defect formation in the early universe was a much more fruitful line of research. An early study of CP-symmetry breaking (Zel'dovich et al. 1975) showed that the resulting inhomogeneity (with energy density concentrated in domain walls) would be far too large to fit observational constraints.[42] But Zel'dovich et al. (1975) also calculated the equation of state for this "cellular medium" (averaged over a volume containing both domain walls and the empty cells), and remarked that evolution dominated by matter in this state might solve the horizon problem.[43] The authors did not highlight this point (it was not mentioned in the introduction, abstract, or conclusion); their main interest was to establish that cosmology rules out discrete symmetry breaking, in itself a remarkable constraint on particle physics.

Later work on the formation of defects in theories with SSB of local gauge symmetries also ran afoul of observational constraints. Tom Kibble, an Indian-born British physicist at Imperial College, established a particularly important result (Kibble 1976): defect formation depends on the topological structure of the vacuum solutions to a particular field theory, and is thus relatively independent of the details of the phase transition. Roughly, defects result from the initial domain structure of the Higgs field, which Kibble argued should be uncorrelated at distances larger than the particle horizon at the time of the phase transition. This complicated domain structure disappears if the Higgs field in different regions becomes "aligned," but in some cases no continuous evolution of the field can eliminate all nonuniformities; topological defects are the resulting persistent structures. Kibble (1976) noted that point-like defects (called monopoles and previously studied by 't Hooft 1974; Polyakov 1974) might form, but thought that they would "not be significant on a cosmic scale." However, given the absence of any natural annihilation mechanism, Zel'dovich and Khlopov (1978); Preskill (1979); Einhorn et al. (1980) established a dramatic conflict between predicted monopole abundance and observations: in Preskill's calculation, monopoles alone would contribute a mass density 10^{14} times greater than the *total* estimated mass density![44]

The resolution of this dramatic conflict would ultimately come from considerations of the third type of effect. Linde, Veltman and Joseph Dreitlein at the University of Colorado independently realized that a non-zero $V(\phi)$ would couple to gravity as an effective Λ term.[45] Linde (1974) argued that although earlier particle physics theories "yielded no information" on the value of Λ (following Zel'dovich, he held that Λ is fixed only up to an arbitrary constant), theories incorporating SSB predicted a tremendous shift – 49 orders of magnitude – in $V(\phi)$ at the critical temperature T_c.[46] However, this dramatic change in the cosmological "constant" would apparently have little impact on the evolution of the universe (Linde 1974, 183):[47]

> To be sure, almost the entire change [of Λ] occurs near $T_c = 10^{15} - 10^{16}$ deg. In this region, the vacuum energy density is lower than the energy density of matter and radiation, and therefore the temperature dependence of Λ does not exert a decisive influence on the initial stage of the evolution of the universe.

Linde implicitly assumed that the phase transition was second-order, characterized by a transition directly from one state to another with no intermediate stage of "mixed" phases.[48] Unlike Linde, Veltman (1974) regarded the idea that an arbitrary constant could be added to the vacuum energy density to yield a current value of $\Lambda \approx 0$ as "ad hoc" and "not very satisfactory." Veltman took the "violent" disagreement with observational constraints on Λ and the value calculated using the electroweak theory as one more indicator that the Higgs mechanism is "a cumbersome and not very appealing burden" (Veltman 1974, 1).[49] Dreitlein (1974) explored one escape route: an *incredibly* small Higgs mass, on the order of $2.4 \times 10^{-27} MeV$, would lead to an effective Λ close enough to 0. Veltman (1975) countered that such a light Higgs particle would mediate long-range interactions that should have already been detected. In sum, these results were thoroughly discouraging: Veltman had highlighted a discrepancy between calculations of the vacuum energy in field theory and cosmological constraints that would come to be called the "cosmological constant problem" (see

Rugh and Zinkernagel 2002). Even for those willing to set aside this issue and focus only on the shift in vacuum energy, there appeared to be "no way cosmologically to discriminate among theories in which the symmetry is spontaneously broken, dynamically broken, or formally identical and unbroken" (to quote Bludman and Ruderman 1977, 255).

By the end of the 1970s several physicists had discovered that this conclusion does not hold if the Higgs field became trapped in a "false vacuum" state (with $V(\phi) \neq 0$). Demosthenes Kazanas, an astrophysicist working at Goddard Space Flight Center, clearly presented the effect of persistent vacuum energy (Kazanas 1980): the usual FLRW dynamics is replaced with a phase of exponential expansion. He also clearly stated an advantage of incoroporating such a phase (L62):

> Such an exponential expansion law occurring in the very early universe can actually allow the size of the causally connected regions to be many orders of magnitude larger than the presently observed part of the universe, thus potentially accounting for its observed isotropy.

But it was not clear how to avoid an undesirable consequence of a first-order phase transition, namely the production of large inhomogeneities due to the formation of "bubbles" of the new "true" vacuum phase immersed in the old phase. Linde and Chibisov (Linde 1979, 433–34) explored the possibility of combining Zel'dovich's "cold universe" idea with a first-order phase transition, but they did not see a way to avoid excessive inhomogeneity.[50] During a stay at NORDITA in Copenhagen, the Japanese astrophysicist Katsuhiko Sato studied first-order phase transitions in considerable detail, focusing on the consequences of a stage of exponential expansion driven by a false vacuum state. Sato (1981) derived constraints on various parameters, such as the rate of bubble formation and coupling constants.[51] Although Sato appears to have been optimistic that these constraints could be met, a slightly later collaborative paper with the University of Michigan theorist Martin Einhorn (Einhorn and Sato 1981) ended on a skeptical note (401):[52]

> We have seen that most of the difficulties with the long, drawn-out phase transition discussed in Section V stems [sic] from the exponential expansion of the universe. This was due to the large cosmological constant. If a theory could be developed in which the vacuum did not gravitate, i.e., a theory of gravity which accounts for the vanishing cosmological constant term in a natural way, then the discussion would be drastically changed. Although scenarios have been developed in which the effect of the cosmological constant term remains small for all times, we would speculate that the problem here is less the choice of GUT but rather reconciling gravity with quantum field theory.

To avoid the unpalatable consequences of a first order phase transition Einhorn and Sato were willing to abandon the starting point of this entire line of thought.[53]

By the time these papers appeared in print, the young American physicist Alan Guth had presented an argument that an "inflationary" stage is a desirable consequence of an early universe phase transition, rather than a source of difficulties. After persistent lobbying from his friend and collaborator Henry Tye, Guth undertook serious study of GUTs in the summer of 1979, focused on production of monopoles in the early universe (Guth 1997, chapter 9). Tye and Guth discovered that a first-order transition could alleviate the monopole problem: within each bubble produced in a first-order transition, the Higgs field is uniform. Monopoles would only be produced at the boundaries between the bubbles as a consequence of bubble wall collisions. Thus the abundance of monopoles ultimately depends upon the nucleation rate of the bubbles. Guth and Tye (1980) argued that reasonable models of the phase transition have a low nucleation rate, leading to a tolerably low production of monopoles. Einhorn and Sato (1981) highlighted various difficulties with this proposal, commenting that "although it is *possible* to meet the necessary requirements, it is unclear whether this scenario is *natural* in the sense that it may require fortuitous relationships between the magnitude of the gauge coupling and the parameters of the Higgs potential" (Einhorn and Sato 1981, 385) and noting the difficulties associated with a phase of exponential expansion. Shortly after Guth and Tye (1980) was submitted, Guth independently discovered that the equation of state for the Higgs field trapped in a "false vacuum" state drives exponential expansion. In short order, he discovered several appealing features of what he called, alluding to economic worries at the end of Carter's presidency, an "inflationary universe."

13.2.4 Guth's "Spectacular Realization"

Guth modestly concluded as follows (Guth 1981, 354):

> In conclusion, the inflationary scenario seems like a natural and simple way to eliminate both the horizon and flatness problems. I am publishing this paper in the hope that it will highlight the existence of these problems and encourage others to find some way to avoid the undesirable features of the inflationary scenario.

To say that Guth's paper (and the series of lectures he gave before and after it appeared) achieved these goals would be a dramatic understatement. This success stemmed not from fundamentally new physics, but from the clear presentation of a rationale for pursuing the idea of inflation. Even those who had been aware of the work discussed above, such as Martin Rees, have commented that they only understood it in light of Guth's paper.[54] Guth's paper significantly upped the explanatory ante for early universe cosmology: he showed that several apparently independent features of the universe could be traced to a common source, an early stage of inflationary expansion. This effectively set a new standard for theory choice in early universe cosmology. The situation resembles several other historical episodes in which a significant success set new standards. Einstein's accurate prediction of the anomaly in Mercury's perihelion motion raised the bar for gravitational theories: although the perihelion motion was not regarded as a decisive check prior to his prediction, it subsequently served as a litmus test for competing theories of gravitation. Similarly, following Guth's paper the ability to solve these problems served as an entrance requirement.

To my knowledge Guth was the first to explicitly recognize the connection between an inflationary stage and a puzzling balance between the initial expansion rate and energy density. Guth's work notebook dated Dec. 7, 1979 begins with the following statement highlighted in a double box: "SPECTACULAR REALIZATION: This kind of supercooling can explain why the universe today is so incredibly flat—and therefore resolve the fine-tuning paradox pointed out by Bob Dicke." [55] Dicke's paradox highlights an odd feature of the density paramter Ω. Using the Friedmann equation, we can write Ω as follows:[56]

$$\Omega := \frac{8\pi G}{3H^2}\rho = \left(1 - \frac{3k}{8\pi G\rho}\right)^{-1}. \tag{13.10}$$

During expansion ρ scales as $\propto a^{-3}$ for normal matter and $\propto a^{-4}$ for radiation. Thus, if the value of Ω initially differs from 1, it evolves rapidly away from 1; the value $\Omega = 1$ is an unstable fixed point under dynamical evolution. For the observed universe to be anywhere close to $\Omega = 1$ (as it appears to be), the early universe must have been *incredibly* close to the "flat" FLRW model ($\Omega = 1, k = 0$). Guth discovered that during exponential expansion Ω is driven rapidly *towards* 1; ρ is a constant for a false vacuum state, so Ω approaches 1 as a^{-2} during inflation. If the universe expands by a factor $Z \geq 10^{29}$, where $Z =: e^{\chi \Delta t}$ and Δt is the duration of the inflationary stage, then $\Omega_0 = 1$ to extremely high precision, for nearly any pre-inflationary "initial value" of Ω.

Unlike the horizon problem, the flatness problem was not widely acknowledged as a legitimate problem prior to Guth's paper. In an appendix added to "convince some skeptics," Guth comments that (Guth 1981, 355):

> In the end, I must admit that questions of plausibility are not logically determinable and depend somewhat on intuition. Thus I am sure that some physicists will remain convinced that there really is no flatness problem. However, I am also sure that many physicists agree with me that the flatness of the universe is a peculiar situation which at some point will admit a physical explanation.

Whether or not this argument swayed many physicists, several of the interviewees in Lightman and Brawer (1990) made remarks similar to Misner's (Lightman and Brawer 1990, 240):

> I didn't come on board thinking that paradox [Dicke's flatness paradox] was serious until the inflationary models came out. [...] The key point for me was that inflation offers an explanation. Even if it's not the right explanation, it shows that finding an explanation is a proper challenge to physics.

The existence of a proposed solution to the flatness problem lent it an air of legitimacy; the universe's flatness had been previously regarded as puzzling (Dicke and Peebles 1979), but following Guth's paper it was widely interpreted as a telling sign of an early inflationary stage.

Several proposals discussed above implied that horizons would disappear, as the horizon distance in equation (13.3) diverges. A transient inflationary phase increases the horizon distance by a factor of Z; for $Z > 5 \times 10^{27}$ the "horizon problem disappears" in the sense that the horizon length at the time of the emission of the background radiation approaches the current visual horizon. Particle horizons don't disappear, but they are stretched enough to encompass the visible universe. Guth stressed the striking difference between initial conditions needed in the inflationary universe and the standard cosmology (Guth 1981, 347): for the standard cosmology, "the initial universe is assumed to be homogeneous, yet it consists of at least $\approx 10^{83}$ separate regions which are causally disconnected." For an inflationary period with sufficiently large Z, a single homogeneous pre-inflationary patch of sub-horizon scale expands to encompass the observed universe.

Despite these successes, Guth's original proposal did not solve the transition problem. As Einhorn and Sato (1981) had argued, bubbles of new phase formed during the phase transition do not percolate, i.e., they do not join together to form large regions of the same phase. The energy released in the course of the phase transition is concentrated in the bubble walls, leading to an energy density far too high near the bubble walls and far too low in the interior. Frequent bubble collisions would be needed to smooth out the distribution of energy so that it is compatible with the smooth beginning of an FLRW model.[57] The phase transition never ends, in the sense that large volumes of space remain "stuck" in the old phase, with vast differences in the energy density between these regions and the bubble walls. In summary, a first-order phase transition appropriate for inflation also produces a universe marred by the massive inhomogeneities due to the formation of bubbles, rather than the smooth early universe required by observations.

The solution to the transition problem led to difficulties with Guth's original identification of the Higgs field of an $SU(5)$ GUT as the source of an inflationary stage. Briefly, Albrecht and Steinhardt (1982); Linde (1982) both developed models of the phase transition based on a Coleman–Weinberg effective potential for the Higgs field. In these new models the inflationary expansion persists long enough that the initial bubble is much, much larger than the observed universe; within this single bubble the matter and radiation density needed for the big bang model is generated via decay of the Higgs field. Within a year theorists had turned to implementing Chibisov's (1981) idea that small fluctuations stretched during inflation would serve as the seeds for galaxy formation. The intense work on structure formation during the Nuffield workshop, a conference held in Cambridge from June 21–July 9, 1982, led to the "death and transfiguration" of inflation (from the title of the conference review in *Nature*, Barrow and Turner 1982). Inflation "died" since detailed calculations of the density perturbations produced during an inflationary era indicated that an $SU(5)$ Higgs field could not drive inflation, as originally thought. The "transfiguration" of the field involved a significant shift in methodology: the focus shifted to implementing inflation successfully rather than treating it as a consequence of independently motivated particle physics. In his recollections of the Nuffield conference, Guth wrote:

> [A] key conclusion of the Nuffield calculations is that the field which drives inflation cannot be the same field that is responsible for symmetry breaking. For the density perturbations to be small, the underlying particle theory must

contain a new field, now often called the *inflaton* field [...], which resembles the Higgs field except that its energy density diagram is much flatter. (Guth 1997, 233–34)

The "inflaton" may resemble the Higgs, but the rules of the game have changed: it is a new fundamental field distinct from any scalar field appearing in particle physics.

The explosion of research interest in inflationary cosmology in the early 1980s attests to its appeal. Inflation allowed theorists to replace several independent features of the initial conditions — overall uniformity, flatness, lack of monopoles and other relics, and the presence of small scale fluctuations — with a theoretical entity they knew how to handle: the effective potential of a fundamental scalar field.[58] The discussion of earlier proposals highlights an important advantage of inflation: the Higgs mechanism is a central component of the Weinberg–Salam model and of GUTs, which provided a rich source of ideas for further refinements of inflation. Starobinsky drew on the more esoteric subject of quantum corrections to the stress-energy tensor in semiclassical quantum gravity, and the other proposals discussed above required a number of bald stipulations. Inflation still has not solved the source problem, in the sense that there is still no canonical identification of the "inflaton" field with a particular scalar field. The fertile link with particle physics has instead produced an embarassment of riches: inflation has been implemented in a wide variety of models, to such an extent that cosmologists have sometimes complained of the difficulty in coining a name for a new model.

In closing, I should emphasize an important difference between inflation and other cases of "upping the explanatory ante." Prior to Einstein's work, astronomers agreed that there was a discrepancy between the observed perihelion motion of Mercury and Newtonian calculations, although this was not seen as a telling failure of Newtonian theory. By way of contrast, several critics of inflation have not been convinced that inflation has cured *genuine* explanatory deficiencies of the standard big bang model.[59] Intellectual descendants of Ludwig Boltzmann such as Roger Penrose (see, in particular Penrose 1979, 1989) *expect* the universe to be in an initially "improbable" state, which is ultimately responsible for the second law of thermodynamics and the arrow of time. Special initial conditions play the crucial role of insuring that the observed universe has an arrow of time; they are not something to be avoided by introducing new dynamics that "washes away" the dependence on an initial state. Two of the proposals above also did not take this approach to "erasing" the singularity: Starobinsky accepted that his proposal would require stipulating that the early universe began in an early de Sitter state, and the Brussels Consortium aimed to develop an account of the creation event itself. In developing theories of the early universe, the methodological strategy exemplified by inflation was by no means mandatory.

13.3 Conclusions

In the epilogue of their recent textbook, Kolb and Turner (1990) contrast the adventurous attitude of their contemporaries with those of earlier cosmologists, commenting that (Kolb and Turner 1990, 498):

Whatever future cosmologists write about cosmology in the decades following the discovery of the CMBR, we can be certain they will not criticize contemporary cosmologists for failure to take their theoretical ideas — and sometimes wild speculations — seriously enough.

Following a story of speculative theories regarding the universe at $t \approx 10^{-35}$ s after the big bang, it is easy to agree with their assessment. As I have described above, various problems and opportunities led cosmologists to develop theories of the early universe. The incredible extrapolations to the early universe allowed theorists to grapple with issues that have no bearing on more directly accessible phenomena, including the creation of particles in strong gravitational fields and the predictions of symmetry restoration at incredibly high temperatures. Many theoretical roads led to the consideration of an early de Sitter phase, and all faced the difficulties of identifying a believable physical source driving the de Sitter expansion and accounting for the transition to customary big bang expansion. Guth's seminal work on inflation did not introduce new physics, and did not solve these problems, but it did provide a rationale that has done much to underwrite the adventurous optimism characterizing the field.

Acknowledgments

I would like to thank John Earman, Al Janis, Michel Janssen, David Kaiser, John Norton, and Laura Ruetsche, for helping in various ways to make this a better paper. My research was supported in part by the NSF under grant SES 0114760.

References

Aitchison, Ian J. R. (1982). *An informal Introduction to Gauge Field Theories*. Cambridge University Press.

Albrecht, Andreas and Steinhardt, Paul (1982). Cosmology for grand unified theories with induced symmetry breaking. *Physical Review Letters* **48**, 1220–1223.

Anderson, Philip W. (1963). Plasmons, gauge invariance, and mass. *Physical Review* **130**, 439–442.

— (1958). Coherent excited states in the theory of superconductivity: Gauge invariance and the Meissner effect. *Physical Review* **110**, 827–835.

Barrow, John D. and Turner, Michael S. (1982). The inflationary universe – birth, death, and transfiguration. *Nature* **298**, 801–805.

Bekenstein, Jacob D. (1975). Nonsingular general-relativistic cosmologies. *Physical Review D* **11** , 2072–2075.

Belinskii, V. A., Khalatnikov, I. M. and Lifshitz, E. M. (1974). General solutions of the equations of general relativity near singularities. In *Confrontation of Cosmological Theories with Observational Data*, M. Longair, ed. No. 63 in IAU Symposium, D. Reidel, Dodrecht, 261–275.

Bernard, Claude W. (1974). Feynman rules for gauge theories at finite temperature. *Physical Review D* **9**, 3312–3320.

Birrell, Neil C. and Davies, Paul C. W. (1982). *Quantum Fields in Curved Space*. Cambridge University Press, Cambridge.

Bludman, Sidney A. and Ruderman, Malvin A. (1977). Induced cosmological constant expected above the phase transition restoring the broken symmetry. *Physical Review Letters* **38**, 255–257.

Brout, R., Englert, F. and Gunzig, E. (1980). Spontaneous symmetry breaking and the origin of the universe. In *Gravitation, Quanta, and the Universe*, A. R. Prasanna, Jayant V. Narlikar and C. V. Vishveshwara, eds., 110–118.

— (1979). The causal universe. *General Relativity and Gravitation* **10**, 1–6.

Brout, Robert, Englert, François and Gunzig, Edgard (1978). The creation of the universe as a quantum phenomenon. *Annals of Physics* **115**, 78–106.

Brown, Laurie M. and Cao, Tian Yu (1991). Spontaneous breakdown of symmetry: its rediscovery and integration into quantum field theory. *Historical Studies in the Physical and Biological Sciences* **21**, 211–235.

Cao, Tian Yu (1997). *Conceptual Developments of 20th Century Field Theories*. Cambridge University Press, Cambridge.

Coleman, Sidney (1985). *Aspects of Symmetry*. Cambridge University Press. Selected Erice lectures.

Criss, Thomas, Matzner, Richard, Ryan, Michael and Shepley, Louis (1975). Modern theoretical and observational cosmology. In *General Relativity and Gravitation*, Shaviv and Rosen, eds. 7, New York: John Wiley & Sons, 33–108.

de Sitter, Willem (1931). The expanding universe. *Scientia* **49**, 1–10.

Dicke, Robert and Peebles, P. J. E. (1979). The big bang cosmology–enigmas and nostrums. In *General relativity: an Einstein centenary survey*, S. W. Hawking and W. Israel, eds., Cambridge University Press, Cambridge, 504–517.

Dolan, Louise and Jackiw, Roman (1974). Symmetry behavior at finite temperature. *Physical Review D* **9**, 3320–3341.

Dreitlein, Joseph (1974). Broken symmetry and the cosmological constant. *Physical Review Letters* **20**, 1243–1244.

Earman, John (1995). *Bangs, Crunches, Whimpers, and Shrieks*. Oxford University Press, Oxford.

— (1999). The Penrose-Hawking singularity theorems: History and implications. In *The Expanding Worlds of General Relativity*, Jürgen Renn, Tilman Sauer and Hubert Gönner, eds. No. 7 in Einstein Studies, Birkhäuser Boston, 235–270.

— (2001). Lambda: The constant that refuses to die. *Archive for the History of the Exact Sciences* **55**, 189–220.

Earman, John and Mosterin, Jesus (1999). A critical analysis of inflationary cosmology. *Philosophy of Science* **66**, 1–49.

Eddington, Arthur S. (1933). *The Expanding Universe*. MacMillan, New York.

Einhorn, Martin B. and Sato, Katsuhiko (1981). Monopole production in the very early universe in a first order phase transition. *Nuclear Physics* **B180**, 385–404.

Einhorn, Martin B., Stein, Daniel L. and Toussaint, Doug (1980). Are grand unified theories compatible with standard cosmology? *Physical Review D* **21**, 3295–3298.

Einstein, Albert (1917). Kosmologische betrachtungen zur allgemeinen relativitätstheorie. *Preussische Akademie der Wissenschaften (Berlin). Sitzungs-*

berichte, 142–152. Reprinted in translation in: *Principle of Relativity*, Lorentz et al. eds., Dover, New York, 1923.

Eisenstaedt, Jean (1989). The early interpretation of the Schwarzschild solution. In *Einstein and the History of General Relativity*, John Stachel and Don Howard, eds., Vol. 1 of *Einstein Studies*. Birkhäuser Boston, 213–33.

Ellis, George F. R. and Rothman, Tony (1993). Lost horizons. *American Journal of Physics* **61**, 883–893.

Ellis, John, Gaillard, Mary K. and Nanopoulos, Dimitri V. (1980). The smoothness of the universe. *Physics Letters B* **90**, 253–257.

Englert, François and Brout, Robert (1964). Broken symmetry and the mass of gauge vector mesons. *Physical Review Letters* **13**, 321–23.

Farhi, Edward and Jackiw, Roman, eds. (1982). *Dynamical Gauge Symmetry Breaking: A collection of reprints*. World Scientific, Singapore.

Gibbons, Gary W. and Hawking, Stephen W. (1977). Cosmological event horizons, thermodynamics, and particle creation. *Physical Review D* **15**, 2738–51.

Gliner, Erast B. (1966). Algebraic properties of the energy-momentum tensor and vacuum-like states of matter. (Translated by W. H. Furry.) *Soviet Physics JETP* **22**, 378–382.

— (1970). The vacuum-like state of a medium and Friedman cosmology. *Soviet Physics Doklady* **15**, 559–561.

Gliner, Erast B. and Dymnikova, Irina G. (1975). A nonsigular Friedmann cosmology. *Soviet Astronomy Letters* **1**, 93–94.

Goldstone, Jeffrey (1961). Field theories with 'superconductor' solutions. *Nuovo Cimento* **19**, 154–164.

Goldstone, Jeffrey, Salam, Abdus and Weinberg, Steven (1962). Broken symmetries. *Physical Review* **127**.

Gott, J. Richard (1982). Creation of open universes from De Sitter space. *Nature* **295**, 304–307.

Grib, Andrej A., Mamayev, S. G. and Mostepanenko, V. M. (1984). Self-consistent treatment of vacuum quantum effects in isotropic cosmology. In Markov and West (1984), 197–212, 197–212. Proceedings of the second Seminar on Quantum Gravity; Moscow, October 13–15, 1981.

Guralnik, G. S., Hagen, C. Richard and Kibble, T. W. B. (1964). Global conservation laws and massless particles. *Physical Review Letters* **13**, 585–587.

Guralnik, Gerald S., Hagen, C. Richard and Kibble, Thomas W. B. (1968). Broken symmetries and the Goldstone theorem. In *Advances in Particle Physics*, R. L. Cool and R. E. Marshak, eds., vol. 2. 567–708.

Gurevich, L. E. (1975). On the origin of the metagalaxy. *Astrophysics and Space Science* **38**, 67–78.

Guth, Alan (1981). Inflationary universe: A possible solution for the horizon and flatness problems. *Physical Review D* **23**, 347–56.

— (1997). *The Inflationary Universe*. Addison-Wesley, Reading, MA.

Guth, Alan and Tye, S.-H. Henry (1980). Phase transitions and magnetic monopole production in the very early universe. *Physical Review Letters* **44**, 631–34.

Guth, Alan H. and Weinberg, Erick J. (1983). Could the universe have recovered from a slow first order phase transition? *Nuclear Physics* **B212**, 321.

Hagedorn, Rolf (1970). Thermodynamics of strong interactions at high energy and its consequences for astrophysics. *Astronomy and Astrophysics* **5**, 184–205.

Hawking, Stephen W. (1970). Conservation of matter in general relativity. *Communications in Mathematical Physics* **18**, 301–306.

Hawking, Stephen W. and Ellis, George F. R. (1968). The cosmic black-body radiation and the existence of singularities in our universe. *Astrophysical Journal* **152**, 25–36.

Higgs, Peter W. (1964). Broken symmetries, massless particles, and gauge fields. *Physical Review Letters* **12**, 132–133.

Janssen, Michel (2002). Explanation and evidence: COI stories from Copernicus to Hockney. *Perspectives in Science* **10**, 457–552.

Kazanas, Demosthenes (1980). Dynamics of the universe and spontaneous symmetry breaking. *Astrophysical Journal Letters* **241**, L59–L63.

Kibble, Thomas W. B. (1976). Topology of cosmic domains and strings. *Journal of Physics* **A9**, 1387–97.

Kirzhnits, David A. (1972). Weinberg model in the hot universe. *JETP Letters* **15**, 529–531.

Kirzhnits, David A. and Linde, Andrei (1972). Macroscopic consequences of the Weinberg model. *Physics Letters B* **42**, 471–474.

Kolb, Edward W. and Turner, Michael S. (1990). *The early universe*, vol. 69 of *Frontiers in Physics*. Addison-Wesley, New York.

Kolb, Edward W. and Wolfram, Stephen (1980). Spontaneous symmetry breaking and the expansion rate of the early universe. *Astrophysical Journal* **239**, 428.

Lapchinksy, V. G., Nekrasov, V. I., Rubakov, V. A. and Veryaskin, A. V. (1984). Quantum field theories with spontaneous symmetry breaking in external gravitational fields of cosmological type. In Markov and West (1984), 213–230, 213–230. Proceedings of the second Seminar on Quantum Gravity; Moscow, October 13–15, 1981.

Lemaître, Georges (1934). Evolution of the expanding universe. *Proceedings of the National Academy of Science* **20**, 12–17.

Lightman, Alan and Brawer, Roberta (1990). *Origins: The Lives and Worlds of Modern Cosmologists*. Harvard University Press, Cambridge.

Linde, Andrei (1974). Is the Lee constant a cosmological constant? *Soviet Physics JETP* **19**, 183–184.

— (1979). Phase transitions in gauge theories and cosmology. *Reports on Progress in Physics* **42**, 389–437.

— (1982). A new inflationary universe scenario: a possible solution of the horizon, flatness, homogeneity, isotropy, and primordial monopole problems. *Physics Letters B* **108**, 389–393.

Lindley, David (1985). The inflationary universe: A brief history. Unpublished manuscript.

Markov, M. A. and West, P. C., eds. (1984). Proceedings of the second Seminar on Quantum Gravity; Moscow, October 13–15, 1981. *Quantum Gravity*. Plenum Press, New York.

Misner, Charles W. (1968). The isotropy of the universe. *Astrophysical Journal* **151**, 431–457.

— (1969). Mixmaster universe. *Physical Review Letters* **22**, 1071–1074.

Misner, Charles W., Thorne, Kip and Wheeler, John Archibald (1973). *Gravitation.* W. H. Freeman & Co., New York.

Mukhanov, Viatcheslav F. and Chibisov, G. V. (1981). Quantum fluctuations and a nonsingular universe. *JETP Letters* **33**, 532–535.

Parker, Leonard and Fulling, Stephen A. (1973). Quantized matter fields and the avoidance of singularities in general relativity. *Physical Review D* **7**, 2357–2374.

Peebles, Phillip James Edward (1971). *Physical Cosmology.* Princeton University Press, Princeton.

Penrose, Roger (1979). Singularities and time-asymmetry. In *General Relativity: An Einstein centenary survey*, Stephen Hawking and Werner Israel, eds. Cambridge University Press, Cambridge, 581–638.

— (1989). Difficulties with inflationary cosmology. *Annals of the New York Academy of Sciences* **271**, 249–264.

Pickering, Andrew (1984). *Constructing Quarks: A sociological history of particle physics.* University of Chicago Press, Chicago.

Polyakov, Alexander M. (1974). Particle spectrum in quantum field theory. *JETP Letters* **20**, 194–195.

Preskill, John P. (1979). Cosmological production of superheavy magnetic monopoles. *Physical Review Letters* **43**, 1365–8. Reprinted in Bernstein and Feinberg, pp. 292–298.

Press, William H. (1980). Spontaneous production of the Zel'dovich spectrum of cosmological fluctuations. *Physica Scripta* **21**, 702–702.

Rindler, Wolfgang (1956). Visual horizons in world models. *Monthly Notices of the Royal Astronomical Society* **116**, 662–677.

Rugh, Svend E. and Zinkernagel, Henrik (2002). The quantum vacuum and the cosmological constant problem. *Studies in the History and Philosophy of Modern Physics* **33**, 663–705.

Ryder, Lewis H. (1996). *Quantum field theory.* 2nd ed. Cambridge University Press, Cambridge.

Sakharov, Andrei D. (1966). The initial state of an expanding universe and the appearance of a nonuniform distribution of matter. *Soviet Physics JETP* **22**, 241–249. Reprinted in *Collected Scientific Works.*

— (1970). A multisheet cosmological model. Preprint, Moscow Institute of Applied Mathematics. Translated in *Collected Scientific Works.*

— (1982). *Collected Scientific Works.* Marcel Dekker, New York.

Sato, Katsuhiko (1981). First-order phase transition of a vacuum and the expansion of the universe. *Monthly Notices of the Royal Astronomical Society* **195**, 467–479.

Starobinsky, Alexei (1978). On a nonsingular isotropic cosmological model. *Soviet Astronomy Letters* **4**, 82–84.

— (1979). Spectrum of relict gravitational radiation and the early state of the universe. *JETP Letters* **30**, 682–685.

— (1980). A new type of isotropic cosmological models without singularity. *Physics Letters B* **91**, 99–102.

— (1984). Nonsingular model of the universe with the quantum-gravitational de Sitter stage and its observational consequences. In Markov and West (1984). Proceedings of the second Seminar on Quantum Gravity; Moscow, October 13–15, 1981.

Steigman, Gary, Schramm, David N. and Gunn, James E. (1977). Cosmological limits to the number of massive leptons. *Physics Letters B* **66**, 202–204.

't Hooft, Gerard (1974). Magnetic monopoles in unified gauge theories. *Nuclear Physics* **B79**, 276–284.

Tolman, Richard C. (1934). *Relativity, Thermodynamics, and Cosmology*. Oxford University Press, Oxford. Reprinted by Dover.

Tryon, Edward (1973). Is the universe a vacuum fluctuation? *Nature* **246**, 396–397.

Veltman, Martinus J. G. (1974). Cosmology and the Higgs mechanism. Rockefeller University Preprint.

— (1975). Cosmology and the Higgs mass. *Physical Review Letters* **34**, 777.

— (2000). Nobel lecture: from weak interactions to gravitation. *Reviews of Modern Physics* **72**, 341–349.

Wald, Robert (1984). *General Relativity*. University of Chicago Press, Chicago.

Weinberg, Steven (1967). A model of leptons. *Physical Review Letters* **19**, 1264–66.

— (1972). *Gravitation and Cosmology*. John Wiley & Sons, New York.

— (1974a). Gauge and global symmetries at high temperature. *Physical Review D* **9**, 3357–3378.

— (1974b). Recent progress in gauge theories of the weak, electromagnetic, and strong interactions. *Reviews of Modern Physics* **46**, 255–277.

— (1977). *The First Three Minutes*. Basic Books, Inc., New York.

— (1980). Conceptual foundations of the unified theory of weak and electromagnetic interactions. *Reviews of Modern Physics* **52**, 515–523.

Yoshimura, Motohiko (1978). Unified gauge theories and the baryon number of the universe. *Physical Review Letters* **41**, 281–284.

Zee, Anthony (1979). Broken-symmetric theory of gravity. *Physical Review Letters* **42**, 417–421.

— (1980). Horizon problem and broken-symmetric theory of gravity. *Physical Review Letters* **44**, 703–706.

— (1982). Calculating Newton's gravitational constant in infrared stable Yang–Mills theories. *Physical Review Letters* **48**, 295–298.

Zel'dovich, Yakov B. (1967). Cosmological constant and elementary particles. *JETP Letters* **6**, 316–317.

— (1968). The cosmological constant and the theory of elementary particles. (Translated by J. G. Adashko.) *Soviet Physics Uspekhi* **11**, 381–393.

— (1981). Vacuum theory - A possible solution to the singularity problem of cosmology. *Soviet Physics Uspekhi* **133**, 479–503.

Zel'dovich, Yakov B. and Khlopov, Maxim Yu. (1978). On the concentration of relic magnetic monopoles in the universe. *Physics Letters B* **79**, 239–41.

Zel'dovich, Yakov B., Kobzarev, Igor Yu. and 'Okun, Lev B. (1975). Cosmological consequences of a spontaneous breakdown of a discrete symmetry. *Soviet Physics JETP* **40**, 1–5.

Zel'dovich, Yakov B. and Pitaevsky, Lev P. (1971). On the possibility of the creation of particles by a classical gravitational field. *Communications in Mathematical Physics* **23**, 185–188.

Notes

[1] A singularity cannot be straightforwardly defined as "the points at which some physical quantities diverge," since the metric field itself diverges; given the usual assumption that this field is defined and differentiable everywhere on the space-time manifold, these points are *ex hypothesi* not in space-time. The subtleties involved in giving a precise definition were more important for disentangling horizons and coordinate effects from genuine singularities in the Schwarzschild and de Sitter solutions; to my knowledge there were no published debates about whether there is a genuine initial singularity in the FLRW models. See Eisenstaedt (1989); Earman (1999) for historical discussions of the Schwarzschild singularity and the singularity theorems (respectively), and Wald (1984); Earman (1995) for more recent treatments of the intricate conceptual and mathematical issues involved.

[2] An incomplete geodesic is inextendible in at least one direction, but does not reach all values of its affine parameter; even though it does not have an endpoint it "runs out" within finite affine length. Loosely speaking, one can think of an incomplete geodesic as corresponding to "missing points" in a manifold; unfortunately, this idea can be made precise for a Riemannian metric but not for a pseudo-Riemannian metric like that used in general relativity.

[3] More precisely, this research program aimed to show that the general solution describes a "bounce"—the matter reaches a maximum density, but then expands rather than continuing to collapse—and that the bounce fails to occur only for specialized initial conditions. This program resulted in detailed studies of the evolution of anisotropic, homogeneous vacuum solutions in the neighborhood of the initial singularity (see Belinskii et al. 1974, and references therein).

[4] Cosmological models that reached a finite limiting temperature at early times were explored during this time (see, e.g. Hagedorn 1970), but were never widely accepted.

[5] Lemaître (1934) appears to have been the first to clearly state this idea in print. See Earman (2001) for an account of Λ's checkered history, and Rugh and Zinkernagel (2002) for a detailed discussion of the relation between Λ and vacuum energy density in QFT.

[6] Gliner noted that he is only concerned with local Poincaré invariance, but does not recognize the difficulties in extending Poincaré invariance to general relativity. As

a result, in general the "vacuum" cannot be uniquely specified by requiring that it is a Poincaré invariant state. I thank John Earman for emphasizing this point to me (cf. Earman 2001, 208–209).

[7]The strong energy condition requires that there are not tensions larger than or equal to the (positive) energy density; more formally, for any time-like vector v, $T_{ab}v^a v^b \geq \frac{1}{2}T_a^a$. In particular, for a diagonalized T_{ab} with principal pressures p_i, this condition requires that $\rho + \sum_{i=1}^3 p_i \geq 0$ and $\rho + p_i \geq 0 (i = 1, 2, 3)$, clearly violated by the vacuum state.

[8]Turning this rough claim into a general theorem requires the machinery used by Penrose and Hawking. Gliner refers to Hawking's work in Gliner (1970), but his argument does not take such finer points into account.

[9]This was formulated more clearly as a "cosmic no hair theorem" by Gibbons and Hawking (1977) and in subsequent work. "No hair" alludes to corresponding results in black hole physics, which show that regardless of all the "hairy" complexities of a collapsing star, the end state can be described as simply as a bald head.

[10]Gliner was not alone in this preference; several other papers in the early 1970s discussed violations of the strong energy condition as a way of avoiding the singularity, as we will see in the next section.

[11]Briefly, Sakharov's multi-sheet model is a cyclic model based on Novikov's suggestion that a true singularity could be avoided in gravitational collapse, allowing continuation of the metric through a stage of contraction to re-expansion. I have been unable to find any discussions of the impact of Sakharov's imaginative work in cosmology or its relation to other lines of research he pursued, especially the attempt to derive gravitational theory as an induced effect of quantum fluctuations, but this is surely a topic worthy of further research.

[12]This point is clearly emphasized by Lindley (1985); although it appears plausible that this line of reasoning motivated Gliner and Dymnikova (1975), they introduce the "gradual transition" without explanation or elaboration.

[13]An alert reader may have noticed the tension between this assumption and vacuum dominance mentioned in the last paragraph: the proposed equation of state rather unnaturally guarantees the opposite of vacuum dominance, namely that the *vacuum* is diluted and the density of normal matter and radiation increases in the course of the transition.

[14]Gliner and Dymnikova (1975) derive this equation by solving for the evolution of the scale factor from the transitional phase to the FLRW phase, with matching conditions at the boundary; see Lindley (1985) for a clearer discussion. The constant $0 < \alpha < 1$ fixes the rate at which the initial vacuum energy decays into energy density of normal matter and radiation. H is the (poorly named) Hubble "constant," defined by $H := \frac{1}{a}\frac{da}{dt}$.

[15]Eddington (1933, 37) and de Sitter (1931, 9-10) both argued that a non-zero Λ was needed for a satisfactory explanation of expansion, despite the fact that the FLRW

models with $\Lambda = 0$ describe expanding models; I thank John Earman for bringing these passages to my attention.

[16] Rindler's classic paper introduced and defined various horizons (Rindler 1956); for a recent discussion see Ellis and Rothman (1993). Here I am following the conventional choice to define horizon distance in terms of the time when the signal is received rather than the time of emission (as signalled by the $a(t_0)$ term).

[17] Sakahrov's equation of state is *not* that for a vacuum dominated state, although it is easy to see that the integral diverges for $p = -\rho$ as well.

[18] Hawking's (1970) theorem showed that a vacuum spacetime would remain empty provided that the dominant energy condition holds. The dominant energy condition requires that the energy density is positive and that the pressure is always less than the energy density; formally, for any timelike vector v, $T_{ab}v^a v^b \geq 0$ and $T_{ab}v^a$ is a spacelike vector.

[19] Bekenstein (1975) also discussed the possibility that scalar fields would allow one to avoid the singularity. Starobinsky's (1978) main criticism is that Parker and Fulling dramatically overestimate the probability that their model will reach a "bounce" stage, even granted that the appropriate scalar field exists: they estimate a probability of .5, whereas Starobinsky finds 10^{-43}!

[20] The expression for the trace anomaly was derived before Starobinsky's work; in addition, it was realized that de Sitter space is a solution of the semi-classical EFE incorporating this anomaly (see, e.g. Birrell and Davies 1982). Starobinsky was the first to consider the implications of these results for early universe cosmology.

[21] In the course of this calculation Starobinsky assumed that initially the quantum fields are all in a vacuum state. In addition, the expression for the one-loop correction includes constants determined by the spins of the quantum fields included in $\langle T_{ab} \rangle$, and these constants must satisfy a number of constraints for the solutions to hold. Finally, Starobinsky argued that if the model includes a large number of gravitationally coupled quantum fields, the quantum corrections of the gravitational field itself will be negligible in comparison.

[22] This extended discussion was clearly motivated by Guth's (1981) discussion of the "flatness problem" (which Starobinsky duly cited), but Starobinsky notably did not endorse Guth's emphasis on the methodological importance of the flatness problem.

[23] Misner (1968) advocated an approach to cosmology that focused on "predicting" various features of the observed universe, in the sense of finding features insensitive to the choice of initial conditions.

[24] Zel'dovich's review does not include any references. He had already discussed the horizon problem in a different context (Zel'dovich et al. 1975), see section 2.3 below.

[25] However, this is more a triumph of approach than actual implementation; a decade after this assessment an account of baryogenesis consistent with all the constraints has yet to be developed.

[26]*Very* roughly, in a renormalizable theory such as QED divergent quantities can be "absorbed" by rescaling a finite number of parameters occuring in the Lagrangian (such as particle masses and coupling constants); these techniques did not carry over to massive Yang–Mills theories (see, e.g., §10.3 of Cao 1997 for an overview).

[27]The three "proofs" of Goldstone's theorem given in Goldstone et al. (1962) hold rigorously for classical but not quantum fields; see, e.g., Guralnik et al. (1968) for a detailed discussion of the subtleties involved.

[28]Quantizing the electromagnetic field in Lortenz gauge leads to photons with four different polarization states: two transverse, one longitudinal, and one "time-like" (or "scalar"). In the Gupta–Bleuler formalism, the contributions of the longitudinal and time-like polarizations states cancel as a result of the Lorentz condition $\partial_\mu A^\mu = 0$, leaving only the two transverse states as true "physical" states. See, e.g., Ryder (1996), section 4.4 for a brief description of the Gupta-Bleuler formalism.

[29]Goldstone's theorem held for Lagrangians invariant under the action of a continuous, "global" gauge transformation of the fields, but not for "local" symmetries or discrete symmetries (such as parity). As Chris Martin has pointed out to me, the terms "local" and "global" suggest a misleading connection with space-time: global gauge groups are finite dimensional Lie groups (such that a specific element of the group can be specified by a finite number of *parameters*), whereas local gauge groups are infinite dimensional Lie groups whose elements are specified via a finite number of *functions*.

[30]This discussion of the Higgs mechanism is by necessity brief; for a clear textbook treatment see, for example, Aitchison (1982).

[31]See, e.g., Coleman (1985, chapter 5) for a concise introduction to the effective potential and arguments that it represents the expectation value of the energy density for a given state.

[32]Englert and Brout (1964) explicitly mentioned the possibility: "The importance of this problem [whether gauge mesons can acquire mass] resides in the possibility that strong-interaction physics originates from massive gauge fields related to a system of conserved currents." The other papers introducing the Higgs mechanism are more directly concerned with exploiting the loophole in Goldstone's theorem.

[33]The number of citations of Weinberg (1967) jumped from 1 in 1970 to 64 in 1972, following 't Hooft and Veltman's proof of renormalizability (Pickering 1984, 172).

[34]In dynamical symmetry breaking, bound states of fermionic fields play the role of Higgs field; see the various papers collected in Farhi and Jackiw (1982) for an overview of this research, which was pursued actively throughout the 1970s and early 1980s.

[35]The Cambridge theorists, including Claude Bernard, Sidney Coleman, Barry Harrington, and Steven Weinberg at Harvard, and Louise Dolan and Roman Jackiw at MIT, seem to have worked fairly closely on this research, based on the acknowledgements and references to personal communication in their papers (Weinberg 1974b;

Dolan and Jackiw 1974; Bernard 1974). See Linde (1979) for a review of this literature.

[36]Conventional QFT treats interactions between fields in otherwise empty space, neglecting possible effects of interactions with a background heat bath. Finite temperature field theory was developed in the 1950s in the study of many-body systems in condensed matter physics.

[37]Weinberg (1974a) gives examples of models with no symmetry restoration and even low-temperature symmetry restoration; symmetry restoration can also be induced by large external fields or high current densities. See Linde (1979) for further discussion and references.

[38]In general relativity a conformal transformation is a map: $g_{ab} \rightarrow \Omega^2 g_{ab}$ where Ω is a smooth, non-zero real function. A field theory is conformally invariant if $\phi' = \Omega^s \phi$ is a solution to the field equations with the metric $\Omega^2 g_{ab}$ whenever ϕ is a solution with the original metric, for a given number s (called the conformal weight) (see, e.g., Wald 1984, Appendix D). A field theory is said to be "conformally coupled" if additional terms are introduced to insure conformal invariance; the conformally coupled Klein–Gordon equation, for example, includes a term, $\frac{1}{6} R$, absent from the "minimally coupled" equation obtained by replacing normal derivatives with covariant derivatives.

[39]I call this an assumption since I cannot understand the argument in favor of it, which invokes Birkhoff's theorem along with the conformal flatness of the FLRW models (see Brout et al. 1978, 78–79).

[40]Zee (1982) described the rationale for this approach in greater detail. The program (partially based on Sakharov's conception of "induced gravity") aimed to formulate a renormalizable, conformally invariant theory in which the gravitational constant is fixed by vacuum fluctuations of the quantum fields.

[41]Kirzhnits and Linde defer the detailed argument for this conclusion to a later paper, which apparently did not appear; in any case it is not clear to me that long range repulsive forces are necessarily incompatible with either a closed or uniform model.

[42]"C" denotes charge conjugation, a transformation implemented by replacing field operators for a given particle with those for its anti-particle; "P" stands for the parity transformation, which (roughly speaking) maps fields into their mirror image.

[43]They comment that "Owing to the peculiar expansion law during the initial (domain) stage it is quite possible that $X_c >> X_p$ [X_c is the causal horizon, X_p is the particle horizon]." The averaged equation of state for the domain stage is $p = -\frac{2}{3}\rho$, leading to $a(t) \propto t^2$ during the "cellular medium"-dominated stage of evolution.

[44]Zel'dovich and Khlopov (1978) calculated the abundance of the lighter monopoles produced in electroweak symmetry breaking, with mass on the order of $10^4 \ GeV$, whereas Preskill (1979) calculated the abundance of monopoles (with mass on the order of $10^{16} \ GeV$) produced during GUT-scale symmetry breaking.

[45]The stress energy tensor for a scalar field is given by $T_{ab} = \nabla_a \phi \nabla_b \phi - \frac{1}{2} g_{ab} g^{cd} \nabla_c \nabla_d \phi - g_{ab} V(\phi)$; if the derivative terms are negligible, $T_{ab} \approx -V(\phi) g_{ab}$.

[46]Linde estimated that before SSB the vacuum energy density should be $10^{21} g/cm^3$, compared to a cosmological upper bound on the total mass density of $10^{-28} g/cm^3$. In an interview with the author, Linde noted that the title of this paper was mistranslated in the English edition (see the bibliography); the correct title is "Is the Cosmological Constant a Constant?"

[47]The radiation density $\rho_{rad} \propto T^4$, which dominates over the vacuum energy density for $T > T_c$; Bludman and Ruderman (1977); Kolb and Wolfram (1980) bolstered Linde's conclusion with more detailed arguments.

[48]This assumption was not unwarranted: Weinberg (1974a) concluded that the electroweak phase transition appeared to be second order since the free energy and other thermodynamic variables were continuous (a defining characteristic of a second-order transition).

[49]Veltman described the idea of "cancellation" of a large vacuum energy density as follows: "If we assume that, before symmetry breaking, space-time is approximately euclidean, then after symmetry breaking ... a curvature of finite but outrageous proportions result [sic]. The reason that no logical difficulty arises is that one can assume that space-time was outrageously "counter curved" before symmetry breaking occurred. And by accident both effects compensate so precisely as to give the very euclidean universe as observed in nature."

[50]In a 1987 interview he commented that "we understood that the universe could exponentially expand, and bubbles would collide, and we saw that it would lead to great inhomogeneities in the universe. As a result, we thought these ideas were bad so there was no reason to publish such garbage" (Lightman and Brawer 1990, 485–86).

[51]Sato apparently hoped that an early phase transition would effectively separate regions of matter and anti-matter, so that observations establishing baryon asymmetry could be reconciled with a baryon-symmetric initial state; he also mentions the possibility that small inhomogeneities could seed galaxy formation.

[52]The original draft of this paper was completed in July 1980, revised in November of 1980 partially in response to comments from Guth and his collaborator, Erick Weinberg. Einhorn and Guth met and discussed phase transitions in November of 1979, but judging from Guth's comments in Guth (1997, 180), Einhorn and Sato hit upon the idea of false-vacuum driven exponential expansion independently.

[53]Einhorn and Sato were not alone in making this suggestion; a year earlier, the Harvard astrophysicist Bill Press had proposed an account of structure formation in which vacuum energy does not couple to gravity. In Press's (1980) scenario, inhomogeneities in the vacuum are converted into fluctuations in the energy density of matter and radiation. This "conversion" only works if the vacuum does not itself gravitate; Press noted the speculative nature of this suggestion, but argued that the other possibility — an incredibly precise cancellation of vacuum energy density — is equally unappealing.

[54]Rees attended talks about the early universe by both Starobinsky and Englert before 1981, but by his own account he did not see the appeal of these ideas until he had read Guth's paper (Lightman and Brawer 1990, 161).

[55]See Guth (1997), chapter 10 for a detailed account (quotation on 179). Guth attended a lecture by Princeton's Bob Dicke, in which he mentioned the flatness problem, on Nov. 13, 1978.

[56]The density parameter is defined as the ratio of the observed density to the critical density, namely the value such that $k = 0$ in the Friedmann equation. The Friedmann equation is given by: $H^2 = \frac{8\pi G}{3}\rho - \frac{k}{a^2(t)}$, where $k = 0$ for a flat model, $k > 0$ for a closed model, and $k < 0$ for an open model.

[57]Guth and Weinberg (1983) later showed that for a wide range of parameters the bubbles do not percolate, and they also do not collide quickly enough to thermalize.

[58]Michel Janssen has recently argued that "common origin inferences" (COIs) play a central role in scientific methodology (Janssen 2002). These inferences license a preference for a theory that traces several apparent coincidences to a common origin. Guth's case for inflation is a particularly clear example of this style of reasoning. I have benefitted from extensive discussions with Janssen regarding whether the case for inflation should be treated as another "COI" story, but I do not have space to explore the issue further here.

[59]For a detailed discussion of the demand for explanatory adequacy see Earman (1995), and for a critical overview of inflationary cosmology see Earman and Mosterin (1999).

14

Hilbert's "World Equations" and His Vision of a Unified Science

U. Majer[1] and T. Sauer[2]

[1] Hilbert-Edition, Universität Göttingen, Papendiek 16, D-37073 Göttingen, Germany; tilman@einstein.caltech.edu
[2] Einstein Papers Project, California Institute of Technology 20-7, 1200 E. California Blvd., Pasadena, CA 91125, U.S.A.; umajer@gwdg.de

14.1 Introduction

In the history of unified field theory, many contributors may be identified, Goldstein and Ritter (2003), Goenner (2004) among them certainly, and perhaps foremost, Einstein. Hilbert's place in the history of the unified field theory program is also well recognized (see, e.g., the discussion of his work in Vizgin's study, Vizgin (1994)). But we tend to view the history of physics in which Einstein was involved through that scholarship which has focussed exclusively, or at least predominantly, on Einstein's work as such. For the case of Einstein's "later journey," we believe that many physicists as well as historians would subscribe to Pais's verdict that "his work on unification was probably all in vain" (Pais 1982, p. 329). The dismissal of Einstein's efforts over three decades is to some extent supported by Einstein's own self-image, in his later years, as the "lonesome outsider" working without real appreciation in his golden Princeton cage. Einstein was an original thinker and an influential voice in the debate, and for this reason understanding Einstein's obsession with the problem of a unified field theory over the last thirty years of his life presents as much of a challenge to the historian as understanding the achievements of his early work.

To this purpose, it helps to free one's mind from preconceptions. We then find Hilbert's insights of great advantage since he was both knowledgeable and had a well-founded and original perspective of his own.

Let us make a distinction right at the beginning in order to disentangle different scientific approaches. The problem of a unified field theory, as of the 1920s, can be seen in a more specific sense as the problem of finding a consistent and satisfactory mathematical unification of the gravitational and electromagnetic fields, be it by modified field equations, by a modification of the space-time geometry, or by increasing the number of space-time dimensions. But there is another aspect to the problem that is, we believe, of both historical and philosophical interest. This aspect concerns the way in which contemporary scientists perceived the technical problem of unification in the wider context of a unified corpus of human knowledge and understanding. In this respect, Hilbert's perspective on the mathematical sciences as an integrated whole can contribute to our modern attempts to come to grips with the philosophical implications of an ever increasing specialization in the natural sciences. Hilbert certainly was not the only one who envisaged a unified science at the time. Many contemporary mathematicians shared this concern. Felix Klein's *History of the Development of Mathematics in the 19th century*, Klein (1979), can also be seen as a most interesting attempt to understand the inner organic unity of the corpus of mathematical knowledge. Other names that come to mind immediately are those of Kaluza and Weyl, but the list certainly does not end here.

Einstein, too, shared this interest in understanding the inner unity of our knowledge of nature, and for him, too, the problem of finding a mathematical representation that would provide a unification of the gravitational and electromagnetic fields was more than just a technical problem. This aspect of his work is expressed most convincingly in Einstein's own account of his lifelong research concerns as given in his 1949 *Autobiographical Notes*, Einstein (1949). Einstein, as we will argue, followed in his later work a path that is not at all very different from Hilbert's. Hilbert himself perceived Einstein as sharing his concern. Of course, there are differences, which we do not deny. But from a broader perspective, both Einstein and Hilbert – and others, one may add – belong to a tradition which attempts to integrate our human knowledge and to perceive an inner unity in science. For today's philosophers, this tradition seems to belong to the 18th and 19th century rather than to the 20th, or to the 21st, for that matter. In this respect, Einstein and Hilbert are akin more to the encyclopedists and enlightenment natural philosophers than to modern puzzle solvers.

14.2 Hilbert's Lectures on Fundamental Questions of Modern Physics of 1923

The document to which we would like to draw attention in this paper is a manuscript extant in the Hilbert archives in Göttingen. It will be published in one of the physics volumes of the Hilbert Edition under the title "Fundamental Questions of Modern Physics." It is a batch of roughly 100 manuscript pages with notes for a trilogy of lectures that Hilbert delivered at the end of the summer semester of 1923 in Hamburg.[1] The three lectures focus on three different topics: the first deals with what Hilbert called the "World Equations," where these equations are introduced; the second part discusses applications and consequences of those equations; and the third lecture contains a discussion of the old problem of theory and experience.

To Hilbert at that time, the epistemological and philosophical implications of recent developments in physics were of central concern. He himself had contributed substantially to modern mathematical physics in the preceding years, most notably through his two Communications to the Göttingen Academy Proceedings on the Foundations of Physics of November 1915 and December 1916, respectively, Hilbert (1915, 1917). By the summer of 1917 at the latest, however, another problem was increasingly occupying Hilbert's mind, namely the problem of an absolute consistency proof of arithmetic that would provide a sound logical foundation for the whole body of mathematics. Just as in Hilbert's work in physics, the roots of this preoccupation date back to his very early work, at least to his "Mathematical Problems" of 1900 Hilbert (1900). This interest resurfaced with a lecture on set theory held in the summer term of 1917.

As a matter of fact, Hilbert's renewed attention to the foundations of mathematics in general, and to a theory of proof in particular, contributed to his taking a broader perspective on the contemporary debates in General Relativity and Field Theory. He had kept an active interest in the development of General Relativity after 1915 but was increasingly concerned with the philosophical implications of the new theories rather than with contributing solutions of some of its outstanding technical problems.[2] He also began to spend a great deal of energy in popularizing these new developments and in acquainting a larger audience with the results of modern physics. It is therefore no accident that when Hilbert spoke on the same topic a few weeks later in Zurich, but in a single lecture, he chose to center on the third of his Hamburg lectures.[3] He used the same manuscript notes for the Zurich lecture, but since he had to cut down the material, he summarized the main points of the first two lectures. This editing of his own manuscript makes it difficult to exactly associate specific phrases with either the Hamburg or Zurich lectures.

14.3 Hilbert's "World Equations" of summer and fall of 1923

Hilbert starts his first lecture by introducing what he calls the "World Equations" or the "World Laws" ("Weltgleichungen" or "Weltgesetze"). The way Hilbert introduces these equations is interesting in itself but for the sake of brevity, we shall only say that these equations basically are the same ones that he had proposed in his First Communication on the Foundations of Physics, Hilbert (1915), considering the fact that Hilbert had, originally, not completely specified the Lagrangian I of the variational integral

$$\int I\sqrt{-g}d\tau, \tag{14.1}$$

where $g = \det(g_{\mu\nu})$ and the integral is over (a domain of) four-dimensional space-time. But both in 1915 and now again in 1923 he pointed out that the fundamental dynamical variables are the ten components $g_{\mu\nu}$ of the metric tensor and the four components φ_l of the electromagnetic potential.[4]

In his Hamburg and Zurich lectures, he takes the Lagrangian to be the sum of a gravitational part K and a matter part L,

$$I = K + L. \tag{14.2}$$

The gravitational part K is understood to be the Riemann curvature scalar. The matter part L is taken to be a sum of a term proportional to the square of the fields, and another term proportional to the square of the potential,

$$L = \alpha\Phi + \beta\varphi, \tag{14.3}$$

where $\Phi \equiv \sum \Phi_{mn} \Phi^{mn}$ with $\Phi_{mn} \equiv \varphi_{[m;n]}$[5] denoting the electromagnetic field, and $\varphi \equiv \varphi_k \varphi^k$.[6] As usual, variation with respect to the components of the metric tensor produces the gravitational field equations,

$$K_{\mu\nu} = -\frac{\partial \sqrt{-g}\, L}{\partial g^{\mu\nu}}, \tag{14.4}$$

and variation with respect to the components of the electromagnetic four-potential produces generalized Maxwell equations of the form

$$\mathcal{D}iv\Phi^{mn} = \frac{\beta}{\alpha}\varphi^m, \tag{14.5}$$

where $\mathcal{D}iv\phi^{mn} \equiv \phi^{mn}_{;u}$. The latter equations are determined by the matter term alone. More specifically, the first term in (14.3) produces the left-hand side of the inhomogeneous Maxwell equations, $\alpha \mathcal{D}iv\Phi^{mn}$, while the second term in (14.3) produces a term proportional to the electromagnetic vector potential, $\propto \varphi^k$, the latter acting as the source of the inhomogeneous Maxwell equations. Following Mie's approach, external currents and charges are not part of the theory. The homogeneous field equations,

$$\Phi_{(mn;k)} = 0, \tag{14.6}$$

follow, in the usual way, from the definition of the field and the fact that the connection was assumed to be the symmetric Levi-Civita connection.

14.4 Hilbert's Comments on Einstein's Recent Work on Affine Field Theory

At this point, Hilbert introduces a remark which at first sight may seem preposterous, or, if you wish, arrogant and self-serving. He claims that Einstein, in his most recent publications, would have arrived, after "a colossal detour," ("kolossaler Umweg") at the very same results and equations that Hilbert had put forward in his first note on the Foundation of Physics of November 1915. But before dismissing this claim as a stubborn and senile insistence of a mathematician who "has left reality behind," let us examine his claim more closely and see whether it is conducive to a more nuanced historical interpretation.

The starting point is Hilbert's claim that the invariance of the action integral allows one to interpret the electromagnetic field equations as implicit in the gravitational field equations. Hilbert here reiterates the claim of his first note that this fact would provide the solution to a problem that he traces back to Riemann, namely the problem of the connection between gravitation and light. He goes on to observe that since then many investigators had tried to arrive at a deeper understanding of this connection by merging the gravitational and electromagnetic potentials into a unity. The one example Hilbert mentions explicitly is Weyl's unification of the two fields in a "unified world metric," as he calls it, by means of Weyl's notion of gauge invariance. Another approach would be Eddington's who proceeded by selecting "certain invariant combinations" as fundamental potentials of the quantities determining the fields. Schouten then had investigated the manifold of possibilities of such combinations and realized that there would be a rich variety of them. At this point, Hilbert inserts his comment on Einstein's recent work. He says explicitly:

> Einstein finally ties up to Eddington in his most recent publications and, just as Weyl did, arrives at a system of very coherent mathematical construction.

But, Hilbert goes on,

> However, the final result of Einstein's latest work amounts to a Hamiltonian principle that is similar to the one that I had originally proposed. Indeed, it might be the case that the content of this latest Einsteinian theory is *completely equivalent* to the theory originally advanced by myself.[7]

It is important to note that Hilbert makes his claim somewhat more specific than that. Looking at the variational principle which he explicitly writes down in the form

$$\delta \int \int \int \int \left\{ K + \alpha \Phi + \beta \varphi \right\} \sqrt{-g} dx_1 dx_2 dx_3 dx_4 = 0, \tag{14.7}$$

he observes that Einstein in his latest note had arrived at the very same Hamiltonian principle

> where φ is defined through $\varphi^m = \text{Div } \Phi^{mn}$ and variation with respect to $g_{\mu\nu}$ and Φ^{mn} produces the eqs. $\Phi_{mn} = \text{Rot } \varphi_m$ instead of my eq. [(14.5)].

Hilbert concludes:

> Hence, nothing else than *an exchange of the two series* [of] Max[well] eq[uations].[8]

The emphasis in the last quote is Hilbert's. He was not only pointing at a vague similarity between his own work and Einstein's. Rather he had identified the differences in their work as being of a purely nominal nature.

14.5 Einstein's "colossal detour"

In view of this remark, let us briefly examine Einstein's post-1915 work in General Relativity, in particular with regard to the problem of unifying gravitation and electromagnetism, see also Vizgin (1994), Goenner (2004).

Until 1923, it is perhaps not too unjust to say that Einstein basically had been reacting to the work of others. He had submitted Kaluza's theory of a five-dimensional metric for publication in the Prussian Academy Proceedings, Kaluza (1921), and had himself done calculations along this approach, partly in collaboration with Jakob Grommer, Einstein and Grommer (1923). He had also published a couple of notes that further elaborated on Weyl's ideas, Einstein (1921), notwithstanding his critical evaluation of its physical viability. Thirdly, he had lately picked up on Eddington's approach of basing the theory on the affine connection rather than on the metric Einstein (1923a,b,c).

In order to evaluate Hilbert's claim, let us take a closer look at Einstein's work along Eddington's approach, as he had published it in those papers of 1923 to which Hilbert refers. Following Eddington,[9] Einstein had taken the components of a real, symmetric affine connection $\Gamma^{\kappa}_{\lambda\mu}$ as the basic quantities of the theory instead of the metric tensor field $g_{\mu\nu}$ which provided the dynamical variables in the original theory. From the symmetric connection he had constructed an asymmetric contracted curvature tensor,

$$R_{kl} = -\Gamma^{\alpha}_{kl,\alpha} + \Gamma^{\alpha}_{k\beta}\Gamma^{\beta}_{l\alpha} + \Gamma^{\alpha}_{k\alpha,l} - \Gamma^{\alpha}_{kl}\Gamma^{\beta}_{\alpha\beta}. \tag{14.8}$$

Since $R_{kl}dx^k dx^l$ is an invariance of the line element, it was tempting to split the Ricci tensor into a symmetric part g_{kl}, to be interpreted as a metric tensor associated with the gravitational field, and an antisymmetric part φ_{kl}, to be associated with the electromagnetic field tensor.

In a first note presented to the Berlin Academy on 15 February 1923, Einstein observed that Eddington had not yet solved the problem of finding the necessary equations that would determine the 40 components of the connection. He therefore set out to provide just such equations. He postulated a Hamiltonian principle,

$$\delta\left\{\int \mathcal{H}d\tau\right\} = 0, \tag{14.9}$$

with a Lagrangian that would depend only on the contracted curvature tensor, $\mathcal{H} = \mathcal{H}(R_{kl})$.[10] More specifically, he proposed a tentative set of field equations for the affine connection based on a Lagrangian proportional to the square root of the determinant of the contracted curvature tensor

$$\mathcal{H} = 2\sqrt{-|R_{kl}|}. \tag{14.10}$$

In his first note, Einstein does not proceed to derive the field equations explicitly from that Lagrangian. Instead, he does the variation for a general Lagrangian $\mathcal{H}$ which gives him

$$\mathfrak{s}^{kl}_{;\alpha} - \frac{1}{2}\delta^k_\alpha \mathfrak{s}^{l\sigma}_{;\sigma} - \frac{1}{2}\delta^l_\alpha \mathfrak{s}^{k\sigma}_{;\sigma} - \frac{1}{2}\delta^k_\alpha \mathfrak{f}^{l\sigma}_{,\sigma} - \frac{1}{2}\delta^l_\alpha \mathfrak{f}^{k\sigma}_{,\sigma} = 0, \tag{14.11}$$

where $\mathfrak{s}^{kl}$ and $\mathfrak{f}^{kl}$ are defined as variations of $\mathcal{H}$ with respect to g_{kl} and ϕ_{kl}, respectively, i.e.,

$$\delta\mathcal{H} = \mathfrak{s}^{kl}\delta g_{kl} + \mathfrak{f}^{kl}\delta\phi_{kl}. \tag{14.12}$$

Solving with respect to $\Gamma^{\lambda}_{\mu\nu}$, he obtains

$$\Gamma^{\alpha}_{kl} = \frac{1}{2}s^{\alpha\beta}\left(s_{k\beta,l} + s_{l,\beta,k} - s_{kl,\beta}\right) - \frac{1}{2}s_{kl}i^{\alpha} + \frac{1}{6}\delta^{\alpha}_k i^l + \frac{1}{6}\delta^{\alpha}_l i^k, \tag{14.13}$$

where $i^l = \sqrt{-|s_{kl}|} i^l = f^{l\sigma}_{,\sigma}$ and indices are raised and lowered by means of s_{kl} and s^{kl} respectively, a fundamental tensor which in turn is defined via $\mathfrak{s}^{kl} = s^{kl} \sqrt{-|s_{kl}|}$ and $s_{\alpha i} s^{\beta i} = \delta^\beta_\alpha$.

Explicit field equations were given by Einstein in a short follow up note to his paper, Einstein (1923b), published on May 15, 1923. In it he briefly recapitulates the basic equations of his previous note, implicitly introducing a change of notation by denoting the Ricci tensor as r_{kl}, and denoting the Ricci tensor formed from the fundamental tensor s^{kl} only as R_{kl}. The field equations were now given as the symmetric and antisymmetric parts of

$$r_{kl} = R_{kl} + \frac{1}{6}\left[\left(i_{k,l} - i_{l,k}\right) + i_k i_l\right]. \tag{14.14}$$

These field equations would not hold up for long. Already two weeks after the publication of the second note, Einstein presented a third note to the Prussian Academy dealing with the affine theory Einstein (1923c), published in the Academy's Proceedings on 28 June. While Einstein in the introductory paragraph of that paper announced that "further considerations" ("Weiteres Nachdenken") had led him to a "perfection" ("Vervollkommnung") of the theory laid out in the previous two notes, he was, in fact, going to present some major revisions, including a new set of field equations.

One change in his understanding is reflected in an implicit overall change of notation. While he had previously regarded the symmetric and antisymmetric parts g_{kl} and ϕ_{kl} of the Ricci tensor $R_{kl} = R_{kl}(\Gamma^\lambda_{\mu\nu})$ as the "metric and electromagnetic field tensors," he now attaches this physical meaning to different quantities. Hence he now denotes the symmetric part of $R_{\mu\nu}$ as $\gamma_{\mu\nu}$ and uses the letter g, resp. $\mathfrak{g}$, to denote the quantities that he had previously denoted by s, resp. $\mathfrak{s}$,

$$\delta\mathcal{H} = \mathfrak{g}^{kl}\delta\gamma_{kl} + \mathfrak{f}^{kl}\delta\phi_{kl}. \tag{14.15}$$

It is the quantities $\mathfrak{g}^{kl}$ and $\mathfrak{f}^{kl}$ that were now "regarded as tensor densities of the metric and electric field." Einstein also pointed out that he no longer would assume the Lagrangian $\mathcal{H}$ to depend on $R_{\mu\nu}$, i.e., only on the sum of $\gamma_{\mu\nu} + \phi_{\mu\nu}$, but would now allow for the possibility that it depend on $\gamma_{\mu\nu}$ and $\phi_{\mu\nu}$ independently.

Thirdly, Einstein does not simply proceed to discuss restrictive conditions or other motivations for a definite choice of $\mathcal{H}$ in order to fix the field equations. Instead, he argues that since, by assumption, (14.15) is a complete differential,

$$\gamma_{\mu\nu}d\mathfrak{g}^{\mu\nu} + \phi_{\mu\nu}d\mathfrak{f}^{\mu\nu} \tag{14.16}$$

is a complete differential of another scalar density $\mathcal{H}^*$ where $\mathcal{H}^*$ is a function of the tensor densities of the metric and electric fields, $\mathcal{H}^* = \mathcal{H}^*(\mathfrak{g}^{\mu\nu}, \mathfrak{f}^{\mu\nu})$. For the choice of a definite $\mathcal{H}^*$ Einstein then gives some arguments. It should be a function of the two invariants of the electromagnetic fields, and specifically, he argues that, "according to our present knowledge, the most natural ansatz" would be[11]

$$\mathcal{H}^* = 2\alpha\sqrt{-g} - \frac{\beta}{2}f_{\mu\nu}\mathfrak{f}^{\mu\nu}. \tag{14.17}$$

The resulting field equations, after a rescaling of the electromagnetic field, read

$$R_{\mu\nu} - \alpha g_{\mu\nu} = -\left[\left(-f_{\mu\sigma}f_{\nu}^{\sigma} + \frac{1}{4}g_{\mu\nu}f_{\sigma\tau}f^{\sigma\tau}\right) + \frac{1}{\beta}i_{\mu}i_{\nu}\right] \tag{14.18}$$

$$-f_{\mu\nu} = \frac{1}{\beta}i_{[\mu,\nu]}. \tag{14.19}$$

For us, the last half-page of his note, following immediately after equations (14.18), (14.19) is most interesting. Einstein observed that the field equations derived along the lines sketched above may also be derived, in fact quite easily, from a different Hamiltonian principle. He conceived of $\mathcal{H}$ as a function of $\mathfrak{g}^{\mu\nu}$ and $\mathfrak{f}^{\mu\nu}$ which Einstein, as was mentioned, in this third note took to be the tensor densities of the metric and electromagnetic fields, $\mathcal{H} = \mathcal{H}(\mathfrak{g}^{\mu\nu}, \mathfrak{f}^{\mu\nu})$. The Lagrangian whose variation with respect to $\mathfrak{g}^{\mu\nu}$ and $\mathfrak{f}^{\mu\nu}$ would produce the field equations (14.18), (14.19) directly then reads

$$\mathcal{H} = \sqrt{-g}\left[R - 2\alpha + \kappa\left(\frac{1}{2}f_{\sigma\tau}f^{\sigma\tau} - \frac{1}{\beta}i_{\sigma}i^{\sigma}\right)\right]. \tag{14.20}$$

Here R denotes the Riemannian curvature scalar formed from the metric tensor $g_{\mu\nu}$. Notwithstanding the cosmological constant term -2α, the Lagrangian already looks familiar. But we need one more little step. In the penultimate paragraph of his paper, Einstein suggests that for a physical interpretation it would be most useful to introduce the "electromagnetic potential"

$$-f_{\mu} = \frac{1}{\beta}i_{\mu}, \tag{14.21}$$

a step that would eventually turn the field equations into those that were identical — up to the sign of the constant β — to field equations proposed by Weyl.

Let us now pause and look at Einstein's result through Hilbert's eyes. If we substitute the electromagnetic potential (14.21) for i_{μ}, we get the variational principle in the form

$$\delta\mathcal{H} = \delta\int\left\{R - 2\alpha + \kappa\left(\frac{1}{2}f_{\sigma\tau}f^{\sigma\tau} - \beta f_{\sigma}f^{\sigma}\right)\right\}\sqrt{-g}d\tau = 0. \tag{14.22}$$

Comparing this variational principle with the variational integral (14.7) given by Hilbert in his lectures, we see that Hilbert's interpretation actually does capture Einstein's result of his third note on the affine theory, provided we make the following identifications. Hilbert's K would be Einstein's $R - 2\alpha$, i.e., Hilbert ignored the cosmological term. However, such a term would fit easily into Hilbert's original scheme. We would also identify Hilbert's $\alpha\Phi$ with Einstein's $\frac{\kappa}{2}f_{\sigma\tau}f^{\sigma\tau}$. Finally, we would identify Hilbert's $\beta\varphi$ with Einstein's $\kappa\beta f_{\sigma}f^{\sigma}$.

One technical difference remains. Hilbert is doing the variation with respect to the electromagnetic potential φ_{μ} whereas Einstein is doing the variation with respect to the electromagnetic tensor density $\mathfrak{f}^{\mu\nu}$. In Hilbert's theory, the electromagnetic field was *defined* as $\Phi_{mn} \equiv \varphi_{[m;n]}$ and the variation *produced* the generalized Maxwell equations (14.5). In Einstein's theory, the variation is done with respect to the field $\mathfrak{f}^{\mu\nu}$. The variation of the term $\beta f_\sigma f^\sigma$ makes use of the *definition* $f^\mu = -\frac{1}{\beta} f^{\mu\nu}_{;\nu}$ and *produces* the relation $f_{\mu\nu} = f_{[\mu;\nu]}$ as an electromagnetic field equation. Taking into account that for symmetric connections the homogeneous Maxwell equations (14.6) follow from the fields being given as the rotation of a vector, we can now see the point of Hilbert's remark.

Regardless of how Einstein had derived his field equations in the first place, he himself had cast them into a form that was technically equivalent to Hilbert's initial framework of 1915. The resulting equations were essentially equivalent to Hilbert's with the only difference that what appeared as a definition and a field equation in one framework turned out to be the resulting field equations and the defining relation in the other. In Hilbert's words, the difference amounted to an "interchange of the two series of Maxwell equations." To be sure, the identification involves some amount of interpretation but essentially we can see why Hilbert rejoiced:

> And if on a colossal detour via Levi-Civita, Weyl, Schouten, Eddington, Einstein returns to this result, then this certainly provides a beautiful confirmation.[12]

It also becomes conceivable that Hilbert's reprint of his 1915 and 1917 notes on the *Grundlagen der Physik* in 1924 as a single paper in the *Mathematische Annalen* was not motivated by his desire to revise his original theory (as has been argued in Renn and Stachel (1999)). His lectures of 1923 in Hamburg and Zurich rather suggest that the true motivation for Hilbert becomes visible on the background of his perception of Einstein's latest work on the affine theory. He saw Einstein's work as a confirmation of his original approach. Hence, there is no reason to assume that Hilbert did not believe what he wrote about his original 1915 theory in the new introduction to the 1924 reprint:

> I firmly believe that the theory which I develop here contains a core that will remain and that it creates a framework that leaves enough room for the future construction of physics along the field theoretic ideal of unity.[13]

14.6 Accessorial Laws of Nature?

As we have seen, Hilbert meant what he said, even though he was deliberately formulating his claim as a hypothesis. Having established that his "world equations" are confirmed, if only by his own perception of a convergence of related research efforts, Hilbert in his second lecture became somewhat more speculative. Of central importance for the argument of his second lecture is the notion of "accessorial laws."[14] While Hilbert does use the term "accessorial" in a contemporary lecture course,[15] we are not aware of any other usage of the term, neither in Hilbert's own Oeuvre nor in any of his contemporaries' writings. Our guess is that Hilbert created a neologism based on the Latin "accedere" — in its meaning "to add."[16] What notion then does Hilbert want to capture by the term "accessorial"? He says:

Anything that needs to be added to the world equations in order to understand the events ("Geschehnisse") of inanimate nature, I will briefly call "accessorial."[17]

An obvious candidate for something "accessorial" with respect to the "world equations" immediately comes to mind. These equations being differential equations, require for the explanation of "events" certainly the determination of initial or boundary conditions. Indeed, Hilbert concedes that initial or boundary conditions are necessary in order to allow for definite solutions of the "world equation,"[18] but, obviously he has something more demanding than "initial conditions" in mind, because he does not qualify them as "accessorial." Hence, the question arises, what else does he want to capture with the term "accessorial." To answer this question, we have to discuss how he proceeds in the second lecture.

Conceding that the world equations are in need of initial or boundary conditions, the main point of Hilbert's second Hamburg lecture is to argue for another and nontrivial meaning of "accessorial." Even with initial conditions, the equations, being differential equations with respect to some time coordinate, would only predict the future from the past, but would they also teach us something about the present which after all, as Hilbert argues, is what we really want? If the answer is no, then we are in need of "accessorial" *laws*, that can tell us something about the present state of nature. Now the interesting point is, as we will see in a moment, that Hilbert argues that no such accessorial laws of nature exist, for the simple reason that precisely that which we want to capture with such laws is either inconsistent with the world equations or already contained in them.

A first argument supporting his claim is a discussion of the irreversibility of thermodynamics. He looks at the example of the mixing of a gas that is initially distributed over two separate halves of a container and emphasizes that the apparent asymmetry with respect to past and future is exclusively a consequence of the choice of the initial states and the initial conditions, and hence that the irreversibility is not one that exists objectively in inanimate nature and its lawfulness but is only an apparent irreversibility, arising from what he called our anthropomorphic point of view.

The argument is interesting in itself, especially with respect to Hilbert's epistemological position.[19] While Hilbert is unambiguous about his claim that there are no accessorial laws introduced in statistical mechanics, he himself brings up an obvious objection. The example of the diffusion of a gas in a container in the theoretical context of kinetic gas theory presupposes the assumption that there exist atoms and molecules, and that these are the fundamental constituents of the diffusing gas. This argument leads him to a discussion of the question whether the principle of atomism is an accessorial law of nature. Hilbert's position on this issue is just as unambiguous as is his position on the issue of irreversibility. He argues that the world equations, possibly after necessary elaborations or corrections, suffice to explain the existence, and even the structure, and properties, of matter. In order to justify this claim, Hilbert refers to Bohr's quantum theory and to the explanation of basic features of the periodic system of elements (such as its periodicity and the chemical stability of the inert gases) on the grounds of the electron orbit model.

Hilbert's conclusion from this discussion is that the field equations and laws of motion suffice to derive the deepest properties of matter including the characteristic details of the chemical elements as particular mathematical integrations of the field equations. It is important to note that in this respect the "world equations" differ fundamentally from Newton's laws, including gravity, because the latter do not imply anything about the existence of atoms and molecules. Of course, Hilbert would take it for granted, among other things, that particle-like solutions of the field equations would exist whose dynamics would then be governed by the field equations as well, rather than by independent equations of motion.

Hilbert's belief that the world equations can tell us something about the present presupposes that we accept only those solutions to the equations that correspond to constant or periodic processes in nature. Hence we have to qualify the assertion about the non-existence of accessorial laws by admitting that there are accessorial *ideas* and *principles*, such as stability and periodicity. But the crucial difference, according to Hilbert, is that these accessorial ideas and principles do not have the character of new equations but are of a more general nature that is connected to our thinking as such and to our attitude towards nature.

It so happened that a number of the assumptions made by Hilbert, both explicitly and implicitly, turned out to be highly problematic, if not false. This is the case, e.g., with the violation of gauge invariance implied by accepting an explicit dependence of the Lagrangian on the electromagnetic potential. But before dismissing Hilbert as a bad speculative physicist, let us take seriously the fact that he himself in a most enthusiastic manner pointed to the rapid development of the natural sciences and the rapid succession of fundamental discoveries. It seems to us that his perhaps premature acceptance of results which had yet to be confirmed appears to us today naïve for a very specific reason. Hilbert's optimism was fuelled by his unwillingness to accept the fact that the modern development of the natural sciences no longer allows for a conceptual unity of knowledge. In this respect, by the way, he was not alone. Indeed, the purpose of the first two lectures of his trilogy was to provide the scientific underpinning for a more philosophical claim that he made in the third lecture.

14.7 Hilbert's Position between Kantian Apriorism and Poincaré's Conventionalism

Let us therefore return now to Hilbert's epistemological position.[20] In his third lecture, Hilbert addresses the ancient question as to the sources of our knowledge, or, in his own words:

> We are dealing here with a decision of an important philosophical problem, namely the old question as to the portion of our knowledge that comes from our thinking, on the one hand, and from experience, on the other hand.[21]

In the remainder of this paper we want to say a few words about Hilbert's answer to the question of the borderline between knowledge a priori and knowledge by experience. Hilbert's position is based on two fundamental presuppositions. The first of these is the distinction between two different domains of the natural sciences, the domain of "inanimate" nature, which is the proper domain of physics in the widest sense, and the domain of living beings including "man as such" which is the domain of biology, including the social and cultural sciences. Even though the distinction seems problematic from a physicalistic point of view, it has not been shown to this day whether the laws of physics, as we know them today, suffice to deduce the phenomena of life, or whether we need in fact some accessorial laws or principles.[22] But for our context, it is sufficient to take this distinction as a warning that the claim that there exist "world equations" in the strong sense, i.e., that we do not need any accessorial laws, is certainly more difficult to establish if the life sciences were included in the claim.

The second fundamental distinction that plays a role here is a distinction between three different levels of experience: (i) a level of every day experience, (ii) a level of scientific experience in the broadest sense of the term, and (iii) a level of totally objective knowledge that is achieved by an emancipation from what Hilbert calls our anthropomorphic point of view. The principle of objectivity that Hilbert had introduced earlier in his first lecture illustrates what Hilbert has in mind by the emancipation from the anthropomorphic point. This principle states

> A sentence about nature, expressed in coordinates, is only then a proposition about the objects in nature, if the sentence has a content which is independent of the coordinates.[23]

According to Hilbert, this emancipation from the coordinate system can be achieved in three different ways that correspond to the three forms of singular, particular, and general judgment: First, by showing or presenting a concrete object, in respect to which the coordinate system has to be fixed; second, in the form of an existential assertion by saying: there exists a coordinate system in which all the formulated relations between the objects considered are valid; third, by formulating the proposition in a form that is valid in every coordinate system. Evidently, this distinction implies that the introduction of coordinates in the first place is a compromise to our human way of looking at nature, and the third way of emancipating from a coordinate system therefore represents the most far-reaching "emancipation from the anthropomorphic point of view." A certain view of the actual and proper development of science is implicit in this latter assumption, and Hilbert's epistemology is indeed a philosophy of graded progress Majer and Sauer (2003).

Hilbert emphatically points out that the Kantian question is ripe for an answer for two reasons, (1) the spectacular progress in the sciences of the time and (2) the advent of the axiomatic method. Much more needs to be said about Hilbert's epistemological position in general and the interrelations of these two moments in particular. But for the sake of brevity, let us here only point to the role of the world equations. Hilbert says:

> If now these world equations, and with them the framework of concepts, would be complete, and we would know that it fits in its totality with reality, then in fact one needs only thinking and conceptual deduction in order to acquire all physical knowledge.[24]

Leaving aside the difficult question concerning completeness of physical theories, we only wish to emphasize that Hilbert, contrary to what one might expect from this quote, by no means wants to take an idealistic position. He emphasizes

> I claim that precisely the world equations can be obtained in no other way than from experience. It may be that in the construction of the framework of physical concepts manifold speculative view points play a role; but whether the proposed axioms and the logical framework erected from them is valid, experience alone can decide this question.[25]

In the sequel to the lecture, Hilbert refined this somewhat crude position by taking issue with Kantian apriorism and with Poincaré's conventionalism. The upshot is

> The opinion advocated by us rejects the absolute Apriorism and the Conventionalism; but nevertheless it does in no way retreat from the question of the precise validity of the laws of nature. I will instead answer this question in the affirmative in the following sense. The individual laws of nature are constituent parts of the total conceptual framework, set up axiomatically from the world-equations. The world-equations are the precipitation of a long, in part very strenuous, experimental inquiry and of experience, often delayed by going astray. In this way we come to the idea that we approximate asymptotically a Definitivum by continued elaboration and completion of the world-equations.[26]

Whatever may be said about this position from a historical and philosophical standpoint, we hope to have at least shown that Hilbert's work along the unified field theory program is embedded in a broader perspective of epistemological and methodological concerns that well deserves to be taken seriously, even on today's philosophical horizon.

Acknowledgments

One of us (T.S.) would like to thank Jim Ritter for illuminating discussions about Einstein and the unified field theory program during our common stay at the Max Planck Institute for the History of Science in Berlin. We also thank Diana Buchwald, Dan Kennefick, and Stephen Speicher for helpful comments on an earlier draft of this paper. Hilbert's lectures are quoted with kind permission of the *Niedersächsisdie Staats—Und Universitätsbibliothek (Handschriftenabaterlung)*.

References

Einstein, Albert (1921). "Über eine naheliegende Ergänzung des Fundamentes der allgemeinen Relativitätstheorie." *Preussische Akademie der Wissenschaften, Phys.-math. Klasse, Sitzungsberichte*, 261–264.

— (1923a). Zur allgemeinen Relativitätstheorie. *Preussische Akademie der Wissenschaften, Phys.-math. Klasse, Sitzungsberichte*, 32–38.

— (1923b). Bemerkung zu meiner Arbeit 'Zur allgemeinen Relativitätstheorie.' *Preussische Akademie der Wissenschaften, Phys.-math. Klasse, Sitzungsberichte*, 76–77.

— (1923c). Zur affinen Feldtheorie. *Preussische Akademie der Wissenschaften, Phys.-math. Klasse, Sitzungsberichte*, 137–140.

— (1949). Autobiographisches—Autobiographical Notes. In *Albert Einstein. Philosopher-Scientist*. Paul Arthur Schilpp, ed. The Library of Living Philosophers, Evanston, Ill., 1–49.

Einstein, Albert and Grommer, Jakob. Beweis der Nichtexistenz eines überall regulären zentrisch symmetrischen Feldes nach der Feld-theorie von Th. Kaluza. In *Scripta Universitatis atque Bibliothecae Hierosolymitanarum: Mathematica et Physica* I. Jerusalem, 1–5.

Goenner, Hubert (2004). On the history of unified field theories. *Living Reviews of Relativity* 7(2004), no. 2 [online article]: http://www.livingreviews.org/lrr-2004.

Goldstein, Catherine and Ritter, Jim (2003). *The Varieties of Unity: Sounding Unified Theories 1920–1030*. In: *Revisiting the Foundations of Physics*. A. Ashtekar et al., eds. Kluwer, Dordrecht, Boston, London, 93–149.

Hilbert, David (1900). Mathematische Probleme. Vortrag, gehalten auf dem internationalen Mathematiker-Kongress zu Paris 1900. *Königliche Gesellschaft der Wissenschaften zu Göttingen. Mathematisch-physikalische Klasse. Nachrichten*, 253–297.

— (1915). Die Grundlagen der Physik. (Erste Mitteilung). ' *Königliche Gesellschaft der Wissenschaften zu Göttingen. Mathematisch-physikalische Klasse. Nachrichten*, 395–407.

— (1917). Die Grundlagen der Physik. (Zweite Mitteilung). *Königliche Gesellschaft der Wissenschaften zu Göttingen. Mathematisch-physikalische Klasse. Nachrichten*, 53–76.

— (1924). Die Grundlagen der Physik. ' *Mathematische Annalen* **92**, 1–32.

Kaluza, Theodor (1921). Zum Unitäts-Problem der Physik. *Königlich Preußische Akademie der Wissenschaften* (Berlin). *Sitzungsberichte*, 966–972.

Klein, Felix (1979). *Vorlesungen über die Entwicklung der Mathematik im 19. Jahrhundert*. Wissenschaftliche Buchgesellschaft, Darmstadt.

Majer, Ulrich (2002). Lassen sich phänomenologische Gesetze "im Prinzip" auf mikrophysikalische Theorien reduzieren? In *Phänomenales Bewußtsein — Rückkehr zur Identitätstheorie?* M. Pauen & A. Stephan, eds. Paderborn: mentis, 2002.

Majer, Ulrich and Sauer, Tilman. Intuition and Axiomatic Method in Hilbert's Foundation of Physics: Hilbert's Idea of a Recursive Epistemology in his Third Hamburg Lecture. To appear in *Intuition in the Mathematical Sciences*, Carson, Emily and Huber, Renate, eds. (in press).

Mastrobisi, Giorgio Jules (2002). Il ⟨⟨ Manoscrito die Singapore ⟩⟩ (1923) di Albert Einstein. Per una Teoria del ⟨⟨ Campo Unificato ⟩⟩ tra possibilità fisica e necessità matematica. ' *Nuncius* **17**, 269–305.

Mie, Gustav (1912). Grundlagen einer Theorie der Materie. Zweite Mitteilung. *Annalen der Physik* **39**, 1–40.

Pais, Abraham (1982). *'Subtle is the Lord...': The Science and Life of Albert Einstein.* Oxford University Press, Oxford.

Renn, Jürgen, and Stachel, John (1999). Hilbert's Foundation of Physics: From a Theory of Everything to a Constituent of General Relativity. Max Planck Institute for the History of Science, Berlin. Preprint 118.

Sauer, Tilman (1999). The Relativity of Discovery: Hilbert's First Note on the Foundations of Physics. *Archive for History of the Exact Sciences* **53**, 529–575.

— (2002). Hopes and Disappointments in Hilbert's Axiomatic Foundation of Physics. In *History of Philosophy of Science*. M. Heidelberger and F. Stadler, eds. Kluwer, Dordrecht, 225–237.

Vladimir P. Vizgin (1994). *Unified Field Theories in the First Third of the 20th Century*, Birkhäuser, Basel, Boston, Berlin.

Notes

[1] The lectures were held in Hamburg on July 26, 27, and 28, 1923. They were announced under the title "Grundsätzliche Fragen der modernen Physik," see "Hamburgische Universität. Verzeichnis der Vorlesungen. Sommersemester 1923," Hamburg 1923, p. 41. The third of the three lectures was delivered a second time, with short summaries of the first two lectures, in a lecture held at the "Physikalische Gesellschaft" in Zürich on October 27, 1923. This lecture was announced under the title "Erkenntnistheoretische Grundfragen der Physik," see "Neue Zürcher Zeitung," Nr. 1473, Erstes Morgenblatt, 27 October, 1923. The manuscript Cod. Ms. Hilbert 596 in the *Handschriftenabteilung* at the *Niedersächsische Staats- und Universitätsbibiliothek* (NSUB) contains the notes for both the Hamburg and Zurich lectures. It will be cited in the following as *Lectures*.

[2] In this respect, we disagree with the claim made by Renn and Stachel, who characterize Hilbert's work in GRT as the transition from a "Theory of Everything to a Constitutent of General Relativity." Renn and Stachel (1999). While their assessment may be true in abstraction of its actors, it is certainly not true for Hilbert himself. Rather than beginning to see his own work as a constituent of General Relativity, his main effort with respect to General Relativity in later years was to emphasize his claim that his approach would provide the basis for a true unification of physics.

[3] The lecture was arranged by Peter Debye following Hilbert's request. "Herr Prof. Hilbert, welcher zur Zeit in der Schweiz weilt, hatte den Wunsch im zusammengefassten Physik. und Mathematischen Kolloquium einen Vortrag zu halten." P. Debye to Robert Gnehm, 22 October 1923. Archiv des Schweizerischen Schulrats, ETH-Bibliothek, Zürich.

[4] As an aside, Hilbert observed in his 1923 lectures that the difference between his own fundamental equations of November 1915 and Einstein's gravitational field

equations pertains to the choice of fundamental variables: "Die Einsteinschen Gravitationsgl[eichungen] sind in dem hier definirten Sinne die Grundgl[eichungen] der
Physik, wenn man darin das Gravitationspotential $g_{\mu\nu}$ und ausserdem den Energie-
tensor als Grundpotentiale nimmt. Ich habe zurselben Zeit Grundgl[eichungen] der
Physik aufgestellt, in denen neben dem Gravitationspotential $g_{\mu\nu}$ nur noch das elektromagnetische Viererpotential ϕ_k als Grundpotential auftritt." *Lectures*, part I, p. 16.

[5]We are closely following Hilbert's and Einstein's notation, with the following
exceptions: for notational brevity, we denote partial (coordinate) derivatives by a subscript index separated by a semicolon (comma), and indicate (anti)symmetrization by
setting the relevant indices in (square) brackets. We also do not use an imaginary x_4-
coordinate, as Hilbert did.

[6]Already in his First Note on the Foundations of Physics, Hilbert had left open the
final choice of a matter term in the Lagrangian. It should be diffeomorphism invariant,
and it should not depend on the derivatives of the metric. But Mie's example of a
term proportional to the sixth power of φ had obviously been unacceptable, and a
different specification of the Lagrangian that would allow for solutions of a reasonable
physical interpretation had not yet been found, see Mie (1912) and also the discussion
in Hilbert's own lecture notes on "Die Grundlagen der Physik," of the summer of 1916,
which are located at the library of the Mathematics Institute of Göttingen University,
see especially §§ 27–30. For further discussion of Hilbert's First Communication, see
Sauer (1999).

[7]"Einstein endlich knüpft in seinen letzten Publikationen an Eddington an und
gelangt ebenso wie Weyl zu einem mathematisch sehr einheitlich aufgebauten System. Indess mündet das Schlussresultat der letzten Einsteinschen Untersuchung wieder
auf ein Hamiltonsches Prinzip, das dem ursprünglich von mir aufgestellten gleicht; ja
es könnte sein, dass diese Einsteinsche Theorie inhaltlich sich mit der von mir ursprünglich aufgestellten Theorie *völlig deckt*." *Lectures*, part I, p. 19 (Hilbert's emphasis).

[8]"... wo φ durch $\varphi^m = \text{Div } \Phi^{mn}$ definiert ist und durch Variation nach $g_{\mu\nu}$ und
Φ^{mn} die Gl. $\Phi_{mn} = \text{Rot } \varphi_m$ an Stelle meiner Gl. (14.5) entstehen. Also Nichts als eine
Vertauschung der beiden Serien [von] Max. Gl. (Hilbert's emphasis).

[9]In this paper, we will not deal with Eddington's own work but only with Einstein's
perception of it.

[10]We are using Einstein's and Hilbert's notation, both of whom referred to the Lagrangian as a Hamiltonian function.

[11]"Der im Sinne unserer bisherigen Kenntnisse natürlichste Ansatz" (Einstein 1923c,
p. 139).

[12]"Und wenn auf dem kollossalen Umweg über Levi-Civita, Weyl, Schouten,
Eddington Einstein zu dem Resultat zurückgelangt, so liegt darin sicher eine schöne
Gewähr." *Lectures*, part I, p. 20.

[13]"Ich glaube sicher, daß die hier von mir entwickelte Theorie einen bleibenden Kern enthält und einen Rahmen schafft, innerhalb dessen für den künftigen Aufbau der Physik im Sinne eines feldtheoretischen Einheitsideals genügender Spielraum da ist." (Hilbert 1924, p. 2).

[14]For another discussion of this concept, see Majer and Sauer (2003).

[15]See lecture notes for course "Über die Einheit in der Naturerkenntnis," held in winter 1923/24. NSUB Cod. Ms. Hilbert 568, p. 247.

[16]We realize that the English word "accessorial" is not a neologism and its meaning of auxiliary, supplementary makes good sense in the present context.

[17]"Ich möchte Alles, was noch zu den Weltgleichungen hinzugefügt werden muss, um die Geschehnisse in der leblosen Natur zu verstehen, kurz *accessorisch* nennen." *Lectures*, part II, p. 1.

[18]For further discussion Hilbert would assume the world to be Euclidean–Newtonian at infinity, but with respect to contemporary cosmological debates Hilbert added a disclaimer to the effect that this choice was only motivated by formal simplicity and was made only to fix the ideas.

[19]For a more detailed discussion of the non-objective but anthropomorphic character of certain apparently irreversible processes in inanimate nature, see Majer (2002).

[20]See Majer and Sauer (2003) for a more extensive discussion.

[21]"Wir stehen da vor der Entscheidung über ein wichtiges philosophisches Problem, nämlich vor der alten Frage nach dem Anteil, den das Denken einerseits und die Erfahrung andererseits an unserer Erkenntnis haben." *Lectures*, part III, p. 1.

[22]For a discussion of this intricate question in connection with the supposed irreversibility of living processes, see Majer (2002).

[23]"Ein in Koordinaten ausgedrückter Satz über die Natur ist nur dann eine Aussage über die Gegenstände in der Natur wenn er von den Koordinaten unabhängig einen Inhalt hat." *Lectures*, part I, p. 3.

[24]"Wenn nun diese Weltgleichungen und damit das Fachwerk vollständig vorläge, und wir wüssten, *dass es* auf die Wirklichkeit in ihrer Gesamtheit passt und dann bedarf es tatsächlich nur des *Denkens* d.h. der begrifflichen *Deduktion*, um alles phys. Wissen zu gewinnen." *Lectures*, part III, pp. 20f. (Hilbert's emphasis).

[25]"... behaupte ich, dass gerade die Weltgesetze auf keine andere Weise zu gewinnen sind, als aus der Erahrung. Mögen bei der Konstruktion des Fachwerkes der phys[Begriffe] mannigfache spekulative Gesichtspunkte mitwirken: *ob* die aufgestellten Axiome und das aus ihnen aufgebaute logische Fachwerk stimmt, das zu *entscheiden*, ist allein die *Erfahrung* im Stande." ibid., p. 21 (Hilbert's emphasis).

[26]"Die von uns vertretene Meinung verwirft den unbedingten Apriorismus und den Konventionalismus; aber sie entzieht sich trotzdem keineswegs der vorhin aufgeworfenen Frage nach der genauen Gültigkeit der Naturgesetze. Ich möchte diese Frage vielmehr bejahen und zwar in folgendem Sinne. Die einzelnen Naturgesetze

sind Bestandteile des Gesamtfachwerkes, das sich aus den Weltgleichungen axiomatisch aufbaut. Und die Weltgleichungen sind der Niederschlag einer langen zum Teil sehr mühsamen und oft durch Irrwege aufgehaltenen experimentellen Forschung und Erfahrung. Wir gelangen dabei zu der Vorstellung, da wir uns durch fortgesetzte Ausgestaltung und Vervollständigung der Weltgleichungen asymptotisch einem Definitivum nähern." ibid., pp. 42f.

15

Einstein, Kaluza, and the Fifth Dimension

Daniela Wünsch

Institute for the History of Science, Papendiek 16, 37073 Göttingen, Germany;
dwuensc@gwdg.de

After grappling intensively with the geometrized unified field theories for three years, Einstein began publishing on this topic in 1921. Kaluza's unified field theory of gravitation and electromagnetism in five dimensions evidently influenced these efforts. Although he had been aware of Kaluza's theory since 1919 it was only between 1919 and 1921 that he became convinced that this method of unification represented a significantly new path in physics. A biographical sketch will answer the question of who Kaluza was. At the same time, it will be shown that Einstein's "reorientation" was also based on a change in his epistemological conception which occurred during this period. In the process of this transformation, geometrical unification took precedence in Einstein's scientific research over the unification as a solution to the quantum problem.

15.1 Introduction

In early April 1919 Einstein received a letter from an unknown privatdocent from Königsberg. The letter included the manuscript of a unified theory of gravitation and electromagnetism in a five-dimensional world. The originality of the theory impressed Einstein to such an extent that he replied as follows on 21 April 1919: "The thought that the electric field quantities are truncated $\{\mu_{\nu\rho}\}$) terms has been a thorn in my side for quite some time. But the idea of achieving this with a five-dimensional cylindrical world never occurred to me and is probably completely new. At first glance I find your idea extraordinary."[1] Indeed Kaluza's theory convinced Einstein to take seriously the idea of a unified view of the physical world. Einstein dedicated the rest of his life, a period of 36 years, to developing unified field theories.

This article will focus on two points:

1. The question of exactly who this unknown lecturer from Königsberg was, whose original idea so impressed Einstein.

2. The question concerning the extent to which Kaluza's theory influenced Einstein's work on the unified theories. It seems that Kaluza's theory had a considerable influence on Einstein and served to convince him that the unification of the physical forces represented an important path in the development of physics. The period from 1919 to 1923 marked a turning point in Einstein's view: from that time on he was convinced of the epistemological cogency of the unification.

15.2 Theodor Kaluza, An Unknown Privatdozent

Kaluza first aroused the interest of theoretical physicists in the mid-1960s when they became aware that his theory might be viewed as one that could lead to a paradigm shift. This interest has gained even more momentum since 1979 when Abdus Salam, one of the creators of the theory of electroweak unification, spoke of the "Kaluza–Klein miracle"[2] in his Nobel Prize acceptance speach. Since then, physicists have set out to answer Steven Weinberg's query: "Who was Kaluza?"[3]

In my more thoroughly researched biography of Kaluza,[4] I have tried to answer this question. In the following article, I present a brief summary of Kaluza's life and work.

15.2.1 Theodor Kaluza's life[5]

Kaluza was born on 9 November 1885 in the Prussian town of Wilhelmstal-Oppeln[6] in Upper Silesia (now Poland). Two years later his father, Max Kaluza, moved the family to Königsberg (East Prussia, today Kalinigrad), where he later became a respected professor of English philology at the University of Königsberg. Kaluza descended from a German family whose origins are documentable back to the 16th century in Ratibor (Upper Silesia).[7]

At the Gymnasium Friedericianum —the same Gymnasium that Immanuel Kant and David Hilbert had attended —Kaluza received a classical education, excelling in mathematics and physics.

In 1902 Kaluza began studying mathematics, physics, and astronomy at the Albertus University in Königsberg, a university particularly renowned for its strong mathematics and physics departments.

Three great scholars, the astronomer Wilhelm Bessel, the mathematician Carl Gustav Jacob Jacobi and the physicist Franz Neumann, considered to be the father of theoretical physics in Germany, founded the "Königsberger Schule." In the late nineteenth century, under the intellectual leadership of David Hilbert, Hermann Minkowski and their mentor Adolf Hurwitz, the mathematics department at Königsberg reached a new acme.

In 1907 Kaluza defended his doctoral dissertation in mathematics on "Tschirnhaus Transformations." Afterwards he worked for two semesters at the observatory in Königsberg, then spent one year in Göttingen, which was, at that time, the most renowned center of mathematics. At the Department of Mathematical Physics in Göttingen, headed by David Hilbert and Hermann Minkowski, Kaluza would witness the genesis of the mathematical model of the theory of special relativity, conceived by Minkowski.

After returning to Königsberg in 1909, Kaluza completed his habilitation and became a privatdocent in Königsberg. In that same year he also married. His scientific activity was interrupted in 1916 by the First World War. Soon after his return from the Western Front, Kaluza developed his unified theory in five dimensions. In early April 1919 he sent his theory to Einstein, who was very impressed. Despite his positive assessment of these ideas, two years elapsed before Einstein submitted Kaluza's theory to the Prussian Academy in Berlin. Upon publication, the theory generated great interest among physicists.

In the years that followed, Kaluza continued to be highly productive academically in spite of his difficult financial situation as privatdocent.

In 1925 Kaluza was forced to move away from research in the field of physics. He was in a very precarious financial situation having, at the age of 40, not yet received a full professorship. In 1929 Kaluza finally received a call to a professorship in mathematics at the University of Kiel. This was a result of his having published intensively in this field. He remained in Kiel until 1935 when he was appointed professor by the University of Göttingen, replacing Richard Courant. In Göttingen he lived through the dark period of the Third Reich as Director of the Institute of Mathematics. He died in Göttingen in 1954.

Kaluza was, as a result of his philosophical convictions, a deeply moral person. Kant's categorical imperative had influenced him greatly and he was a proponent of Albert Schweitzer's ethical ideas. These inner convictions manifested themselves in Kaluza's resolute stand against the pseudo-philosophy and ideology of National Socialism.

His best friends were the Jewish mathematicians Gabor Szegö, Willy Feller, and Werner Rogosinski, all forced to emigrate as a result of Nazi oppression. Another friend, the mathematician Kurt Reidemeister, was dismissed from the faculty of the University of Königsberg in 1933 because of political differences with the Nazi party.

Kaluza himself never was a member of the National Socialist German Workers Party and was ideologically opposed to all forms of violence. Thus he eschewed the official ideology of power and violence, facing the Nazis in a characteristically reserved but determined manner, refusing to take part opportunistically in military research. In 1944 the president of the University of Göttingen, Hans Drexler, a professor of classics with National Socialist convictions, placed Kaluza's name on a list of undesirable persons to be removed from the university. This never came to pass, however, as the uproar resulting from the assassination attempt on Hitler a few month later proved too distracting.

Theodor Kaluza's work and philosophical views form an inseparable entity. We can best describe Kaluza's philosophical understanding as Kantian: nature, in its overarching harmony, can be understood in mathematical language. This conviction was also the basis of Kaluza's belief that all natural interactions originate from one force.

15.2.2 Scientific works

In all, Theodor Kaluza published four papers on physics: "On the Theory of Relativity"(I)[8] (1910) in which he extended the principle of relativity to accelerated motion of a rigid rotating disk;[9] "On the Unity of Physics"[10] (1921), his theory of the unification of gravity and electromagnetism in five dimensions; "On the Structure and Energy Content of Nuclei"[11] (1922), where he anticipated the diagram of the nucleus' binding energy, developed much later and now familiar to all nuclear physicists; and "On the Theory of Relativity"[12] (II) (1924), where he tried to prove that the definition of the simultaneity of two separate events can be derived "without any reference to the principle of the constant velocity of light in vacuum."[13] The impressive originality and significance of these works alone are grounds enough to arouse the interest of any physicist.

Although Kaluza's publishing activity in physics stopped in 1924 —despite Einstein's encouragement[14]—there is clear evidence[15] that in 1952 Kaluza intended to continue his work on his unified field theory and to connect it to Weyl's theory. His papers on mathematics, 13 in all, two of them born of the discussion about the foundations of mathematics at the end of the 19th century, never attained the same importance as his work in physics. Nevertheless Kaluza is still known in mathematics for his contributions to analysis.

15.3 Einstein and Kaluza's Idea

15.3.1 Introduction

Vizgin remarks in his book[16] that Einstein's conceptual interpretation of physics underwent a "reorientation" in 1919. Up to that point, Einstein had been interested in "obtaining particles and quanta on the basis of a field theory,"[17] and had rejected as not at all promising the theory Weyl developed in 1918 proposing a geometrized unification of gravitation and electromagnetism. In spring 1919, however, Einstein reoriented his views and from this time on, he was no longer unreceptive to the idea of the geometrization of the fields. Vizgin conjectures that Einstein was moved to this "reorientation" by Kaluza's theory:

> It is entirely possible that Kaluza's investigations already had a strong influence on Einstein's change in position with regard to the possibility of a geometrical synthesis of gravitation and electromagnetism.[18]

I concur with Vizgin's thesis with one exception — I believe that Einstein's "reorientation" was a process which took place between 1919 and 1921. I would also like to examine this topic from a different point of view: Einstein's reorientation is reflected not only in his physical theory but also in his epistemology. It is my thesis that Einstein's epistemological conception changed at the time he began to examine the geometrized unified field theories. His new philosophy was dominated much more by rationalist elements. Einstein's physical theory is closely connected to his epistemology and only by considering both of these aspects is it possible to understand the complex phenomenon which was taking place at that time, i.e., the development of the unified field theories as a new path in physics.

In the first section I will analyse the historical development of the events. In the process, it will become clear that the historical sources, considerable as they are, do not suffice to explain this development fully. Only an epistemological analysis will allow clarification of the events which served to convince Einstein that the unified theories represented a physical path worth pursuing as a logical and conceptual extension of the theory of relativity. The second section deals with the development of Einstein's epistemology within the framework of his work on the unified field theories and will show that Einstein developed a new "philosophical picture" in which the concept of "mathematical simplicity" played an important role in the construction of physical theories.

15.3.2 Historical development

In the following I will summarize the historical course of events.[19] In 1918 Einstein examined closely Weyl's unified theory "Gravitation and Electricity."[20] Weyl had attempted to unify electromagnetism and gravitation by a gauge transformation. His theory contained not only a metric tensor $g_{\mu\nu}$ but also a metric fundamental vector Φ_μ meant to represent the electromagnetic potential. Weyl had perfected the world metric; however, the two interacting forces continued to have separate origins.

Throughout 1918, Weyl tried unsuccessfully to convince Einstein of the correctness of his theory. Einstein considered Weyl's theory to be an admirable mathematical construction, nevertheless, it violated the criterion of empirical verification. It would have the physical consequence that, for example, the spectral lines of hydrogen atoms would change depending on the path they had travelled, a phenomenon as yet never observed. In his letter of 18 September 1918 to Walter Dällenbach, Einstein wrote: "Weyl's theory of electricity is a wonderful concept, it has a touch of genius. But I am convinced that the laws of nature are different. [...] all in all: It is my firm opinion that this theory does not correspond to reality."[21]

At that time Einstein was quite skeptical of Weyl's geometrized synthesis and believed that the unification could be attained by reducing the particles to a field. This he called the "solution of the quantum problem"[22] which indicates that Einstein's concept of unification went deeper. Whereas Weyl's theory was based exclusively on the field concept, Einstein, in his article, tried to solve the field-particle dualism by reducing the particle concept to a field. This explains why in April 1919 Einstein, after having rejected Weyl's theory, published his article "Do Gravitational Fields play a Role in the Structure of the Elementary Particles of Matter?,"[23] in which he tried to find a possible fundamental connection between the general theory of relativity and the structure of matter. By trying to show that the elementary electric particles were held together by the forces of gravitation, Einstein intended to demonstrate "the possibility of a theoretical construction of matter out of the gravitational and electromagnetic fields."[24]

Thus Kaluza's theory reached Einstein at a time when Einstein himself was having serious doubts about the idea of the unification as proposed by Weyl. Between 21 April and 29 May 1919 Einstein wrote four letters and a postcard to Kaluza. In these he expressed his admiration for Kaluza's theory: On 29 May 1919 he wrote: "I have great respect for the beauty and the boldness of your thought."[25] What led Einstein to this statement? Kaluza wrote Einstein's gravitational equations in a five-dimensional space with a Riemannian metric such that all physical quantities were explicitly independent of the fifth coordinate. This he described as the "cylinder condition" (Kaluza 1921, 967). What were the advantages of this theory?

1. The two forces no longer appear separately, as in Weyl's theory, but rather originate from a unique five-dimensional $F_{\mu\nu}$ gravitation-like world tensor which generates a universal field. In four-dimensional space it splits into a gravitational and an electromagnetic field.

2. Kaluza was able to prove that the fifth dimension influences only electromagnetic interactions but not gravitational force, the $g_{\mu 5}(\mu = 1, 2, 3, 4)$ components of the five-dimensional metric being identified with the Maxwellian field A_μ. In this manner he was able to define electromagnetism as a geometrical manifestation of the fifth dimension.

3. The fifth dimension is such that in the transition to normal four-dimensional space, one obtains, on the one hand, Einstein's gravitational equations and, on the other, Maxwellian equations, thus allowing a coherent reduction of five-dimensional space so as to fit into the four-dimensional space perceivable to us. The idea that two forces appearing with a different structure in our four-dimensional space may be perfectly unified to only one, if we add an additional dimension, goes back to Kaluza's theory.[26]

Unlike Weyl's theory which contradicted the known empirical facts and thus did not fulfill the criterion of "external corroboration," Kaluza's theory had no such deficits. Einstein, while checking its "inner coherence," was impressed by the perfection of the theory. In his first letter to Kaluza of 21 April 1919, in which Einstein expressed his admiration, he also emphasized the physical advantages of Kaluza's theory compared to Weyl's theory:

> From the physical standpoint it seems to me to be definitely more promising than Weyl's very *mathematical* theory because your theory sets out from the electric field and not from, in my opinion, physically insignificant four-potential. Moreover your theory leaves the four-dimensional geodetic line (the path of the uncharged mass point) intact, which is very nice too.[27]

The latter point referred to the fact that Kaluza's theory could embed the general theory of relativity without changing it. For Einstein, this might very well be an argument in favor of Kaluza's theory.

Einstein also expressed his "physical criticism:"

> The main issue is whether your idea can stand up to physical criticism. I would like to mention two points:

1) It has to be compatible with the existence of different positive and negative elementary quanta. [...]
2) It should lead to the correct solution of the cosmological problem.[28]

Thus Einstein required that the theory solve the more important conceptional problems of physics and that it be compatible with the quantum problem.

In his letter of 5 May 1919 Einstein proposed that he present Kaluza's work to the Prussian Academy in Berlin for its *Sitzungsberichte*. However, in this letter he also mentioned his objections:

From a general standpoint only *one* thing bothers me. The theory requires
1) general covariance in R_5
2) and in combination with it, the relation $\partial/\partial w_5 = 0$ not be covariant in R_5.
Naturally this is not very satisfying.[29]

Through the cylinder condition, an asymmetry of the fifth axis occurs, as the physical quantities do not depend on the fifth dimension. Kaluza set this condition in order to explain why we do not perceive any consequences of the fifth dimension in our four-dimensional space. In this way, the covariance of the equations is restricted, as the fifth dimension plays a particular role.[30] This is also most probably the reason why Einstein hesitated to publish Kaluza's theory. But in the same letter Einstein once again expressed his admiration: "But on the other hand, the formal unity of your theory is astonishing."[31] Despite this positive assessment, in his letter of 29 May 1919 Einstein retracted his offer to recommend Kaluza's theory to the Academy. This did not happen until October 1921 when he finally supported the publication of Kaluza's theory.

The period between May 1919 when he rejected Kaluza's theory and 3 March 1921 when Einstein submitted his article on Weyl's theory represented a break in Einstein's publications on unified theories. It was during this period that Einstein tackled the problem. The historical sources (his letters, speeches and popular science articles) indicate that Einstein was engrossed in the topic. Ultimately, Einstein considered geometrized unified field theories to be a promising path in theoretical physics. But before reaching this conclusion, Einstein explored the possibility of attaining the unification by reducing particles to fields. In his letter of 27 January 1920 to Max Born, Einstein expressed his opinion about the quantum problem:

I do not believe that one has to give up the continuum in order to solve the quantum problem. [32]

Einstein viewed the continuum theory as the most promising approach to solving the more important physical problems. The quantum-like phenomena were to be explained by "over-determination," i.e., through more equations than unknowns: "I still believe that one must look for the over-determination by differential equations, such that the *solutions* no longer have continuum character. But how?"[33] Einstein pursued this idea throughout 1920. This is reflected in some of his letters to Ehrenfest. On 2 February 1920 Einstein wrote him:

I haven't been successful in the theory of relativity since then. The electromagnetic field refuses to work out, despite all my efforts and an over-determination by differential equations which take the "horizontal" regularities into account. But I am convinced that this will be the way to genuine internal progress. [34]

In Einstein's letters to Ehrenfest of 9 March 1920,[35] 7 April 1920[36] and 26 November 1920[37] he likewise mentioned his efforts to use the method of over-determination, to attain what he described as his "favorite scientific goal" —the unification.[38]

Contrary to Vizgin's contentions, at that time Einstein's position regarding Weyl's theory had not yet changed.[39] Both in his letter of 2 February 1920 to Ehrenfest as well as in his speech in Leiden "Ether and Relativity"[40] on 27 October 1920 he once again expressed his doubts concerning Weyl's theory:

In the latest edition of his textbook, Weyl has unfortunately added his electromagnetic theory so that this —albeit ingenious —nonsense will engrain itself into the brains of the readers. But I comfort myself with the thought that the sieve of time will also do its work here.[41]

In his talk in Leiden as well as in his talk of 27 January 1921 "Geometry and Experience"[42] Einstein considered the unification from the conceptual standpoint of reducing the particles to fields. In his Leiden talk he discussed the unification of "space and matter" within the framework of his view that "the nature of the elementary particles of matter is nothing but a concentration of the electromagnetic field"[43] and in "Geometry and Experience" he examined the question of whether the theory of relativity was applicable to the molecular scale.[44]

In his article in *Nature*, "A Brief Outline of the Development of the Theory of Relativity" published on 17 February 1921, he briefly mentions the unified theories as one "of the important questions which are awaiting solution at the present time."[45] Einstein formulated the question about the role the gravitational fields play in "the constitution of matter." For the first time Einstein called the unification the "formal unit": "Are electrical and gravitational fields really so different in character that there is no *formal unit* to which they can be reduced?"[46] This formulation might be connected with Einstein's idea of improving Weyl's theory. In his letter of 1 March 1921 to Ehrenfest, Einstein mentioned his "lucky insights:"

The mathematical apparatus is relatively simple. Whether this is physically acceptable remains to be seen. This will be clarified within a relatively short time.[47]

Einstein submitted Weyl's article on 3 March 1921 but just six days later he expressed his skepticism in a letter to Sommerfeld of 9 March:

I have my doubts about the physical correctness of the theory. The Lord makes the rules—and isn't impressed by our theories.[48]

Einstein's article, "On a Natural Addition to the Foundation of the General Theory of Relativity" appeared on 17 March. In it he tried to improve Weyl's theory by assuming that the condition of the "existence of transportable clocks and rods" has to be abandoned. By replacing the interval ds^2 as the fundamental characteristic of the space-time continuum by the "invariant meaning of the equation $ds^2 = g_{\mu\nu}dx^\mu dx^\nu = 0$," he hoped to be able to eliminate from the theory the concept of distance and of rods and clocks (Einstein 1921c, 262). Thus the theory would no longer contradict experience.

In May 1921 Einstein held a series of lectures in Princeton "Four Lectures on the Theory of Relativity."[49] Here he talked about Kaluza's theory for the first time, before it was even published:

A theory in which the gravitational field and the electromagnetic field do not enter as logically distinct structures would be much preferable. H. Weyl, and recently, Th. Kaluza, have put forward ingenious ideas along this direction; but concerning them, I am convinced that they do not bring us nearer to the true solution of the fundamental problem.[50]

Despite this negative appraisal, in October 1921 Einstein began working in collaboration with Jakob Grommer on an article about Kaluza's theory.[51] In a letter of 14 October 1921 to Kaluza, Einstein offered to have his paper published. Once again he considered Kaluza's theory better than that of Weyl's:

Your approach definitely seems to me to be better than that of H. Weyl. If you like, I will present your article to the *Academy*.[52]

Einstein's "reorientation" could be explained by the fact that by September 1921 he found little hope in other attempts at unification. Thus, on 5 September 1921, Einstein wrote to Weyl:

Bach's work is quite nice but physically insignificant, Eddington's work is likewise. Moreover it is imprecise. And I still consider your interpretation of the electric field to be incorrect.[53]

Kaluza's article appeared on 8 December 1921 but only one month later, on 10 January 1922, Einstein and Grommer submitted their article about Kaluza's theory as well. On 9 December 1921 Einstein wrote to Kaluza:

Your theory is really captivating. There must be some truth to it.[54]

In the introduction to his article with Jakob Grommer, Einstein emphasized the significance of the new unification program. He claimed that the question about the "fundamental unity of the gravitational and the electromagnetic field" is "now the most important question of general relativity."[55] Furthermore Einstein enumerated his "considerable doubts" about Weyl's theory. He saw three disadvantages to Weyl's approach:

1. It contradicts empirical facts: "The invariance of roads, clocks and atoms with regard to their history is not taken into account" (Einstein and Grommer 1923, 1).
2. Conceptual and logical deficiencies by not eliminating the gravitational-electromagnetic dualism. "Moreover the theory does not eliminate that dualism because the Hamilton function is composed of two additive parts, an electromagnetic and a gravitational part which are —logically seen– independent of each other."[56] Weyl's theory essentially did not answer the question of "the fundamental unity of the two fields."
3. Deficiencies in the mathematical argumentation: "Furthermore this theory leads to differential equations of the fourth order while up to now we have no reason to assume that equations of the second order wouldn't suffice."[57]

Einstein emphasized that Kaluza's theory, in contrast to Weyl's theory, avoids "all of these evils."[58] Repeating what he said in his first letter to Kaluza, Einstein stated that Kaluza's theory exhibited "an astonishingly formal simplicity."[59] Einstein attached considerable importance to Kaluza's theory. First of all its logical-conceptual unity is remarkable, as both fields derive from a unique field which is what Einstein expressed in his article using the Hamilton function:

> Kaluza's essential hypothesis consists of the assumption that the natural laws in this five-dimensional world should generally be covariant. [...] This provides us with the possibility of constructing the physical world using a unified Hamilton function which does not contain heterogeneous terms connected artificially by addition
> $$H = g_{\nu\mu}\Gamma^{\alpha}_{\mu\beta}\Gamma^{\beta}_{\nu\alpha} \quad {}^{60}$$

Secondly Kaluza's theory, unlike Weyl's theory, did not contradict empirical facts. Einstein also enumerated the "weak points" of the theory. The fifth dimension is purely "abstract," and is thus unable to provide any measurable quantities. The construction of the five-dimensional world — no matter how perfect its formal unity is — was much too remote from the empirical world, even if it did not contradict it. The second objection refers to the preferred position of the fifth dimension arising from the "cylinder condition:"

> The cylinder property leads to a disturbing asymmetry based on giving preference to one dimension and at the same time requiring that in the construction of the equations all five dimensions play the same role."[61]

In addition, Einstein proved that Kaluza's theory did not provide a centrally symmetric solution "which could be interpreted as a (non-singular) electron."[62]

Nevertheless Einstein, despite his criticism, did adopt one element from the structure of Kaluza's theory. In his three-part article of 1923,[63] in which he proposed an extension of the Eddington–Weyl theory, Einstein borrowed one element of Kaluza's theory: The Hamilton function contains a single tensor to describe the unified field. The two independent fields (gravitational and electromagnetic) could thus be unified to one field. In his article "On the General Theory of Relativity" he succeeded in expressing the Hamilton function H with the help of one single tensor R_{kl}:

$$H = 2\sqrt{-|R_{kl}|},$$

which "is formed without splitting it into the symmetrical and anti-symmetrical part." Thus he emphasized that now "the Hamilton function is definitely a unified one whereas before it consisted of summands which were logically independent of each other" (Einstein 1923b, 35). Even though Einstein intended in his article to pursue "the Weyl–Eddington path to the end," he expressed doubts whether this approach was "true."[64]

As we have seen, after rejecting Kaluza's work in May 1919, Einstein struggled with the quantum problem until early 1921. He had hoped to solve it by applying the method of over-determination by diffential equations. Then, in three articles, he examined the geometrical unified theories in the forms elaborated by Weyl, Kaluza and Weyl–Eddington. While he was rather sceptical about his own Weyl treatment, he praised the formal unity of Kaluza's theory and in his three-part treatment on Weyl–Eddington, he used Kaluza's idea of unifying both fields in a unique field tensor.

Einstein continued, however, to pursue the over-determination approach. By 1923 he returned again to the quantum problem and, on 13 December, in the session of the Prussian Academy in Berlin, Einstein presented his paper "Does the Field Theory Offer Possibilities for Solving the Quantum Problem?" (Einstein 1923e). Einstein's "final goal" was to derive matter from the field variables, as he expressed it in his article with Grommer: "[to construct] a pure field theory such that the field variables [...] represent the electrical elementary particles which constitute matter." (Einstein and Grommer 1923, 2). Einstein also pursued this goal in connection with the geometrized field theories by trying unsuccessfully in his articles on the theories of Weyl, Kaluza and Weyl–Eddington to derive the existence of non-singular solutions.[65]

The foregoing analysis of the historical events raises a number of questions: Why did Einstein—in 1921—suddenly develop such interest in the geometrized unification, which he then persistently pursued? What conceptual path led him to change his mind so radically? Exactly how much was he influenced by Kaluza's theory?

15.3.3 Epistemology and unified field theories

Numerous studies have shown that in the course of his scientific activity Einstein changed his philosophical stance more than once, adapting it to his physical theories.[66] Some historical studies[67] have demonstrated that Einstein, while working on special relativity, was a "skeptical empiricist like Mach"[68] whereas, while working on general relativity, he had to contradict Mach. Kanitscheider showed that even Einstein's early philosophy contained rationalistic elements which led him to apply principles of symmetry (which are not derivable empirically) to special relativity (Kanitscheider 1979, 151). Friedman demonstrated convincingly that Einstein's general relativity contains *a priori* elements (Friedman 2001).

Einstein's many physical articles underline the importance he placed on his philosophy: it constitutes a part of his scientific method. Whereas detailed studies exist of Einstein's philosophical conceptions during the years he was occupied with special and general relativity, there are none which examine Einstein's philosophical approach while researching unified theory.[69] Such a study may reveal new epistemological connections Einstein discovered within the context of the unified theories, otherwise not explicable within the system of his former philosophy.

Einstein's work on the unified theories is likewise characterized by a change in his philosophical stance. The thought process he underwent while working on the unified theories touches on several points.

Einstein's new epistemology

1. Einstein realized that fundamental theories conceivable as attempts at unification constitute important stages in the development of physics. The unification idea is thus construed as a necessary logical path toward unifying gravitation and electromagnetism. In his lecture in Leiden in 1920, Einstein mentioned the unification of the two fields as an idea one had to take seriously. It was the first time that Einstein talked about this idea in public, describing it as promising. He went on to emphasize the continuity of the unification idea in theoretical physics, pointing out that Faraday and Maxwell had also pursued it.

The next stage toward unification Einstein mentioned was the general theory of relativity, which drew "geometry, kinematics and gravitational theory" together within a unified system. A unification theory of both fields would transform "all of physics" into a "closed conceptual system" similar to the one formed by the general theory of relativity. Thus Einstein concluded:

> Naturally it would represent considerable progress if we succeeded in regarding the fields of gravitation and electromagnetism as a single unified structure (Einstein 1920, 14).

In this speech Einstein portrayed unification as a geometrized synthesis of the two fields, mentioning Weyl's theory in this connection as an unsatisfactory attempt. On the other hand, the field-matter dualism, which he expressed in the "aether-matter" or "space-matter opposition" was to be solved by the unification. But it was not until 1923 in his Gothenburg speech that geometrized unification took the front stage, with the special and general theories of relativity constituting intermediary steps along the way.

If the idea of unification featured in the speech at Leiden merely as a central theme of physics, it matured as a consistent logical approach in Einstein's articles from 1923 and from 1929. In his Nobel lecture, held on 11 July 1923 in Gothenburg, "Fundamental Ideas and Problems of the Theory of Relativity," Einstein viewed the concept of unification as a successful heuristic program in historical developments. He mentioned the special theory of relativity as a high point, since it "reconciled" mechanics with electrodynamics.

This speech already outlines the epistemological substance of the unification method. On the one hand, "it reduced the number of logically independent hypotheses." On the other hand, it necessitated "clarification of the fundamental concepts in epistemological terms" and drew together concepts like mass and energy (Einstein 1923a, 4).

The general theory of relativity Einstein considered to be the next highpoint along the way to unification, "combining the hitherto separate concepts of inertia and gravitation [...] into a logical unit." (Einstein 1929, 127).

As special relativity and general relativity had proved to be successful as unification attempts, one could feel confident continuing along this path of unification. Based on this past experience in constructing physical theories, further unification could be expected to follow as a logical consequence.

The fact that Einstein viewed relativity theory as a unification theory shows that he placed physics within a new framework of thought grounded on a new epistemology. It is obvious that at the time of his Gothenburg speech, Einstein was already perfectly aware of the significance of the new epistemological content of the unification program.

Much later, in a letter of mid-June 1952 to Carl Seelig, Einstein gave a keener formulation of the path to unification and described it as a programm of three stages. The unified field theories were the third stage in the development of theoretical physics which began with the special theory of relativity. The second stage was the general theory of relativity, which Einstein did not consider to be concluded as yet: "[It] does, however, allow considerable leeway for the theoretical representation of the electromagnetic field." (Seelig 1960, 141–142) Here, too, Einstein viewed the unified theories as a logical consequence and extension of the general theory of relativity.

2. In constructing and verifying physical theories, Einstein differentiated between empirical and epistemological criteria. The priority he gave to these criteria changed, however, in the period between 1919 and 1921.

In his letter of 4 December 1919 to Ehrenfest, Einstein drew a clear distinction between the empirical and the epistemological bases leading to the theory of relativity:

> I understand your difficulties in describing the development of relativity. They are simply due to the fact that you want to put the innovation of 1905 on an epistemological basis (the non-existence of the ether at rest) instead of on an empirical basis (equivalence of all inertial systems with regard to light).
> The epistemological need did not emerge until 1907. Why should relativity prevail with respect to uniform motion alone, especially in view of the fact that the equality of inertial and ponderable mass suggests an extension of the principle of relativity?[70]

Thus Einstein confirmed that epistemological reasons emerged in 1907 (besides empirical ones) to play a decisive role in the construction of physical theories (the general theory of relativity). However, in the same letter, in connection with his rejection of Weyl's theory, Einstein emphasized the empirical criterion and considered it more important than epistemological considerations:

> I simply cannot comprehend why Weyl himself and everyone else do not realize to what an enormous extent the underlying idea of his theory contradicts experience.[71]

The about-face which occurred in Einstein's verification criteria while evaluating Weyl's theory is remarkable. He now accepted Weyl's theory even though the theory contradicted empirical fact, and in March 1921 proposed a further development of it. This evidently arose out of basic epistemological reasoning which had changed in the interim. "Mathematical simplicity," which we will discuss in the next section, played a central role in this.

3. Having considered mathematics a mere tool, devoid of any epistemological purpose in physical research, Einstein realized between 1919 and 1921 that its role was more profound than he had thought. If he meant to work on geometrized unification, Einstein had first to accept a new epistemological criterion in theoretical verification: the criterion of "mathematical simplicity."

In his letter of 15 December 1917 to Felix Klein, Einstein expressed his doubts about applying mathematical formalism as an epistemological method:

> It does seem to me that you are very much overestimating the value of purely formal approaches. The latter are certainly valuable when it is a question of formulating definitively an already established truth, but they almost always fail as a heuristic aid.[72]

In his letter to Weyl of 16 December 1918 it is also quite clear that at that time Einstein was very skeptical about mathematical argument and that he did not view mathematics to be a source of important physical knowledge:

> Everyone I have spoken to talks about your theories with the deepest respect, from the *mathematical point of view*, and [...] I also admire it as an edifice of ideas. [...] genuine admiration but disbelief, those are my feelings toward the subject.[73]

This relationship to mathematics changed in the period between 1919 and 1921. In his *Nature* article Einstein already uses the terms "formal unity," with "formal" unambiguously referring to the mathematical structure of the theory. His change of mind comes out more clearly in his Nobel lecture of 11 July 1923. There Einstein laid out in detail the program of unifying gravitation and electromagnetism for the first time, describing the unification as "mathematically unified field theory." Thus Einstein indicated that unification was based on a mathematical structure:

> A mathematically unified field theory is being sought in which the gravitational field and the electromagnetic field are interpreted merely as different components or manifestations of the same unified field (Einstein 1923a, 9).

Einstein also went into what the logical structure of the unified field should look like: The field equations should not consist of "logically independent summands."[74] Whereas Weyl's theory contained these sums, thus not satisfying Einstein's requirement, Kaluza's theory achieved unification of both fields by means of a unique field tensor: "A unique potential tensor generates a universal field which under ordinary conditions splits into a gravitational part and an electrical part." (Kaluza 1921, 971)

In the same speech Einstein elevated the principle of "mathematical simplicity" to a verification criterion of physical theories. Raising the argument that no empirical verification criterion existed, he emphasized:

> Unfortunately we are unable here to base ourselves on empirical facts as when deriving the gravitational theory (equality of the inertial and heavy mass) but we are restricted to the criterion of mathematical simplicity which is not free of arbitrariness (Einstein 1923a, 9).

When Einstein's former position on mathematical formalism is taken into account, then it becomes apparent that a major shift has occurred in his epistemological stance. This change did not arise *ad hoc*. It was the result of long and searching epistemological inner reflection before Einstein could allow himself to lay his trust in this criterion. A more thorough examination and analysis of his philosophical statements is required to reveal the significance and content Einstein assigned to the criterion of "mathematical simplicity" toward verifying a physical theory.

It was only in 1929 in his article "On the Present State of the Field Theories" that Einstein propounded the fruits of his reflections on the unification method. Einstein described the goals of a unified theory as follows:

> 1. To include, to the greatest extent possible, all phenomena and their relations (completeness).
> 2. to attain this using as few logically independent concepts and arbitrarily set relations between them (fundamental laws resp. axioms) as possible. I will call this goal that of "logical uniformity" (Einstein 1929, 126).

The concept of "logical uniformity" referred to the logical connection between terms and axioms. Did Einstein restrict this criterion to "the creation of connections between terms and axioms."?[75] In his article of 1938, he denied it:

> This aspect which is most difficult to formulate in precise terms, has always played an important role in the selection and judgement of theories. It is not simply a kind of counting of the logically independent premises (if such a counting were clearly possible at all) but a kind of reciprocal weighing up of the advantages and disadvantages of incommensurable qualities (Einstein 1938, 8).

In his 1929 article, other terms also appear, such as "logical simplicity," "logical unity," "logical uniformity" and "formal simplicity." These terms all seem to have a similar meaning and Einstein did not really differentiate among them. These concepts—as well as the concept of "mathematical simplicity"— are not mathematically definable (there are no mathematical criteria, laws or axioms for verifying the mathematical simplicity of a system of equations). Einstein's choice of words in describing them reveals how much this concept is beyond the semantic limits of mathematics, and underscores his conviction in its importance in the verification of a theory.[76]

This is the *promethean element* of scientific experience, which is encapsulated in the above expression "logical unity." For me, this is where the genuine *enchantment of scientific pondering* is always to be found; it is, so to speak, the *religious basis* of scientific thought; it is, so to speak, the *religious basis* of scientific effort."[77]

Indeed, what role did "logical simplicity" play in Einstein's research methods? In his article "Autobiographical notes"[78] in which he presented his "epistemological credo," he analysed a decisive component of his scientific method which is based on the correspondence between theory and experience:

Concepts and axioms acquire "meaning" or "content" only as they relate to sense experiences. The connection between the latter and the former is purely intuitive, not even of a logical nature. The degree of certainty with which this relationship resp. intuitive association can be drawn—and nothing else—distinguishes between figments of imagination from scientific truth' (Einstein 1938, 4).

The criterion of "logical simplicity" goes a step further than logical relations: It must secure reasonable connections to sensory perceptions, which are "purely intuitively" drawn. In doing so, it must offer a high degree of certainty that the connection between concepts and axioms on the one hand and sense-experiences on the other is correct.

This should explain why the concept of "logical uniformity" appeared to Einstein to be so important in constructing theories which were still empirically unverifiable. Particulary for such theories, the certainty that the connection to experience has been satisfactorily established must be guaranteed by a criterion based on a mathematical internal structure of the theory.

In a later letter to Solovine[79] Einstein retrospectively described in detail his epistemological method of research.[80] A statement in this letter shows that Einstein was able to conclude from the certainty of a given "correspondence" between the axioms (A) of a theory and actual experience (E), that the "logical machinery" for "comprehending reality" is reliable—under special circumstances, namely, when the theory exhibits internal perfection:

This relationship of S to E is, however (more pragmatically) far less uncertain than the relationship of A to E.[81] Were such correspondence not attainable with great certainty, (although not logically graspable), then the logical machinery for "the comprehensibility of reality" would be completely worthless."[82]

It is within this context that the criterion of "mathematical simplicity" and "logical unity" finds its place. It is a criterion which makes it possible to pre-verify whether a theory has excellent prospects of attaining agreement with experience. There are indications elsewhere as well among Einstein's writings from this period about the profound transformation his epistemology underwent. Let us look at his statements about the "speculative method" in constructing physical theories. In a letter to Besso of 28 August 1918, Einstein warned against embarking on the "speculative path:" "A genuinely usable and profound theory has never really been found through pure speculation."[83] By 1929, however, he no longer saw this as a disadvantage:

> The success of this attempt to derive subtle natural laws from the conviction of the formal simplicity of the structure of reality through pure thinking encourages us to continue along this speculative path (the dangers of which anyone who dares to take this path must always be aware of [84] (Einstein 1929, 127).

It is, in fact, this path which can now enable the physicist to construct the theory capable of fulfilling the criterion of empirical verification:

> Experience, so it seems, does not provide us with a clue as to how to solve this problem [of the unification of the fields of gravitation and electromagnetism]; but we may hope that the results of a finished theory developed by speculative means will also include such results as allow verification through experience (Einstein 1929, 128).

Einstein did not reject the empirical verification criterion but was convinced that in the case of the unified theories a different criterion—that of mathematical simplicity—had priority.

Einstein's new epistemological conception crystallized during the next few years. In his Herbert Spencer Lecture, "On the method of theoretical physics" held in Oxford on 10 June 1933 Einstein emphasized that mathematics played an epistemological role in the construction of physical theories:

> Our experience hitherto justifies us in believing that nature is the realization of the simplest conceivable mathematical ideas. I am convinced that we can discover by means of purely mathematical constructions the concepts and the laws connecting them with each other, which furnish the key to the understanding of natural phenomena (Einstein 1933, 274).

Einstein then underlined the creative role mathematics plays in the construction of physical theories:

> Experience remains, of course, the sole criterion of the physical utility of a mathematical construction. But the creative principle resides in mathematics (Einstein 1933, 274).

If one compares Einstein's statement about the role of mathematics before 1919 with this admission, one observes a clear transformation in his stance. Based on his experience in constructing physical theories he has come to the conviction that mathematics certainly does play an epistemological role.

In his letter of 15 February 1954 to Louis de Broglie, Einstein retrospectively summarized the epistemological transformation that he had to undergo while working on unified theory. Einstein specifically recognized that a "purely formal principle" had played a decisive role in his development of the general theory of relativity. This physico-mathematical principle (of general covariance) was based on a broader episte-mological principle that had accompanied his research: the principle of greatest logical simplicity in the laws of nature:

> The gravitational equations could only be found on the basis of a purely for-mal principle (general covariance) i.e., on the basis of the belief in the greatest conceivable logical simplicity of natural laws.[85]

Furthermore Einstein called the path to geometrized unified field theories "the logical way," which had to be thought through to the end. He had recognized that the gravitation theory was only a initial step along the path to unification:

> Since it was clear that the theory of gravitation was just a first step towards finding the simplest possible general field laws, it seemed to me that this logi-cal way must first be thought through to the end before one could hope to find a solution to the quantum problem.[86]

On his quest for unification, Einstein also realized that geometrical unified the-ory was an extension of the general theory of relativity, unlike the quantum problem, which could not be reconciled with the field concept. This realization convinced him to concentrate first on unified theory and then to solve the quantum problem within this context.

In this section we saw that Einstein came to this conclusion while working on the unified theory in the period between 1919 and 1921, when he identified the principle of "logical simplicity" as a criterion for verifying a theory. In his letter to Louis de Broglie he concluded: "That is how I became a fanatical believer in the method of logical simplicity."[87]

Einstein's epistemological transformation in the course of constructing unified the-ory proved to be so complex that one might speak of a radical "epistemological adap-tation." It gave him the philosophical grounding upon which he could accommodate the physical findings.

How did Kaluza's theory influence Einstein's epistemology?

Weyl's theory had been unacceptable to Einstein between 1918 and 1920 because it did not agree with known physical facts. Kaluza's theory, by contrast, did not violate experience. Thus it was the first unified theory that he took seriously. Its "formal sim-plicity" impressed him and showed him *how* the idea of unification could be realized as an extension of the general theory of relativity. Kaluza's theory fit best within this conceptual framework because, unlike the theories of Weyl and Eddington, it retained the Riemannian metric of the general relativity. Kaluza's unification was achieved with great mathematical simplicity. This may well have convinced Einstein of the useful-ness of new criteria for verifying the physical theories, namely that of logical unity and mathematical simplicity.

In the period 1919–1921 he seemed to have recognized in Kaluza's theory "the greatest conceivable logical simplicity of the natural laws." This is the only way to ex-plain the fact that, despite his conviction that Weyl's theory was wrong, Einstein, from this point on, no longer seemed to consider the thought of a unification of gravitation and electromagnetism unacceptable. Thus it becomes understandable why Einstein stopped rejecting Weyl's theory in March 1921, turning his attention instead toward improving it. The signficance of its lack of agreement with empirical findings receded into the background. The criterion of "mathematical simplicity" was elevated to a means of verifying newly postulated theories.

The formal perfection of Kaluza's theory was closely bound with the realization of the unification of both fields which were fused into a unique "universal field tensor." The mathematical content of Kaluza's theory was no longer empty formalism but came closer to reality.[88] As shown in the previous chapter, in 1923, Einstein used the fusing of the two fields to one unique field tensor also for the structure of his extension of the Weyl–Eddington theory.[89] This also inspired Einstein's appraisal in his letter of 6 June 1922 to Weyl, where he attributed more reality to Kaluza's theory than to Weyl's and Eddington's:

> His [Kaluza's theory] smells more like reality even though he too fails to provide the singularity-free electron.[90]

Einstein emphasized the conceptual advantage of Kaluza's theory later again, in a letter of 22 September 1932 to Abraham Fraenkel:

> I too hold Kaluza in great esteem. His idea of interpreting the electromagnetic field in the framework of general relativity is the basis of all later attempts to work out a uniform interpretation of the field.[91]

Here Einstein conveys what influence Kaluza's idea had on him. Kaluza's conceptual interpretation of the electromagnetic field was the basis for all later unification attempts. It seems very credible that Kaluza's idea was — besides a variety of other factors — what primarily convinced Einstein of the significance and the epistemological value of the unification idea. Einstein stressed his confidence in Kaluza's theory many times in his correspondence. To Lorentz he wrote on 16 February 1927: "It turns out that the unification of gravitation and Maxwell's theory is achieved to complete satisfaction by the 5-dimensional theory (Kaluza–Klein–Fock)."[92]

Above and beyond its epistemological significance, Einstein also considered Kaluza's theory a serious possibility for achieving unification. Einstein wrote no fewer than eight articles on extensions of it. Since publishing his joint paper with Grommer in 1922, Einstein continued to work on the five-dimensional theory during several periods: in 1927 when hopes were high among physicists of establishing deep connections between the fifth dimension and quantum mechanics; in 1931–1932 together with Walther Mayer, implementing the projective formalism developed in the 1930s by Veblen and Hoffman; and finally, from 1938 to 1943 together with Peter Bergmann, Valentin Bargmann and Wolfgang Pauli.

Einstein's objections to Kaluza's theory were related to two issues which he found problematical: 1. The physical meaning of the fifth dimension and 2. the confrontation with the atomistic structure of matter against which all field theories failed.

1. Einstein considered the hypothesis of the fifth dimension "arbitrary" (Einstein 1936, 312). In 1938, in the article he wrote with Bergmann he emphasized: "This

shows distinctly how vividly our physical intuition resists the introduction of the fifth dimension" (Einstein and Bergmann 1938, 688). This is also the only article in which Einstein argues in favor of assigning a physical meaning to the fifth dimension: "Furthermore it is much more satisfactory to introduce the fifth dimension not only formally, but to assign to it some physical meaning (Einstein and Bergmann 1938, 696).

2. The second point concerned the incompatibility of field theories with the particle concept. In the Herbert Spencer lecture of 1933, Einstein mentioned the decisive difficulty of the unified field theories "in the understanding of the atomistic structure of matter and energy" (Einstein 1953, 155).

In 1943 Einstein finally dropped the concept of the fifth dimension after having shown in his paper with Pauli that the particle concept is not compatible with the five-dimensional field theory (Einstein and Pauli 1943).

15.4 Concluding Remarks

This article argues that the influence Kaluza's theory had on Einstein was mainly epistemological. The resulting profound change, between 1919 and 1921, in Einstein's epistemology led him to regard geometrized unified field theories as the most promising route toward progress in theoretical physics. Thus his program in unified field theory emerged. His new epistemology was based on the concept of mathematical simplicity, a pivotal criterion in the construction of physical theories. Its purpose is to secure agreement between theory and experience.

Above and beyond the epistemological components, Einstein considered Kaluza's theory a serious contender for achieving real physical unification of electromagnetism and gravitation. In this conviction, he labored on constructing theories based on Kaluza's model until 1943. Einstein's main objection to Kaluza's unifying concept seems to have been the nonexistence of a fifth dimension—an objection which, owing to the continuing impossibility of empirical proof, still stands today.

References

Bergia, S. (1987). Explication d'une analogie formelle entre la théorie einsteinienne de la gravitation pour les champs stationnaires et la theorie unitaire de Kaluza–Klein. *Annales de la Fondation Louis de Broglie*, **12**, 349–362.

Dongen, Jeroen van (2002). Einstein and the Kaluza-Klein Particle. *Studies in History and Philosophy of Science B*, **33**, 185–210.

Einstein, Albert (1919). Spielen Gravitationsfelder im Aufbau der materiellen Elementarteilchen eine wesentliche Rolle? *Königlich Preussische Akademie der Wissenschaften* (Berlin). *Sitzungsberichte*, 349–356.

— (1920). *Äther und Relativitäts-Theorie*. Julius Springer, Berlin.

— (1921a). *Geometrie und Erfahrung*. Julius Springer, Berlin.

— (1921b). A Brief Outline of the Development of the Theory of Relativity, *Nature* **17**, Feb. 1921.

— (1921c). Über eine naheliegende Ergänzung des Fundamentes der allgemeinen Relativitätstheorie. *Königlich Preussische Akademie der Wissenschaften* (Berlin). *Sitzungsberichte*, 261–264.

— (1922). *The Meaning of Relativity*, 6th ed. Methuen, London.

— (1923a). Grundgedanken und Probleme der Relativitätstheorie. Vortrag gehalten an der Nordischen Naturforscherversammlung in Gotenburg den 11 Juli 1923. In: *Les Prix Nobel en 1921–1923*, Imprimerie Royale, P.A. Norstedt & Söner, Stockholm, 1–10.

— (1923b). Zur allgemeinen Relativitätstheorie. Königlich Preussische Akademie der Wissenschaften (Berlin). *Sitzungsberichte*, 32–38.

— (1923c). Bemerkung zu meiner Arbeit Zur allgemeinen Relativitätstheorie'. *Königlich Preussische Akademie der Wissenschaften* (Berlin). *Sitzungsberichte*, 76–77.

— (1923d). Zur affinen Feldtheorie. *Königlich Preussische Akademie der Wissenschaften* (Berlin). *Sitzungsberichte*, 137–140.

— (1923e). Bietet die Feldtheorie Möglichkeiten für die Lösung des Quantenproblems? *Königlich Preussische Akademie der Wissenschaften* (Berlin). *Sitzungsberichte*, 359–364.

— (1929). Über den gegenwärtigen Stand der Feld-Theorie. In Festschrift für Prof. Dr. A. Stodola zum 60. Geburtstag. Orell Füssli, Zürich and Leipzig, 126–132.

— (1933). On the Method of Theoretical Physics. The Herbert Spencer lecture, delivered at Oxford, June 10, 1933. In *Albert Einstein. Ideas and Opinions*. Three Rivers Press, New York, 1982, 270–276.

— (1936). Physics and Reality. In *Albert Einstein. Ideas and Opinions*. Three Rivers Press, New York, 1982, 290–323.

— (1938). Autobiographisches. In *Albert Einstein als Philosoph und Naturforscher*, Paul Arthur Schilpp, ed., W. Kohlhammer, Braunschweig and Wiesbaden, 1979.

— (1953). *Mein Weltbild*. Carl Seelig, ed. 2nd ed. Zürich, Europa, Stuttgart and Wien.

— (1956). *Briefe an Maurice Solovine*. Gauthier-Villars, Paris.

— (1998a). *The Collected Papers of Albert Einstein*, vol. 8A, Robert Schulmann, A.J. Kox, Michael Janssen and Jozsef Illy, eds. Princeton University Press, Princeton, N.J.

— (1998b). *The Collected Papers of Albert Einstein*, vol. 8B, Robert Schulmann, A.J. Kox, Michael Janssen and Jozsef Illy eds., Princeton University Press, Princeton, N.J.

— (1998c). *The Collected Papers of Albert Einstein. The Berlin Years: Correspondence, 1914–1918*; transl. by Ann M. Hentschel, vol. 8. Robert Schulmann, A.J. Kox, Michael Janssen and Jozsef Illy eds. Princeton University Press, Princeton, N.J.

Einstein, Albert and Bergmann, Peter (1938). On a Generalisation of Kaluza's Theory of Electricity. *Annals of Mathematics* **39** , 683–701.

Einstein, Albert and Pauli, Wolfgang (1943). On the Non-Existence of Regular Stationary Solutions of relativistic Field Equations. *Annals of Mathematics* **44**, 131–137.

Einstein, Albert and Grommer, Jakob (1923). Beweis der Nichtexistenz eines überall regulären zentrisch symmetrischen Feldes nach der Feld-Theorie von Th. Kaluza. *Scripta Universitatis Atque Bibliothecae Hierosolymitanarum*, 1–5.

Einstein, Albert and Besso, Michele (1972): *Correspondance. 1903–1955*. Hermann, Paris.

Einstein, Albert and Born, Max (1991). *Briefwechsel 1916–1955*. 2nd ed., Nymphenburger, München.

Elkana, Yehuda (1982). The Myth of Simplicity. In *Albert Einstein. Historical and Cultural Perspectives*. Gerald Holton and Yehuda Elkana ed., Princeton University Press, Princeton, N.J., 205–251.

Fölsing, Albrecht (1995). *Albert Einstein*. Suhrkamp, Frankfurt am Main.

Friedman, Michael (2001). *Dynamics of Reason*. CSLI Publications, Stanford.

Hermann, Armin, ed. (1968). Albert Einstein-Arnold Sommerfeld. Briefwechsel. Schwabe, Basel and Stuttgart.

Holton, Gerald, (1970). Mach, Einstein, and the Search for Reality. In Ernst Mach, Physicist and Philosopher. R. S. Cohn, R. J. Seeger, eds., D. Reidel, Dodrecht, 165–199.

— (1979). Einsteins Methode zur Theorienbildung. In Albert Einstein. Sein Einfluß auf Physik, Philosophie und Politik. Peter C Aichelburg, Roman U Sexl, Hrsg. Braunschweig and Wiesbaden: Friedr. Vieweg & Sohn.

Kaluza, Theodor (1910). Zur Relativitätstheorie. (I). *Physikalische Zeitschrift* **11**, 977–978.

— (1921). Zum Unitätsproblem der Physik. *Königlich Preussische Akademie der Wissenschaften* (Berlin). *Sitzungsberichte*, 966–972.

— (1922). Über Bau und Energieinhalt der Atomkerne. *Physikalische Zeitschrift* **23**, 474–476.

— (1924). Zur Relativitätstheorie. (II). *Physikalische Zeitschrift* **25**, 604–606.

Kanitscheider, Bernulf (1979). Einsteins Behandlung theoretischer Größen. In *Albert Einstein. Sein Einfluß auf Physik, Philosophie und Politik*. Peter C. Aichelburg and Roman U. Sexl, ed., Friedr. Vieweg & Sohn, Braunschweig and Wiesbaden.

Mehra, Jagdish (1999). *Einstein, Physics and Reality*. World Scientific, Singapore, New Jersey and London.

Renn, Jürgen (1994). *The Third Way to General Relativity*. Preprint 9, Berlin: Max Planck Institut für die Wissenschaftsgeschichte.

Salam, Abdus (1980). Gauge Unification of Fundamental Forces. *Reviews of Modern Physics* **52**, 525–538.

Seelig, Carl (1960). *Albert Einstein, Leben und Werk eines Genies unserer Zeit*. 2nd ed., Europa, Zürich.

Stachel, John (1986). The Rigidly Rotating Disk as the Missing Link. In *Einstein and the History of General Relativity*, Don Howard and John Stachel eds., Birkhäuser, Boston, Berlin and Basel, 966–972.

Tonnelat, Marie-Antoinette (1959). *Les principes de la théorie électromagnétique et de la relativité*. Masson et Cie, Paris.

Vizgin, Vladimir P. (1994). *Unified Field Theories in the First Third of the 20th Century*. Birkhäuser, Basel, Boston, Berlin.

Weyl, Hermann (1918). Gravitation und Elektrizität. *Königlich Preussische Akademie der Wissenschaften* (Berlin). *Sitzungsberichte*, 465–473.

Wolters, Gereon (1987). *Mach I, Mach II, Einstein und die Relativitätstheorie*. Walter de Gruyter, Berlin and New York.

Wünsch, Daniela (2000): Theodor Kaluza. Dissertation, Stuttgart, Leben und Werk.

Wünsch, Daniela (2003): The fifth dimension: Theodor Kaluza's ground-breaking idea. *Annalen der Physik* **12**, 519–542.

UNPUBLISHED SOURCES

Weinberg, Steven: Kaluza–Klein-Theory and the Great Unified Theory. Lecture in Oxford, 1983.

Wünsch, Daniela (2003): Thinking about Unification: Einstein between 1919 and 1923, forthcoming.

ARCHIVES

Albert Einstein Archiv Jerusalem (EA).

Archives Louis de Broglie, Academie des Sciences de Paris (AB).

Notes

[1] Einstein to Kaluza on 21.4.1919, (EA 14-249.2).

[2] Salam 1980.

[3] Weinberg 1983.

[4] Wünsch 2000 and Wünsch 2003.

[5] For the historical references in the following section, see Wünsch 2000.

[6] In most of the papers on Kaluza and physics, his place of birth is erroneously given as Ratibor.

[7] In many papers, Kaluza is erroneously described as a Pole.

[8] Kaluza 1910.

[9] See also Stachel 1986.

[10] Kaluza 1921.

[11] Kaluza 1922.

[12] Kaluza 1924.

[13] Kaluza 1924, p. 602.

[14] See Einstein to Kaluza on 27.2.1925, (EA 65-731).

[15] See Wünsch 2000.

[16] Vizgin 1994.

[17] Vizgin 1994, p. 169.

[18] Vizgin 1994, pp. 167–168.

[19] For a detailed description see Vizgin 1994 and Wünsch 2003.

[20] Weyl 1918.

[21] Einstein to Dällenbach on 18.9.1918, in Einstein 1998b, p. 803.

[22] Einstein 1923e.

[23] Einstein 1919.

[24] Einstein 1919, p. 355.

[25] Einstein to Kaluza on 29.5.1919 (EA 65-728).

[26] In modern physics it was generalised for four forces and an eleven-dimensional space. Another advantage of the theory—which was only recognized later—was that the structure first introduced in his theory of a five-dimensional cylinder world, made it possible to connect Kaluza's theory with quantum mechanics, as was done in 1926 by Oskar Klein, by the compactification of the fifth dimension.

[27] Einstein to Kaluza on 21.4.1919, (EA 14-249.2).

[28] Ibid.

[29] Einstein to Kaluza on 5.5.1919, (EA 14-249.4).

[30] See Tonnelat 1959, p. 329.

[31] Einstein to Kaluza on 5.5.1919, (EA14-249.4).

[32] Einstein to Born on 27.1.1920, in Einstein and Born 1991, p. 42.

[33] Ibid.

[34] Einstein to Ehrenfest on 2.2.1920, (EA 9-470.2). Vizgin cited an excerpt from this letter as quoted in Seeling and translated it incorrectly by confusing "Überbestimmung" (over-determination) with "Übereinstimmung" (consistency). It is quite clear that here Einstein was referring to his theory of over-determination by differential equations, which was to solve the quantum problem. In Vizgin 1994, p. 169.

[35] Einstein to Ehrenfest on 9.3.1920, (EA 9-473).

[36] Einstein to Ehrenfest on 7.4.1920, (EA 9-477).

[37] Einstein to Ehrenfest on 26.11.1920, (EA 9-531).

[38] Einstein to Ehrenfest on 26.11.1920, (EA 9-531).

[39] See Vizgin 1994, p. 168.

[40] Einstein 1920.

[41] Einstein to Ehrenfest on 2.2.1920, (EA 9-469).

[42] Einstein 1921a.

[43] Einstein 1920, p. 149.

[44]Einstein 1921a, p. 10.

[45] Einstein 1921b, p. 784.

[46]Einstein 1921b, p. 784.

[47]Einstein to Ehrenfest on 1.3.1921, (EA 9-553).

[48]Einstein to Sommerfeld on 9.3.1921, in Hermann 1968, p. 78.

[49]Einstein 1922.

[50]Einstein 1922, p. 94.

[51]See Grommer to Einstein on 25.10.1921, (EA 11-411.1).

[52]Einstein to Kaluza on 14.10.1921, (EA 65-729).

[53]Einstein to Weyl on 5.9.1921, (EA 24-092).

[54]Einstein to Kaluza on 5.12.1921, (EA 65-730).

[55]Einstein and Grommer 1923, p. 1.

[56]Ibid.

[57]Ibid.

[58]Ibid.

[59]Ibid.

[60]Ibid, p. 2.

[61]Ibid, p. 3.

[62]Ibid, p. 5.

[63]Einstein 1923b, 1923 c, 1923d.

[64]Einstein to Ehrenfest on 20.6.1923, (EA 10-074).

[65]See van Dongen 2002.

[66]See Renn 1994; Wolters 1987; Kanitscheider 1979; Holton 1970, 1979 and Mehra 1999.

[67]See Renn 1994 and Mehra 1999.

[68]Einstein to Lanczos on 24.1.1938, cited in Fölsing 1995, p. 638.

[69]See Mehra 1999 and Holton 1970.

[70]Einstein to Ehrenfest on 4.12.1919, (EA 9-452.1).

[71]Ibid.

[72]Einstein to Felix Klein on 15.12.1917, in Einstein 1998c, p. 418.

[73]Einstein to Weyl on 16.12.1918, in Einstein 1998c, p. 711.

[74]Einstein, 1923a, p. 9.

[75]Einstein 1938, p. 4.

[76] See Elkana 1982.

[77] Einstein 1929, p. 127. Italics from D. Wünsch.

[78] Einstein 1938.

[79] Einstein to Solovine on 7.5.1952, in Einstein and Solovine 1956, pp. 119–121.

[80] For the analysis, see Holton 1979.

[81] With (S) Einstein was referring to predictions and other consequences of a theory.

[82] Einstein to Solovine on 7 May 1952 in Einstein 1956, p. 120.

[83] Einstein to Besso on 28.8.1918, in Einstein and Besso 1972, p. 1938.

[84] See on this also Weyl to Einstein on 18.5.1923, (AE 24-076.3).

[85] Ibid.

[86] Ibid.

[87] Ibid.

[88] There are, however, historians who see this influence between Kaluza's theory and Einstein's unification attempts also in a structural context. Bergia, for example, discerns a "formal analogy" between the five–dimensional theory of Kaluza and Einstein's four-dimensional unified theories in the case of stationary fields (Bergia 1987, 349–350).

[89] In answer to the legitimate question of what Eddington's role had been in Einstein's unification efforts, I would like to note that Einstein considered Eddington's theory as a mere extension of Weyl's theory. No comparable expressions of admiration for the theory can be found in their correspondence. Nor can one speak of any comparable epistemological influence on the part of Eddington as had been the case with Kaluza. See Einstein to Weyl on 23.6.1923, (Ea 24-081), Einstein to Holland on 25.6.46, EA (9-304.1), Einstein to Douglas in March 1953, EA (9-310.1).

[90] Einstein to Weyl on 6.6.1922, (EA 24-072.2).

[91] Einstein to Fraenkel on 22.9.1932, (EA 37-049).

[92] Einstein to Lorentz on 16.2.1927, (EA, 16-612).

16

Unified Field Theory: Early History and Interplay Between Mathematics and Physics

Hubert F. M. Goenner

University of Göttingen, Institute for Theoretical Physics, Bunsenstr. 9, 37073 Göttingen, Germany; goenner@theorie.physik.uni-goettingen.de

16.1 Introduction

The following is part of a future extended review;[1] I try to sketch, more or less chronologically, and by trailing Einstein's path, the history of attempts at unifying what are now called the *fundamental interactions* during the period from ca. 1915–1930. Until the 1940s the only known fundamental interactions were the electromagnetic and the gravitational, plus, tentatively, something like the "mesonic" or "nuclear" interaction. The physical fields considered in the framework of "unified field theory" including, after the advent of quantum mechanics, the wave function satisfying either Schrödinger's or Dirac's equation, were all assumed to be *classical* fields. Due to the slow acceptance of the statistical interpretation of quantum mechanics by those working in or close to general relativity, the quantum mechanical wave function often was taken to represent a matter field in space time, the field of the electron.[2] For us in the 21st century, the concept of "unified field theory" extends into two directions: (1) the inclusion of the weak and strong interactions and, (2) the necessary approach to unification in the framework of *quantum field theory*.

We must also remember that the unified field theory of the period considered here included two aspects closely related, i.e.,

- a geometrized theory of matter in the sense of a removal of the energy-momentum tensor of matter in favour of intrinsic *geometrical* structures. In the period discussed, the representation of matter oscillated between the point-particle concept (particles as singularities of a field) to everywhere regular fields (solitonic type solutions).
- the development of a unified field theory *more geometrico* for electromagnetism and gravitation.

General relativity's doing away with *forces* in exchange for a richer (and more complicated) geometry of space and time than the Euclidean remained the guiding principle throughout. Of course, in today's unified field theories which appear in the form of *gauge* theories, the dichotomy between matter and fields in the sense of a dualism is minimized. Matter is represented by operator-valued spin half quantum fields (fermions) while the "forces" mediated by "exchange particles" are embodied in quantum fields of integer spin (bosons).

In the following, a multitude of geometrical concepts (affine, conformal, projective spaces etc.) available for unified field theories, on the one side, and their use as tools for a description of the dynamics of the electromagnetic and gravitational field on the other, will be used. Then, we look at the very first steps towards a unified field theory taken by Foerster (alias Bach), Weyl, Eddington and Einstein. This includes Weyl's generalization of Riemannian geometry to a conformal one and the subsequent extensions of Riemannian to affine geometry by Eddington, Einstein and others. Einstein's treatment of a special case, distant parallelism, set off an avalanche of research papers. To this, Kaluza's idea concerning a geometrization of the electromagnetic and gravitational fields within a five-dimensional manifold will be added. Attention is also given to the mutual influence exerted on each other by the Princeton (Eisenhart, Veblen), French (E. Cartan), and the Dutch (Schouten, Struik) schools of mathematicians, and the work of physicists such as Eddington, Einstein, their collaborators, and others on the basis of their published papers and books.[3]

16.2 The possibilities of generalizing general relativity: a brief overview

Of the main avenues extending general relativity, I shall follow the generalization of geometry and of the number of dimensions. Within the first, the two fundamental independent structural objects are a metric (first fundamental form) and the connection, a device for establishing the comparison of vectors in different points of a manifold. The various possibilities for affine spaces with symmetric or asymmetric *connection*, metric-affine spaces with, in addition, symmetric or asymmetric *metric* (mixed geometry) will appear. In affine geometry, the metric is derived from the connection, while in Riemannian geometry the connection is derived from the metric.

Within a particular geometry, usually various options for the dynamics (field equations; in particular as following from a Lagrangian) exist as well as different possibilities for the identification of physical observables with the mathematical objects of the formalism. These identifications were made on internal, structural reasons as no link-up to empirical data was possible. As an example, we take the identification of the electromagnetic field tensor with either the skew part of the metric, in a "mixed geometry," or the skew part of the Ricci tensor in metric-affine theory, to list only two possibilities. The latter choice obtains likewise in a purely affine theory in which the metric is a derived secondary concept. In this case, among the many possible choices for the metric, one may take it proportional to the variational derivative of the Lagrangian with respect to the symmetric part of the Ricci tensor. This does neither guarantee the proper signature of the metric nor its full rank. Several identifications for the electromagnetic 4-potential and the electric current vector density have also been suggested (Cf. Goenner 1984).

16.3 Early attempts at a unified theory

16.3.1 First steps in the development of unified field theories.

Even before (or simultaneously with) the generalization of the concept of covari-ant derivative by Hessenberg (1917), Levi-Civita (1917), Schouten (1918), Weyl (1918), and König (1919), the introduction of an *asymmetric* metric was suggested (Rudolf Förster 1917).[4] In his letter to Einstein of 11. 11. 1917, he writes "Vielleicht findet sich ein kovarianter Sechservektor der das Auftreten der Elektrizität erklärt und ungezwungen aus den $g_{\mu\nu}$ herauskommt, nicht als fremdes Element herangetragen wird." (ECP **8A**, Doc. 398, 552)[5]. Einstein replied: " Das Ziel, Gravitation und Elek-tromagnetismus einheitlich zu behandeln, indem man beide Phänomengruppen auf die $g_{\mu\nu}$ zurückführt, hat mir schon viele erfolglose Bemühungen gekostet. Vielleicht sind Sie glücklicher im Suchen. Ich bin fest überzeugt, dass letzten Endes alle Feldgrössen sich als wesensgleich herausstellen werden. Aber leichter ist ahnen als finden." (16. 11. 1917, ECP **8A**, Doc. 400, 557)[6]. In his next letter, Förster gave results of his calculations with an asymmetric $g_{\mu\nu} = s_{\mu\nu} + a_{\mu\nu}$, introduced an asymmetric "three-index-symbol" and a possible generalization of the Riemannian curvature tensor as well as tentative Maxwell's equations and interpretations for the 4-potential A_μ, and special solutions. (28. 12. 1917, ECP **8A**, Doc. 420, 581–587). Einstein's answer is skeptical: "Das Ausgehen von einem nichtsymmetrischen $g_{\mu\nu}$ hat mich auch schon lange beschäftigt; ich habe aber die Hoffnung aufgegeben, auf diese Weise hinter das Geheimnis der Einheit (Gravitation, Elektromagnetismus) zu kommen. Verschiedene Gründe flössen da schwere Bedenken ein: [...] Ihre übrigen Bemerkungen sind eben-falls an sich interessant und mir neu."[7] (17. 1. 1918, ECP **8B**, Doc. 439, 610-611.)

Einstein's remarks concerning his previous efforts must be seen under the aspect of the attempts at formulating a unified field theory including matter by Gustav Mie, and of the unified field theory of gravitation and electromagnetism proposed by David Hilbert. "In Folge eines allgem. math. Satzes erscheinen die elektrody. Gl. (verallge-meinerte Maxwellsche) als math. Folge der Gravitationsgl., so dass Gravitation und Elektrodynamik eigentlich garnicht verschiedenes sind."[8] Cf. Letter of D. H. to A. E. of 13 November 1915 (ECP **8A**, Doc. 140, 195.) The result is contained in (Hilbert 1915, p. 397).[9]

Einstein's answer to Hilbert shows that he had been also busy along such lines: "Ihre Untersuchung interessiert mich gewaltig, zumal ich mir oft schon das Gehirn zer-martert habe, um eine Brücke zwischen Gravitation und Elektromagnetik zu schlagen. Die Andeutungen, welche Sie auf Ihren Karten geben, lassen das Grösste erwarten."[10] (15. 11. 1915, ECP **8A**, Doc. 144, 199.)

Even before Förster/Bach corresponded with Einstein, an early bird in the attempt at unifying gravitation and electromagnetism had published two papers in 1917: Ernst Reichenbächer, a mathematics teacher at a Gymnasium in Northern Germany (Re-ichenbächer 1917a,b). His paper amounts to a scalar theory of gravitation with field equation $R = 0$ instead of Einstein's $R_{ab} = 0$ outside the electrons. The electron is considered as an extended volume in the sense of Lorentz–Poincaré and described by a metric joined continuously to the outside metric.

$$ds^2 = dr^2 + r^2\,d\phi^2 + r^2\,cos^2\phi\,d\psi^2 + (1 - \alpha/r)^2\,dx_0^2.^{11} \qquad (16.1)$$

Reichenbächer has only a limited understanding of general relativity. I agree with Weyl's remark in Raum-Zeit-Materie (Weyl 1919, p. 267, footnote 30), i.e., that his reasoning is hard to understand.

We must also keep in mind that the generalization of the metric tensor toward asymmetry or complex values was more or less synchroneous with the development of Finsler geometry (Finsler 1918). In fact, also Finsler geometry was also used in attempts at unifying gravitation and electromagnetism (Reichenbächer 1926, Newman 1927).

16.3.2 Early disagreement about program of explaining elementary particles by field theory

As to the program for building the constituents of matter from the fields the source of which they are, Pauli's remark after Weyl's lecture in Bad Nauheim (86. Natur-forscherversammlung, 19.-25. 9. 1920, Pauli 1920) showed that not everybody believed in it. He held that in bodies smaller than those carrying the elementary charge (electrons), an electric field could not be measured. There was no point in creating the "interior" of such bodies with the help of an electric field. Einstein's answer is tentative and evasive: we just don't know yet:

"Mit fortschreitender Verfeinerung des wissenschaftlichen Begriffssystems wird die Art und Weise der Zuordnung der Begriffe von den Erlebnissen immer komplizierter. Hat man in einem gewissen Studium der Wissenschaft gesehen, dass einem Begriff ein bestimmtes Erlebnis nicht mehr zugeordnet werden kann, so hat man die Wahl, ob man den Begriff fallen lassen oder ihn beibehalten will; in letzterem Fall ist man aber gezwungen, das System der Zuordnung der Begriffe zu den Erlebnissen durch ein komplizierteres zu ersetzen. Vor dieser Alternative sind wir auch hinsichtlich der Begriffe der zeitlichen und räumlichen Entfernung gestellt. Die Antwort kann nach meiner Ansicht nur nach Zweckmässigkeitsgründen gegeben werden; wie sie ausfallen wird, erscheint mir zweifelhaft.[12]

In the same discussion, Gustav Mie came back to Förster's idea of an asymmetric metric but did not like it: " [...] dass man dem symmetrischen Tensor des Gravitationspotentials einen antisymmetrischen Tensor hinzufügte, der den Sechservektor des elektromagnetischen Feldes repräsentierte. Aber eine genauere Überlegung zeigt, dass man so zu keiner vernünftigen Weltfunktion kommt."[13]

Since the twenties Einstein had changed his mind; he now looked for solutions of his field equations which were everywhere *regular* to represent matter particles.

16.4 Mathematicians and physicists at work ca 1915–1933

16.4.1 The unification game: Competition among ideas in 1918–1923

As we are aware, after 1915 Einstein was first busy with extracting important consequences from general relativity such as e.g., cosmology and gravitational waves, gravitational radiation, and in following up the mathematical and physical consequences of Weyl's theory. Although he kept thinking about gravitation and elementary particles (Einstein 1919) and looked closer into Weyl's theory (Einstein 1921), he only *reacted* to the new ideas concerning unified field theory as advanced by others. The first such idea after Förster's, of course, was Weyl's conformal approach to gravitation and electromagnetism, unacceptable to Einstein and to Pauli for physical reasons.[14] Next came Kaluza's 5-dimensional unification of gravitation and electromagnetism.

Kaluza's idea of using four spatial and one time dimension originated in or before 1919; by then he had communicated it to Einstein. Cf. the letter of Einstein to Kaluza of April 21, 1919: "The idea of achieving [a unified field theory] by means of a five-dimensional cylinder world never dawned on me.[...] At first glance I like your idea enormously." This remark is surprising because Nordström had suggested a five-dimensional unification of his scalar gravitational theory with electromagnetism five years earlier (Nordström 1914) by embedding space-time into a five-dimensional world in quite the same way as Kaluza did. Einstein should have known Nordström's work. In the same year, 1914, he and Fokker had given a covariant formulation of Nordström's pure (scalar) theory of gravitation (Einstein and Fokker 1914). In a subsequent letter to Kaluza of May 5, 1919, Einstein still was impressed: "The formal unity of your theory is startling." Kaluza's paper was communicated by Einstein to the Prussian Academy, but for unknown reasons was published only in 1921. (Kaluza 1921; cf. Pais 1982, p. 330.)

The third main idea which emerged was Eddington's suggestion to forgo the metric as a fundamental concept and start right away with a (general) connection which he then restricted to a symmetric one (Eddington 1921, p. 104). His motivation went toward a theory of matter: "In passing beyond Euclidean geometry, gravitation makes its appearance; in passing beyond Riemannian geometry, electromagnetic force appears; what remains to be gained by further generalisation? Clearly, the non-Maxwellian binding forces which hold together an electron. But the problem of the electron must be difficult, and I cannot say whether the present generalisation succeeds in providing the material for its solution." Eddington's main goal was to include matter as an inherent geometrical structure: "What we have sought is not the geometry of actual space and time, but the geometry of the world-structure which is the common basis of space and time and things." (Eddington 1921, p. 121) Eddington's publication early in 1921 generalizing on Einstein's and Weyl's theories, started a new direction of research both in physics and mathematics. At first, Einstein was attracted by his idea and tried to make it work as a physical theory (Eddington had not given field equations):

"I must absolutely publish since Eddington's idea must be thought through to the end."

(Letter of A. Einstein to H. Weyl of May 23, 1923; cf. Pais 1982, p. 343.) And indeed Einstein published fast, even while still on the steamer returning from Japan: the paper of February 1923 in Sitzungsberichte carries, as location of the sender, the ship "Haruna Maru." In two publications following the letters to Weyl (in May and June), Einstein elaborated on the theory, but found it unsatisfactory, because, although the theory for every solution with *positive* charge offered also a solution with *negative* charge, the masses in the two cases were the same. At the time, however, the known particle with positive charge (and what is now the proton) had a mass *greatly different* from the particle with negative charge, the electron. (Einstein 1923a,b)

While, in the meantime, mathematicians had taken over the conceptual development of affine theory, some other physicists including the perpetual "pièce de resistance" Pauli kept a negative attitude (W. Pauli to Eddington, September 20, 1923 (W. Pauli 1979, 115–119):

> "[...] Die Grössen $\Gamma^{\mu}_{\nu\,\alpha}$ können nicht direkt gemessen werden, sondern müssen aus den direkt gemessenen Grössen erst durch komplizierte Rechenoperationen gewonnen werden. Niemand kann empirisch einen affinen Zusammenhang zwischen Vektoren in benachbarten Punkten feststellen, wenn er nicht vorher bereits das Linienelement ermittelt hat. Deswegen halte ich im Gegensatz zu Ihnen und Einstein die Erfindung der Mathematiker, dass man auch ohne Linienelement auf einen affinen Zusammenhang eine Geometrie gründen kann, zunächst für die Physik bedeutungslos."[15]

Also Weyl, in the 5th edition of RZM, appendix 4, in discussing "world-geometric extensions of Einstein's theory," found Eddington's theory not convincing. He critizised a theory which keeps only the connection as a fundamental building block for its lack of a guarantee that it would also house the *conformal* structure (light cone structure). This is needed for special relativity to be incorporated in some sense.

16.4.2 Differential geometry's high tide

In the introduction to his book, Dirk Struik distinguishes three directions in the development of the theory of linear connections (Struik 1934):
(1) The generalization of parallel transport in the sense of Levi-Civita and Weyl. Schouten is the leading figure in this approach (Schouten 1924).
(2) The "geometry of paths" considering the lines of constant direction for a connection — with the proponents Veblen, Eisenhart and T. Y. Thomas. Here, only symmetric connections can appear.
(3) The idea of mapping a manifold at one point to a manifold at a neighboring point is central (affine, conformal, projective mappings). The names of König and Cartan are connected with this program.

In his assessment, Eisenhart (Eisenhart 1927) adds to this all the spaces whose metric is "based upon an integral whose integrand is homogeneous of the first degree in the differentials. Developments of this theory have been made by Finsler, Berwald, Synge and J. H. Taylor. In this geometry the paths are the shortest lines, and in that sense are a generalization of geodesics. Affine properties of these spaces are obtained from a natural generalization of the definition of Levi-Civita for Riemannian spaces."(Eisenhart 1927, p. v.)

In fact, already in 1922 Jan Arnoldus Schouten in Delft had classified all possible connections in two papers (Schouten 1922a,b; paper a accepted 14. 5.1921). He lists 18 different linear connections and classifies them invariantly. The most general connection is characterized by two fields of third degree, one tensor field of second degree, and a vector field. These fields are the torsion tensor $S_{\lambda\mu}^{v}$, the tensor of nonmetricity $Q_{\lambda\mu}^{v}$, the metric, and a vector C_μ, which follows from $C_{\lambda\mu}^{v} = C_\mu \, \delta_v^\mu$ while $C_{\lambda\mu}^{v} = \Gamma_{\lambda\mu}^{v} + \Gamma_{\lambda\mu}^{\prime v}$, if Γ stands for the connection for tangent vectors and Γ' for the connection for linear forms. Torsion is defined by $S_{\lambda\mu}^{v} = 1/2(\Gamma_{\lambda\mu}^{v} - \Gamma_{\mu\lambda}^{v})$, non-metricity by $\nabla_\mu g^{\lambda v} = Q_\mu^{\lambda v}$. Furthermore, on page 57 we find: "The general connection for n = 4 at least theoretically opens the door for an extension of Weyl's theory. For such an extension an invariant affixation of the connection is needed, because a physical phenomenon can correspond only to an invariant expression."

At the end of the paper we find the confirmation that during the proofreading Schouten received Eddingtons paper (Eddington 1921, accepted 19. 2. 1921). Thus, while Einstein and Weyl influenced Eddington, Schouten apparently did his research without knowing of Eddingtons idea. Einstein, perhaps, got to know Schouten's work only later through the German translation of Eddingtons book where it is mentioned (Eddington 1925), or, directly, through Schouten's book on the Ricci calculus (Schouten 1924).[16] On the other hand, Einstein's papers following Eddington's (Einstein 1923a,b) induced Schouten to publish on a theory with vector torsion (Schouten 1923, Friedmann and Schouten 1924) which tried to remedy a problem Einstein had noted in his paper, i.e., that no electromagnetic field could be present in regions of vanishing electric current density.

A similar, but less general, classification of connections has also been given by Cartan. He relied on the curvature-, torsion- and homothetic curvature 2-forms. (Cartan 1923, Chap. III.)

Other mathematicians were also stimulated by Einstein's use of differential geometry in his general relativity and, particularly, by the idea of unified field theory. Examples are Luther Pfahler Eisenhart and O. Veblen, both at Princeton, who developed the "geometry of paths"[17] under the influence of papers by Weyl, Eddington and Einstein (Eisenhart and Veblen 1922, (Eisenhart 1922/23, Veblen and Thomas 1923). In Eisenhart's paper, we can read that

> "Einstein has said (in *Meaning of Relativity*) that 'a theory of relativity in which the gravitational field and the electromagnetic field enter as an essental unity' is desirable and recently has proposed such a theory." (p. 367–368) And " His geometry also is included in the one now proposed and it may be that the latter, because of its greater generality and adaptability will serve better as the basis for the mathematical formulation of the results of physical experiments." (p. 369)

16.4.3 What Einstein did with his collaborators

Mixed geometry

After Kaluza's paper had appeared, Einstein set his calculational aide Grommer to work on regular spherically symmetric solutions of this theory. This resulted in a joint publication (Einstein and Grommer 1923), the negative result of which (no regular, statical s.s.s. exist) led him to abandon Kaluza's idea for the time being.

Instead, in July 1925, Einstein modified Eddington's approach to the extent that he now took both a non-symmetric connection and a non-symmetric metric, i.e., dealt with a mixed geometry (Einstein 1925a):

> "[...] Auch von meiner in diesen Sitzungsberichten (Nr. 17, p. 137 1923) er-schienenen Abhandlung, welche ganz auf Eddingtons Grundgedanke basiert war, bin ich der Ansicht, dass sie die wahre Lösung des Problems nicht gibt. Nach unablässigem Suchen in den letzten zwei Jahren glaube ich nun die wahre Lösung gefunden zu haben."[18]

But also this novel approach did not convince him. Eisenhart commented on it and pointed to some difficulties: when identification of the components of the antisymmetric part of the metric with the electromagnetic field is made in first order "they are not the components of the curl of a vector as in the classical theory, unless an additional condition is added." (Eisenhart 1926, p. 129; communicated 16 Dec. 1925).

Due to its intrinsic difficulties — e.g. the condition of metric compatibility did not have the physical meaning of the conservation of the norm of and angle between vectors by parallel transport, and much of the formalism was hard to handle — essential work along this line was done only much later in the 1940s and 1950s (Einstein, Einstein and Strauss, Schrödinger, Lichnerowicz, Hlavaty, Tonnelat etc).

Kaluza's idea taken up again

Einstein became interested in Kaluza's theory again by Oskar Klein's paper concerning a relation between "quantum theory and relativity in five dimensions" (Klein 1926, received by the journal on 28 April 1926a). He wrote to Paul Ehrenfest on Aug. 23, 1926: "Subject Kaluza, Schrödinger, general relativity," and, again on Sept. 3, 1926: "Klein's paper is beautiful and impressive, but I find Kaluza's principle too unnatural." However, less than half a year later he had completely reversed his opinion: (Einstein to H. A. Lorentz, Feb. 16, 1927): "It appears that the union of gravitation and Maxwell's theory is achieved in a completely satisfactory way by the five-dimensional theory (Kaluza–Klein–Fock)." On the next day (17. Feb. 1927), and ten days later Einstein was to give papers of his own before the Prussian Academy in which the Einstein-Maxwell equations were derived exactly — not just in first order as Kaluza had done, but only after Klein had done the same (Einstein 1927).

Thus, concerning what now is known as Kaluza–Klein theory, Einstein himself acknowledged indirectly that his two notes in the Sitzungsberichte did not contain any new material. In his second communication, he says "Herr Mandel macht mich

darauf aufmerksam, dass die von mir hier mitgeteilten Ergebnisse nicht neu sind."
He then refers to papers of Klein (Klein 1926) and "Fochs Arbeit" which is (Fock
1926) submitted 3 months later than Klein's paper. [19] H. Mandel of Leningrad had
rediscovered some of Oskar Klein's results (Mandel 1926). This was acknowledged
by Klein in his second paper received on 22. 10. 1927 where he also gave further
references on work done in the meantime but remained silent on Einstein's papers
(Klein 1928).

Einstein did not comment on Klein's new idea of "dimensional reduction" as it
is now called and which justifies Klein's name in the "Kaluza–Klein" — theories
of our time. By this, the reduction of 5-dimensional equations — as e.g., the 5-
dimensional wave equation — to 4-dimensional equations by Fourier decomposition
with respect to the 5th coordinate x^5, taken as periodic with period L, is understood:
$\psi(x, x^5) = \frac{1}{\sqrt{L}} \Sigma_n \psi_n(x) e^{inx^5/L})$, n integer. The 5th dimension is assumed to be a
circle, topologically, and thus gets a finite linear scale. By adding to this the idea of de
Broglie waves, Klein brought in Planck's constant and determined this linear scale to
be unmeasurably small. From this, the possibility of "forgetting" the unobserved fifth
dimension arises.

Four years later, Einstein returned to Kaluza's idea in the form of a projective
four-dimensional theory (Einstein and Mayer 1931b). After this paper Einstein wrote
to Ehrenfest in a letter of Sept. 17, 1931 that this theory: "in my opinion definitively
solves the problem in the macroscopic domain." (Pais 1982, p. 333.) Now, Veblen
had worked on projective connections for a couple of years (Veblen 1928) and, with
his student Banesh Hoffmann, suggested an application to physics equivalent to the
Kaluza–Klein theory (Veblen and Hoffmann 1930, Hoffmann 1930). At about the
same time as Einstein and Mayer, van Dantzig worked on projective geometry (van
Dantzig 1932 a,b,c,d). Together with him, Schouten wrote a series of papers on projective geometry as the basis of unified field theories (Schouten and van Dantzig 1932
a,b,c,d). Both the Einstein–Mayer theory and Veblen and Hoffmann's approach turned
out to be subcases of the more general scheme of Schouten and van Dantzig intending
"to give a unification of general relativity not only with Maxwell's electromagnetic
theory but also with Schrödinger's and Dirac's theory of material waves." (Schouten
and van Dantzig 1932d, p. 271.) In this paper (Schouten and van Dantzig 1932d, p.
311, fig. 2 we find an early graphical representation of the parametrized set of all
possible *theories* of a kind.

Schouten and van Dantzig also used a geometry built on *complex* numbers, i.e., on
hermitian forms: "[...] we were able to show that the metric geometry used by Einstein
in his most recent approach to relativity theory [(Einstein 1928a,b)] coincides with
the geometry of a hermitian tensor of highest rank, which is real on the real axis and
satisfies certain differential equations." (Schouten and van Dantzig 1930, p. 319.)

Cartan wrote a paper on this theory as well (Cartan ca 1934) in which he showed
that the Einstein–Mayer theory could be interpreted as a five-dimensional flat geometry with torsion in which space time is embedded as a totally geodesic subspace
(Cartan 1934).

Fernparallelismus (Teleparallelism)

The next geometry Einstein took as a fundament for unified field theory was a geometry with Riemannian metric, vanishing curvature and non-vanishing torsion, named "distant parallelism" or "Fernparallelismus." The contributions from the Levi-Civita connection and from contorsion[20] in the curvature tensor cancel. In place of the metric, tetrads are introduced as the basic variables. As in Euclidean space, these 4-beins can be parallely translated to retain the same fixed directions everywhere. Thus, again, a degree of rigidity is re-introduced into geometry in contrast to Weyl's first attempt at unification.

Now, as concerns "Fernparallelism," it is a special case of a space with Euclidean connexion introduced by Cartan in 1922/23 (Cartan 1922a,b; 1923). When Einstein published his contributions in June 1928, Cartan had to remind him that a paper of his introducing the concept of torsion had

> "[...] parue au moment où vous faisiez vos conférences au Collége de France; je me rappelle même avoire, chez M. Hadamard, essayé de vous donner l'exemple le plus simple d' un espace de Riemann avec Fernparallelismus en prenant une sphère et en regardand commes paralléles deux vecteurs faisant le même angle avec les méridiennes qui passent par leurs deux origines: les géodésiques correspondantes sont les loxodromies."[21]

(Letter of E. Cartan to A. Einstein, 8. 5. 1929; cf. Debever 1979, p. 4.) This remark refers to Einstein's visit in Paris in March/April 1922. In his response (A. E. to E. C., 10. 5. 1929, Debever 1979, p. 10), Einstein admitted Cartan's priority and referred also to Eisenhart's book of 1927 and to Weitzenböck's paper (Weitzenböck 1928). He excused himself by Weitzenböck's likewise omittance of Cartan's papers among his 14 references. The embarassing situation was solved by Einstein's suggestion that he had submitted a comprehensive paper on the subject to *Zeitschrift für Physik*, and he invited Cartan to add his description of the historical record in another paper (AE to EC 10. 5. 1929). After Cartan had sent his historical review to Einstein (24. 5. 1929), the latter answered only three months later: "I am now writing up the work for the Mathematische Annalen and should like to add yours [...]. The publication should appear in the Mathematische Annalen because, for the present, only the mathematical implications are explored and not their applications to physics." (AE to E.C. 25. 8. 1929) (Cartan 1930, Einstein 1930a)[22] Cartan made it very clear that it was not Weitzenböck who had introduced the concept of distant parallelism, as valuable as his results were after the concept became known. Also, he took Einstein's treatment of Fernparallelism as a special case of his more general considerations. Einstein explained:

> "Insbesondere durch die Herren Weitzenböck und Cartan erfuhr ich, dass die Behandlung von Kontinua der hier in Betracht kommenden Gattung an sich nicht neu sei. [...] Was an der vorliegenden Abhandlung das Wichtigste und jedenfalls neu ist, das ist die Auffindung der einfachsten Feldgesetze, welche eine Riemannsche Mannigfaltigkeit mit Fernparallelismus unterworfen werden kann."[23] (Einstein 1930a, p. 685.)

For Einstein, the attraction of his theory consisted in "its uniformity (Einheitlichkeit), and in the highly overdetermined field variables." The split, in first approximation of the tetrad field h_{ab} according to $h_{ab} = \eta_{ab} + \bar{h}_{ab}$ lead to homogeneous wave equations and divergence relations for both the symmetric and the antisymmetric part identified as metric and electromagnetic field tensors, respectively. The equations were seen as corresponding to "the Newton–Poisson theory of gravitation and the Maxwell theory of electromagnetism." (Einstein 1930a, p. 697.)

Pauli as usual was less than enthusiastic:

> "Ich danke Ihnen vielmals dafür, dass Sie die Korrekturen Ihrer neuen Arbeit aus den mathematischen Annalen (Einstein 1930a) an mich senden liessen, die eine so bequeme und schöne Übersicht über die mathematischen Eigenschaften eines Kontinuums mit Riemann-Metrik und Fernparallelismus enthält. [...] Entgegen dem, was ich im Frühjahr zu Ihnen sagte, lässt sich von Standpunkt der Quantentheorie nunmehr kein Argument zu Gunsten des Fernparallelismus mehr vorbringen. [...] Es bleibt [...] nur übrig, Ihnen zu gratulieren (oder soll ich lieber sagen: zu kondolieren?), dass Sie zu den reinen Mathematikern übergegangen sind. [...] Aber ich würde jede Wette mit Ihnen eingehen, dass Sie spätestens nach einem Jahr den ganzen Fernparallelismus aufgegeben haben werden, so wie Sie früher die Affintheorie aufgegeben haben. [...]."[24]

(Letter to Einstein of 19. Dez. 1929; W. Pauli 1979, 526–527).

Einstein's answer of Dec. 24, 1929, (ibid. p. 582):

> "Ihr Brief ist recht amüsant, aber Ihre Stellungnahme scheint mir doch etwas oberflächlich. So dürfte nur einer schreiben, der sicher ist, die Einheit der Naturkräfte vom richtigen Standpunkt aus zu überblicken. [...] Bevor die mathematischen Konsequenzen richtig durchgedacht sind, ist es keineswegs gerechtfertigt, darüber wegwerfend zu urteilen. [...] Dass das von mir aufgestellte Gleichungssystem zu der zugrundegelegten Raumstruktur in einer zwangsläufigen Beziehung steht, würden Sie bei tieferem Studium bestimmt einsehn, zumal der Kompatibilitätsbeweis der Gleichungen sich unterdessen noch hat vereinfachen lassen."[25]

The question of the compatibility of the field equations played a very important role because Einstein, hoped to gain, eventually, the quantum laws from an overdetermined system of equations. (Cf. his extended correspondence on the subject with Cartan. (Debever 1979)

Einstein really seemed to have believed that he was on a good track because, in 1929 and 1930 he published at least 9 articles on distant parallelism and unified field theory before switching his interest.

Pauli's expressed his discontent also in a letter to Hermann Weyl of 26. August (Pauli 1979, p. 518-519):

> "Zuerst will ich diejenige Seite der Sache hervorheben, bei der ich voll und ganz mit Ihnen übereinstimme: Ihr Ansatz zur Einordnung der Gravitation

in die Diracsche Theorie des Spinelektrons. [...] Ich bin nämlich dem Fernparallelismus ebenso feindlich gesinnt wie Sie, [...]. Nun ist die Stunde der
Rache für Sie gekommen; jetzt hat Einstein den Bock des Fernparallelismus
geschossen, der auch nur reine Mathematik ist und nichts mit Physik zu tun
hat, und Sie können schimpfen!)"[26]

Pauli an Ehrenfest 29. Sept. 1929 (Pauli 1979, p. 524):

"Jetzt glaube ich übrigens vom Fernparallelismus keine Silbe mehr, den Einstein scheint der liebe Gott jetzt völlig verlassen zu haben."[27]

That Pauli had been right (except for the time span envisaged) was expressly admitted by Einstein when he gave up his unified field theory based on distant parallelism in 1931. (letter of A. Einstein to W. Pauli, January 22, 1932; cf. Pais 1982,
p. 347.) Nevertheless, before Einstein dropped the subject many more papers were
written by physicists as e.g. Proca in Romania (Proca 1929, 1930), Zaycoff of Sofia
(Zaycoff 1929a,b,c,d,e,f), Tamm and collaborator in Moscow (Tamm 1929a,b, Tamm
and Leontowitsch 1929a,b) and N. Wiener, M. S. Vallarta from MIT (Wiener and
Vallarta 1929a,b,c ; Rosen and Valarta 1930), and others.

The quantum mechanical wave equations as an additional ingredient of unified field theory

Einstein's papers on distant parallelism nevertheless had a shortlived impact on theoretical physicists, in particular in connection with the discussion of Dirac's equation
for the electron. For the time span between 1926 and 1929, there seemed to be some
hope to come to a unified field theory for gravitation, electromagnetism, and the "electron field." This was caused by a poor understanding of the new quantum theory in
Schrödinger's version: the new complex wave function was interpreted in the spirit
of de Broglie's "onde pilote," i.e., as a classical matter-wave, not — as it should have
been — as a probability amplitude for an ensemble of indistinguishable electrons. One
of the essential features of quantum mechanics, the non-commutability of conjugated
observables like space and momentum, nowhere entered this approach. A little more
than one year after his first paper on Kaluza's idea in which he had hoped to gain
some hold on quantum mechanics, Klein wrote: "Particularly, I no longer think it to
be possible to do justice to the deviations from the classical description of space and
time necessitated by quantum theory through the introduction of a fifth dimension."
(Klein 1928, p. 191 footnote.)

Unlike Klein, H. Mandel of Petersburg tried to interpret the wave function as a new
coordinate. He linked the two components of a (Weyl-) spinor to positive and negative
charge; thus, in his 5-dimensional space the fifth coordinate, as a "charge"-coordinate,
assumed only 2 discrete values $\pm e$. (Mandel 1930) His point of view was expressly
introduced to bring both space time geometry and the Hilbert space of quantum mechanics into a close relationship. Other researchers also found the task of an amalga-
mation of relativity and quantum theory attractive: "It is the purpose of the present
paper to develop a form of the theory of relativity which shall contain the theory of

quanta, as embodied in Schrödingers wave mechanics, not merely as an afterthought, but as an essential and intrinsic part." (Struik and Wiener 1927). The reports given by D. Ivanenko and V. Fock from the talks given at the conference in Charkow (19–25 Mai 1929) clearly picture the situation between unified field theory as a geometrical theory and the final quantum theory (Ivanenko and Fock 1929).

A further motivation for the hope to include some aspects of quantum mechanics into unified field theory resulted from teleparallelism and from the new concept of spin. E. Wigner claimed that the Dirac equation could be written only in a Lorentz-covariant form (Wigner 1929), i.e., in flat space time. Now, the Lorentz group is brought in naturally by distant parallelism. However, it very soon became clear that Dirac's equation could be written covariantly in an arbitrarily curved space time. (Schrödinger 1932) What remained in the end was the conviction that the quantum mechanical "wave equations" could be brought into a covariant form but that quantum mechanics, spin, and gravitation were independent subjects

16.5 Mutual influences among mathematicians and physicists?

A most interesting task far beyond this talk would be to reconstruct, *in detail*, the mutual influences in the development of the various strands of unified field theory. It seems safe to say that the mathematical development of differential geometry in the direction of affine and metric-affine geometry received its original impetus from Einstein's general relativity and Weyl's extension of it (statements by Hessenberg 1917, Schouten 1922, Cartan 1922). Weyl, although a mathematician, understood some of his work to be research in physics proper. In this, he was much criticised by Pauli who gave in only when Weyl (after London's remarks) shifted his gauge idea from coupling electromagnetism to gravitation to coupling electromagnetism to the quantum mechanical state function for an electron. Weyl's influence was prominent among both parties, mathematicians (Cartan, Schouten, Struik, Eisenhart, Hlavaty, N. Wiener etc) and physicists Eddington, Einstein, E. Reichenbächer, H. Mandel, V. Fock, Zaycoff etc. Of course, the interaction in terms of co-authored papers both inside the group of mathematicians (e.g., between Delft and MIT, Delft and Prague, Delft and Leningrad, Princeton and Zürich, Weyl–Bach) and inside the group of physicists (Einstein–Pauli–Eddington, Einstein–Reichenbächer, Einstein–Mandel) was more intensive than the interaction between mathematicians and physicists in the form of journal publications (Weyl-Einstein, Einstein–Cartan, Eddington–Schouten, Kaluza–Einstein, Weyl–Pauli).[28] Mathematicians often used unified field theory as a motivation for their research. Within the communications-net of mathematicians and theoretical physicists contributing to unified field theory, J. A. Schouten played a prominent role. Schouten published also in a *physics journal*, i.e., Zeitschrift für Physik. From the mutual references to their papers, among mathematicians Weyl, Cartan, Schouten, Eisenhart, Veblen, T. Y. Thomas, J. M. Thomas, Levi Civita, Berwald, Weitzenböck, and later Hlavatý and Vranceanu stand out.

The development of projective geometry did profit from mathematician Kaluza's idea of a 5-dimensional space as the arena for unified field theory. It enticed physicists such as W. Pauli, O. Klein, H. Mandel, V. Fock, L. Infeld, and inspired such mathematicians as O. Veblen, J. A. Schouten, D. van Dantzig, E. Cartan and others.

16.6 Conclusion

Even a superficial survey as the one made here[29] shows clearly the dense net of mathematicians and theoretical physicists involved in the building of unified field theory and of the geometrical structures underlying it. Mathematician Grossmann introduced physicist Einstein into Ricci's calculus; Einstein influenced many mathematicians such as e.g., Hessenberg, Weyl, Schouten, Struik, Cartan, Eisenhart, Veblen, to name a few. In return, some ideas very influential on Einstein's path within unified field theories came from these mathematicians: Förster's asymmetric metric[30], Cartan's distant parallelism, Kaluza's 5-dimensional space, Weyl's and Schouten's completely general concept of connection.

My greatest surprise was to learn that, in the period considered here, in the area of unified field theories, Einstein did *not* assume the role of *conceptual* leader he had played when creating general relativity. In fact, in the area of unified field theories, he tended to re-invent or improve on developments made by others. The ideas most fruitful for physics in the long run came from Weyl ("Gauge concept"), and Klein ("Dimensional reduction").[31]

Einstein's importance consists in having been the central *missionary* figure in a scientific enterprise within theoretical physics which, without his weight, fame and obstinacy, would have been reduced to an interesting specialty in differential geometry, and become a dead end for physicists. It is interesting how his zig-zagging path through the wealth of constructive possibilities was followed by the body of researchers in the field. His world fame is as strong as to induce people to continue his endeavor with only slightly changed methods even today despite the predictable failure of their theories in bringing progress for the understanding of nature.

It might be an interesting task to confront the methodology which helped Einstein to arrive at general relativity with the one used by him within unified field theory. (Cf. the contribution of J. Renn.) It is no longer exactly the same as J. van Dongen points out in his contribution; the conception of "the mathematically most natural equations" now appears with its wide spread of possible interpretations.

If the situation in the decades looked at, in the continued attempt at a unification of the fundamental interactions, is compared with today's, it looks similar in the first instance: mathematicians take up new concepts from theoretical physicists (quantum field theory, elementary particle theory, string theory, membrane theory, quantum gravity) and develop them according to their own interests (supermanifolds, knot theory, non-commutative geometry, etc). Physicists then absorb some of the mathematicians' new concepts and methods. Now like then, on the side of physics, the game belongs to what I call "extrapolational physics," i.e., follows formal mathematical lines due to the lack of an empirical basis.[32]

16.7 Acknowledgement

My sincere thanks go to the organizers of this conference and to Mr. J. van Dongen for his helpful remarks.

References

Bach, Rudolph (1921). Zur Weylschen Relativitätstheorie und der Weylschen Erweiterung des Krümmungsbegriffs. *Mathematische Zeitschrift* **9**, 11–35.

Cartan, Eli (1922a). Sur les varié tés à connexion affine courbure de Riemann et les espaces à torsion." *Comptes rendues de l' Academie des Sciences, Paris* **174**, 593–595.

— (1922b). Sur les espaces généralisés et la théorie de la relativité." *Comptes rendues de l' Academie des Sciences, Paris* **174**, 734–737.

— (1923). Sur les variétés à connexion affine et la théorie de la relativité généralisée. *Annales de l'École Normale* **40**, 325–412.

— (1924) Les récentes généralisations de la notion d'espace. *Bulletin. Sci. Math.* **48**, 294–320.

— (1924/25). La théorie des groupes et les recherches récentes de géometrie différentielle. *L'enseignement mathématique*, 1–18.

— (1927). La théorie des groupes et la Géometrie. *L'enseignement mathématique*, 201–225.

— (1930). Notice historique sur la notion de parallélisme absolu." *Mathematische Annalen* **102**, 698–706.

— (1934). La théorie unitaire d'Einstein-Mayer. (Manuscrit datant de 1934 environ.) Oeuvres Complètes III, **2**, 1863–1875.

— (1931) Le parallelisme absolu et la théorie unitaire du champs. *Revue de Métaphysique et de Morale* **38**, 13–28. Oeuvres Complètes III, **2**, 1167–1185.

Debever, R., ed. (1979). *Elie Cartan – Albert Einstein: Lettres sur le parallelisme absolu 1929-1932*. Academie Royale de Belgique, Bruxelles, and Princeton University Press.

Eddington, Arthur S. (1921). A generalisation of Weyl's theory of the electromagnetic and gravitational fields. *Proceedings of the Royal Society of London* A99, 104–122.

Eddington, A.S. (1925). *Relativitätstheorie in Mathematischen Behandlung*. Springer, Berlin.

Einstein, Albert and Fokker, Adriaan D. (1914). Nordströms Gravitationstheorie vom Standpunkt des allgemeinen Differentialkalküls." *Annalen der Physik* **44**, 321–328.

Einstein, Albert (1919). Spielen Gravitationsfelder im Aufbau der materiellen Elementarteilchen eine wesentliche Rolle? *Sitzungsberichte der Preussischen Akademie der Wissenschaften*, 349–356.

— (1921). Über eine naheliegende Ergänzung des Fundamentes der allgemeinen Relativitätstheorie. *Sitzungsberichte der Preussischen Akademie der Wissenschaften*, 261–264.

— (1923a). Zur allgemeinen Relativitätstheorie, *Sitzungsberichte der Preussischen Akademie der Wissenschaften*, 32–38.

— (1923b). Bemerkungen zu meiner Arbeit 'Zur allgemeinen Relativitätstheorie', *Sitzungsberichte der Preussischen Akademie der Wissenschaften*, 76–77.

— (1923c). Zur affinen Feldtheorie. *Sitzungsberichte der Preussischen Akademie der Wissenschaften*, 137–140.

Einstein, Albert and Grommer, Jakob (1923). Beweis der Nichtexistenz eines überall regulären zentrisch symmetrischen Feldes nach der Feldtheorie von Kaluza. *Scripta Jerusalem University* 1, No. 7, 5 S.

Einstein, Albert (1925a). Einheitliche Feldtheorie von Gravitation und Elektrizität. *Sitzungsberichte der Preussischen Akademie der Wissenschaften*, 414–419.

— (1925b). Elektron und allgemeine Relativitätstheorie." *Physica* 5, 330–334.

— (1927). Zu Kaluzas Theorie des Zusammenhangs von Gravitation und Elektrizität. *Sitzungsberichte der Preussischen Akademie der Wissenschaften*, 23–25 (1. Mitteilung), 26–30 (2. Mitteilung).

— (1928a). Riemann-Geometrie mit Aufrechterhaltung des Begriffs des Fernparallelismus. *Sitzungsberichte der Preussischen Akademie der Wissenschaften*, 217–221.

— (1928b). Neue Möglichkeit für eine einheitliche Feldtheorie von Gravitation und Elektrizität. *Sitzungsberichte der Preussischen Akademie der Wissenschaften*, 224–227.

— (1929a). Zur einheitlichen Feldtheorie. *Sitzungsberichte der Preussischen Akademie der Wissenschaften*, 3–8.

— (1929b). Einheitliche Feldtheorie und Hamiltonsches Prinzip. *Sitzungsberichte der Preussischen Akademie der Wissenschaften*, 156–159.

— (1929c). New Field Theory. I, II. *Observatory* 52, 82–87, 114–118.

— (1929d). Über den gegenwärtigen Stand der Feld-Theorie. In *Festschrift Prof. A Stodola zum Geburtstag*, Füssli Verlag, Zürich und Leipzig, pp. 126–132.

— (1930a). Auf die Riemann-Metrik und den Fernparallelismus gegründete einheitliche Feldtheorie. *Mathematische Annalen* 102, 685–697.

—(1930b). Zur Theorie der Räume mit Riemann-Metrik und Fernparallelismus. *Sitzungsberichte der Preussischen Akademie der Wissenschaften*, 1–2.

Einstein, Albert and Mayer, Walter (1930). Zwei strenge statische Lösungen der Feldgleichungen der einheitlichen Feldtheorie. *Sitzungsberichte der Preussischen Akademie der Wissenschaften*, 110–120.

— (1931a). Systematische Untersuchung über kompatible Feldgleichungen, welche in einem Riemannschen Raume mit Fernparallelismus gesetzt werden können. '*Sitzungsberichte der Preussischen Akademie der Wissenschaften*, 257–265.

— (1931b). Einheitliche Theorie von Gravitation und Elektrizität. *Sitzungsberichte der Preussischen Akademie der Wissenschaften*, 541–557.

Einstein, Albert (1931). Gravitational and electrical fields. *Science* 74, 438–39.

Einstein, Albert and Mayer, Walter (1932a) Einheitliche Theorie von Gravitation und Elektrizität. (Zweite Abhandlung) *Sitzungsberichte der Preussischen Akademie der Wissenschaften*, 130–137.

— (1932b). 'Semi-Vektoren und Spinoren.' *Sitzungsberichte der Preussischen Akademie der Wissenschaften*, 522–550.

Einstein-ECP (1998). *The Collected Papers of Albert Einstein*, Vol. 8, Part A, B, R. Schulmann, A. J. Kox, Michel Janssen, and József Illy, eds., Princeton University Press, Princeton.

Eisenhart, L.P. (1926). Einstein's Recent Theory of Gravitation and Electromagnetism. *Proceedings of the National Academy of Science, USA* **12**, 125–129.

Eisenhart, L.P. and Veblen, O. (1922). The Riemann Geometry and Its Generalizations. *Proceedings of the National Academy of Science, USA* **8**, 19–23.

Eisenhart, L.P. (1922/23). The Geometry of Paths and General Relativity. *Annals of Mathematics* **24**, 367–392.

— (1927). *Non-Riemannian Geometry*. American Mathematical society Colloquium Publications, vol. 8., American Mathematical Society, Providence, RI.

Finsler, P. (1918). *Über Kurven un Flächen in allgemeinen Räumen*, vol. 11 of Lehrbücher und Monographien aus dem gebeit der exalten wissenschaften: Mathematische Reihe. (Nachrichte 1918), Birkhäuser, Basel, 1951.

Fock, V. (1926). Über die invariante Form der Wellen- und der Bewegungsgleichung für einen geladenen Massenpunkt. *Zeitschrift für Physik* **39**, 226–232.

Förster, Rudolf (1908). Beiträge zur spezielleren Theorie der Riemannschen P-Funktion III. Teubner, Leipzig.

Friedmann, A. and Schouten, J.A. (1924). Über die Geometrie der halb-symmetrischen Übertragungen. *Mathematische Zeitschrift* **21**, 211–223.

Goenner, Hubert (1984). Unified field theories from Eddington and Einstein up to now. In: *Proceedings of the A. Eddington Centennial Symposium*, vol. 1, Relativistic Astrophysics and Cosmology. V. De Sabbata and T. M. Karade, eds., World Scientific, Singapore.

Goenner, Hubert and Havas, Peter (1980). Spherically-Symmetric Space-Times with vanishing Curvature Scalar. *Journal of Mathematical Physics* **21**, 1159–1167.

Goldstein, Catherine and Ritter, Jim (2000). *The Varieties of Unity; Sounding Unified Theories 1920–1930*. Preprint series Max-Planck-Institut für Wissenschaftsgeschichte, Berlin, preprint No. 148.

Hessenberg, Gerhard (1917). Vektorielle Begründung der Differentialgeometrie. *Mathematische Annalen* **78**, 187–217. Breslau.

Hilbert, David (1915). Die Grundlagen der Physik. (Erste Mitteilung). *Königliche Gesellschaft der Wissenschaften zu Göttingen. Mathematisch-physikalische Klasse. Nachrichten*, 395–407.

Hoffmann, Banesh (1930). Projective Relativity and the Quantum Field. *Physical Review* **37**, 88–89.

Ivanenko, D. and Fock, V. (1929). Vorträge und Diskussionen der Theoretisch-Physikalischen Konferenz in Charkow (19.-25. Mai 1929) *Physikalische Zeitschrift* **30**, 645–655, 700–717.

Kaluza, Th. (1921). Zum Unitätsproblem in der Physik. *Sitzungsberichte der Preussischen Akademie der Wissenschaften*, 966–972.

Klein, Oskar (1926a). Quanten-Theorie und 5-dimensionale Relativitätstheorie. *Zeitschrift für Physik* **37**, 895–906.

— (1926b). The Atomicity of Electricity as a Quantum Law. *Nature* **118**, 516.

— (1928). Zur fünfdimensionalen Darstellung der Relativitätstheorie. *Zeitschrift für Physik* **46**, 188–208.

König, Robert (1919). Beiträge zu einer allgemeinen linearen Mannigfaltigkeitslehre. *Jahresberichte der Deutschen Mathematikervereiningung* **28**, 213–228.

König, Matthias (2000). Herleitung und Untersuchung der Feld- und Bewegungsgleichungen von an das Gravitationsfeld gekoppelten Tensorfeld-Theorien. Institute for Theoretical Physics, University of Göttingen, Diplomarbeit (unpublished).

Levi-Civita, Tullio (1917). Nozione di parallelismo in una varietà qualunque e conseguente specificazione geometrica della curvatura Riemanniana. *Circulo matematico di Palermo. Rendiconti* **42**, 173–205.

Mandel, Heinrich (1926). Zur Herleitung der Feldgleichungen in der allgemeinen Relativitätstheorie. Erste Mitteilung *Zeitschrift für Physik* **39**, 136–145.

Newman, M.A.H. (1927). A Gauge-Invariant Tensor Calculus. *Proceedings of the Royal Society of London* A 116, 603–623.

Nordström, G. (1914). Über die Möglichkeit, das elektromagnetische feld und das gravitationsfeld zu vereinigen, *Physikalische Zeitschrift*, **15**, 504–506.

Pais, Abraham (1982). *'Subtle is the Lord': The Science and Life of Albert Einstein*. University Press, Oxford.

Pauli, Wolfgang (1920). Diskussionsbemerkungen. *Physikalische Zeitschrift* **21**, 650–651.

Pauli, Wolfgang (1921). Relativitätstheorie in: *Enzyklopädie der Mathematischen Wissenschaften*, vol. 5, part II. Teubner, Leipzig and Berlin.

Pauli, Wolfgang (1979). Wissenschaftlicher Briefwechsel mit Bohr, Einstein, Heisenberg, u.a. Band 1: 1919–1929. Springer, New York.

Proca, A. (1929). La nouvelle théorie d'Einstein. *Bull. Math. Phys. Bucarest* **1**, 170–176.

— (1930). La nouvelle théorie d'Einstein. II. *Bull. Math. Phys. Bucarest* **2**, 15–22.

O'Raifeartaigh, L. and Straumann, N. (2000). Gauge theory: Historical Origins and Some Modern Developments. *Reviews of Modern Physics* **72**, 1–24.

Reichenbächer, Ernst (1917a). Grundzüge zu einer Theorie der Elektrizität und der Gravitation. *Annalen der Physik* **52**, 134–173.

— (1917b). 3. Nachtrag zu der Arbeit Grundzüge zu einer Theorie der Elektrizität und der Gravitation. *Annalen der Physik* **52**, 174–178.

— (1926). Der Elektromagnetismus in der Weltgeometrie. *Physikalische Zeitschrift* **27**, 741–745.

Rosen, Nathan and Vallarta, M.S. (1930). The Spherically Symmetrical Field in the Unified Theory. *Physical Review* **36**, 110–120.

Scholz, Erhard, ed. (2001) *Hermann Weyl's Raum-Zeit-Materie and a General Introduction to His Scientific Work*. Birkhäuser, Basel, Boston, Berlin.

Schouten, J.A. (1922a) 'Über die verschiedenen Arten der Übertragung in einer *n* dimensionalen Mannigfaltigkeit, die einer Differentialgeometrie zugrundegelegt werden kann.' *Mathematische Zeitschrift* **13**, 56–81.

— (1922b). Nachtrag zur Arbeit 'Über die verschiedenen Arten der Übertragung in einer n-dimensionalen Mannigfaltigkeit, die einer Differentialgeometrie zugrundegelegt werden kann.' *Mathematische Zeitschrift* **15**, 168.

— (1923). On a Non-Symmetrical Affine Field Theory. *Proc. Kon. Akad. v. Wetenschappen, Amsterdam* **26**, 850–857.

— (1924) *Raum, Zeit und Relativitätsprinzip*. Teubner, Leipzig, Berlin.

Schouten, J.A. and van Dantzig, D. (1930) Über unitäre Geometrie. *Mathematische Annalen* **103**, 319–346.

— (1932a). Zum Unifizierungsproblem der Physik; Skizze einer generellen Feldtheorie. (GF I) *Proc. Kon. Akad. v. Wetenschappen, Amsterdam* **35**, 642–655.

— (1932b). Zur generellen Feldtheorie; Diracsche Gleichung und Hamiltonsche Funktion. (GF II) *Proc. Kon. Akad. v. Wetenschappen, Amsterdam* **35**, 642–655.

— (1932c). Generelle Feldtheorie. (GF III) *Zeitschrift für Physik* **78**, 639–667.

— (1932d). On Projective Connections and Their Application to the General Field Theory. (GF VI) *Annals of Mathematics* **34**, 271–312.

Schrödinger, Erwin (1932). Diracsches Elektron im Schwerefeld. I. *Sitzungsberichte der Preussischen Akademie der Wissenschaften*, 105–128.

Struik, Dirk J. (1934). *Theory of Linear Connections*. Ergebnisse der Mathematik und ihrer Grenzgebiete, Bd. 3., Springer, Berlin.

Struik, Dirk J. and Wiener, Norbert (1927). A relativistic theory of Quanta. *Journal of Mathematics and Physics* **7**, 1–23.

Tamm, Igor (1929a). Über den Zusammenhang der Einsteinschen einheitlichen Feldtheorie mit der Quantentheorie. ' *Proceedings of the Royal Academy, Amsterdam* **32**, 288–291.

— (1929b). Die Einsteinsche einheitliche Feldtheorie und die Quantentheorie. *Physikalische Zeitschrift* **30**, 652–654.

Tamm, Igor and Leontowitsch, M. (1929a) Über die Lösung einiger Probleme in der neuen Feldtheorie. *Physikalische Zeitschrift* **30**, 648.

— (1929b). Bemerkungen zur Einsteinschen einheitlichen Feldtheorie. *Zeitschrift für Physik* **57**, 354–366.

van Dantzig, David (1926). Die Wiederholung des Michelson Versuch's und die Relativitätstheorie. *Mathematische Annalen* **96**, 261–228.

— (1932a). Zur allgemeinen projektiven Differentialgeometrie. I. Einordnung der Affingeometrie. *Proc. Kon. Akad. v. Wetenschappen, Amsterdam* **35**, 524–534.

— (1932b). Zur allgemeinen projektiven Differentialgeometrie. II X_{n+1} mit eingliedriger Gruppe.

— (1932c) *Proc. Kon. Akad. v. Wetenschappen, Amsterdam* **35**, 525–542.

— (1932d). Theorie des projektiven Zusammenhangs n-dimensionaler Räume. *Mathematische Annalen* **106**, 400–454.

Veblen, O. and Thomas, T.Y. (1923). The geometry of paths. *Transactions of the American Mathematical Society* **25**, 551–608.

Veblen, O. (1928). Projective tensors and connections. *Proceedings of the National Academy of Science, USA* **14**, 154–166.

— (1933). *Projektive Relativitätstheorie*. Ergebnisse der Mathematik und ihrer Grenzgebiete, Bd. 2, Heft 1. Springer, Berlin.

Veblen, O. and Hoffmann, Banesh (1930). Projective relativity, *Physical Review* **36**, 810–822.

Vizgin, Vladimir P. (1994). *Unified Field Theories in the First Third of the 20th Century*. Birkhäuser, Basel.

Weitzenböck, R. (1928) Differentialinvarianten in der Einsteinschen Theorie des Fernparallelismus. *Sitzungsberichte der Preussischen Akademie der Wissenschaften*, 466–474.

Weyl, Hermann (1919). *Raum-Zeit-Materie*, 3rd edition. Springer, Berlin.

— (1929a). Gravitation and the Electron. *Proceedings of the National Academy of Science, USA* **15**, 323–334.

— (1929b). Elektron und Gravitation. I. *Zeitschrift für Physik* **56**, 330–352.

Wiener, Norbert and Vallarta, Manuel S. (1929a). On the Spherically Symmetric Statical Field in Einstein's Unified Theory of Electromagnetism and Gravitation. *Proceedings of the National Academy of Science, USA*, 353–356.

— (1929b). On the Spherically Symmetric Statical Field in Einstein's Unified Theory of Electromagnetism and Gravitation. *Proceedings of the National Academy of Science, USA* **15**, 802–804.

— (1929c). Unified Field Theory With Electricity and Gravitation. *Nature* **123**, 317.

Wigner, Eugene (1929). Eine Bemerkung zu Einsteins neuer Formulierung des allgemeinen Relativitätsprinzips. *Zeitschrift für Physik* **53**, 592–596.

Zaycoff, R. (1929a). Zur Begründung einer neuen Feldtheorie von A. Einstein. *Zeitschrift für Physik* **53**, 719–728.

— (1929b). Zur Begründung einer neuen Feldtheorie von A. Einstein. II *Zeitschrift für Physik* **54**, 590–593.

— (1929c). Zur Begründung einer neuen Feldtheorie von A. Einstein. III *Zeitschrift für Physik* **54**, 738–740.

— (1929d). Zu der neuesten Formulierung der Einsteinschen einheitlichen Feldtheorie. *Zeitschrift für Physik* **56**, 717–726.

— (1929e). Fernparallelismus und Wellenmechanik. *Zeitschrift für Physik* **58**, 833–840.

— (1929f). Fernparallelismus und Wellenmechanik. II *Zeitschrift für Physik* **59**, 110–113.

Notes

[1] The review has appeared in LivingReviews in relativity cf.
http://relativity.livingreviews.org/Articles/lrr-2004-2/index.html

[2] Nevertheless, the construction of quantum field theory had begun already around 1927. Cf. P. A. M. Dirac, Proc. Roy. Soc. London **A 114**, 234 (1927); P. Jordan u. O. Klein, Zeitschr. f. Physik **45**, 751 (1927); P. Jordan, Zeitschr. f. Physik **44**, 766 (1927) (fields with Bose statistics); P. Jordan, Zeitschr. f. Physik **44**, 473 (1927); P. Jordan u. E. Wigner, Zeitschr. f. Physik **47**, 631 (1928) (fields with Fermi statistics).

[3]Unpublished correspondence has not yet been included. The Einstein correspondence is refered to by ECP and listed in the references under Einstein-ECP.

[4]Förster published under a nom de plume "R. Bach." He wrote also about Weyl's theory (Bach 1921)

[5]"Perhaps, there exists a covariant 6-vector by which the appearance of electricity is explained and which springs lightly from the $g_{\mu\nu}$, not forced into it as an alien element."

[6]"The aim of dealing with graviation and electricity on the same footing by reducing both groups of phenomena to $g_{\mu\nu}$ has already caused me many disappointments. Perhaps, you are luckier in the search. I am fully convinced that in the end all field quantities will show up as alike in essence. But it is easier to suspect something than to dicover it."

[7]"Since long, I also was busy by starting from a non-symmetric $g_{\mu\nu}$; however, I lost hope to get behind the secret of unity (graviation, electromagnetism) in this way. Various reasons instilled in me strong reservations: [...] your other remarks are interesting in themselves and new to me."

[8]"According to a general mathematical theorem, the electromagnetic equations (generalized Maxwell eqs.) appear as a consequence of the gravitational equations such that gravitation and electrodynamics are not really different."

[9]Cf. also Diplomarbeit König, Göttingen April 2000. It is shown there that from the divergence relation $T^{\mu\nu}{}_{;\nu} = 0$ and the most general Lagrangian L(u, v) with $u = F_{\mu\nu}F^{\mu\nu}$; $v = {}^*F_{\mu\nu}F^{\mu\nu}$ the field equations follow for the generic case of full rank of the electromagnetic field tensor $F_{\mu\nu}$.

[10]"Your investigation is of great interest to me because I have often tortured my mind in order to bridge the gap between gravitation and electromagnetism. The hints dropped by you on your postcards bring me to expect the greatest."

[11]Reichenbächer's solution is a special case of a huge number of spherically symmetric solutions of $R = 0$ given in Goenner and Havas (1980). Reichenbächer published 24 papers between 1917 and 1930.

[11]Reichenbächer's solution is a special case of a huge number of spherically symmetric solutions of $R = 0$ given in Goenner and Havas (1980). Reichenbächer published 24 papers between 1917 and 1930.

[12]"With the progressing refinement of scientific concepts, the manner by which concepts are related to (physical) events becomes ever more complicated. If, in a certain stage of scientific investigation, it is seen that a concept can no longer be linked with a certain event, there is a choice to let the concept go, or to keep it; in the latter case, we are forced to replace the system of relations among concepts and events by a more complicated one. The same alternative obtains with respect to the concepts of time- and space-distances. In my opinion, an answer can be given only under the aspect of usability; the outcome appears dubious to me."

[13][...] that an antisymmetric tensor was added to the symmetric tensor of the gravitational potential, who represented the six-vector of the electromagnetic field. But a more precise reasoning shows that in this way no reasonable worldfunction is obtained."

[14](Pauli 1921); cf. also the volume about Weyl edited by E. Scholz (Scholz 2001).

[15]" [...] The quantities $\Gamma^\mu_{\nu\alpha}$ cannot be measured directly, but must be won from the directly measured quantities by complicated calculational operations. Nobody can determine empirically an affine connection for vectors in neighboring points if he has not obtained the line element before. Therefore, unlike you and Einstein, I deem the mathematician's discovery of the possibility to found a geometry on an affine connection without a metric as meaningless for physics, on first sight."

[16]Die Grundlehren der Mathematischen Wissenschaften in Einzeldarstellungen.

[17]The geometry of paths involves a change of connection which preserves the geodesics when vectors are displaced along themselves.

[18]"Also, my opinion about my paper which appeared in these reports [i.e., Sitzungsberichte of the Prussian Academy], and which was based on Eddington's fundamental idea, is such that it does not present the true solution of the problem. After an uninterrupted search during the past two years I now believe to have found the true solution."

[19]Pais, in his book (Pais 1982) expresses his lack of understanding as to why Einstein published his papers at all. It is known that Einstein did not follow the literature closely, including the reprints he received from their respective authors.

[20]A linear combination of torsion appearing in the connection besides the metric contribution.

[21]"[...] appeared at the moment at which you gave your talks at the Collège de France. I even remember having tried, at Hadamar's place, to give you the most simple example of a Riemannian space with Fernparallelismus by taking a sphere and by treating as parallels two vectors forming the same angle with the meridians going through their two origins: the corresponding geodesics are the rhumb lines."

[22]Mathematische Annalen was a journal edited by David Hilbert with co-editors O. Blumenthal and G. Hecke which physicists usually would not read. The editor of Zeitschrift für Physik was Karl Scheel.

[23]"In particular, I learned from gentlemen Weitzenböck and Cartan that the treatment of continua of the species which is of import here, is not really new.[...] In any case, what is most important in the paper, and new in any case, is the discovery of the simplest field laws which can be imposed on a Riemannian manifold with Fernparallelismus."

[24]"I thank you so much for letting be sent to me your new paper from the Mathematische Annalen which gives such a comfortable and beautiful review of the mathematical properties of a continuum with Riemannian metric and distant parallelism [...]. Unlike what I told you in spring, from the point of view of quantum theory, now an argument in favor of distant parallelism can no longer be put forward [...]. It just

remains [...] to congratulate you (or should I rather say condole you) that you have passed over to the mathematicians. But I would bet with you that, at the latest after one year, you will have given up the whole instant parallelism in the same way as you have given up the affine theory earlier."

[25]"Your letter is quite amusing, but your statement seems rather superficial to me. In this way, only someone ought to write who is certain of seeing through the unity of natural forces in the right way. Before the mathematical consequences have not been thought through properly, is is not at all justified to make a negative judgement. [...] That the system of equations established by myself forms a consequential relationship with the space structure taken, you would probably accept by a deeper study - more so because, in the meantime, the proof of the compatibility of the equations could be simplified."

[26]"First let me emphasize that side of the matter about which I fully agree with you: your approach for incorporating gravitation into Dirac's theory of the spinning electron [...] I am as adverse with regard to Fernparallelismus as you are [...]. Now the hour of revenge has come for you, now Einstein has made the blunder of distant parallelism which is nothing but mathematics unrelated to physics, now you may scold [him]."

[27]"By the way, I now do no longer believe one syllable of teleparallelism; Einstein seems to have been abandoned by the dear Lord."

[28]If correspondence is taken into account this can no longer be said.

[29]I did neglect to discuss the different possibilities for Lagrangians used, and the extension of the geometrical concepts from the real to the complex domain.

[30]Förster wrote his thesis in mathematics (Förster 1908)

[31]For the historical development of gauge theory from the point of view of physics cf. (O'Raifeartaigh and Straumann 2000). Note Vizgin's differing statement that Einstein, around 1930, "became the recognized leader of the investigations [in unified field theory], taking over, as it were, the baton from Weyl, who had been the leading authority for the previous five years." (Vizgin 1994, p. 183.)

[32]"All major theoretical developments of the last twenty years, such as grand unification, supergravity, and supersymmetric string theory, are almost completely separated from experience. There is a great danger that theoreticians may get lost in pure speculations." (O'Raifeartaigh and Straumann 2000, p. 45.)

Is Quantum Gravity Necessary? *

James Mattingly

Georgetown University, Washington DC, U.S.A.; jmm67@georgetown.edu

17.1 Introduction

Quantum gravity presents something of a unique puzzle for the philosophy of science. For in a very real sense, there is no such thing as quantum gravity. Despite near unanimous agreement among physicists that a quantum theory of gravitation is needed to reconcile the contradictions between general relativity and quantum mechanics, there are no pressing empirical issues that require this resolution—the regime in which one would expect to observe a conflict between the claims of general relativity and quantum mechanics is at the Planck scale. Thus the question naturally arises "Why quantize gravity?" Are there other issues that compel us to seek a quantum theory of gravity?

The standard response is intimately connected with a desire for theoretical unification. Quantum field theory successfully describes the physical world on small length scales at low "particle" density. General relativity is a successful theory of large length scales where individual features of particular objects are swamped by their mass-energy properties. It is natural to seek a unified theory that captures these successful features and yet is somehow a "fundamental" theory of both regimes. But why should the resulting theory involve a quantized gravitational field? There is clearly *something* wrong with the general relativistic treatment of matter fields as classical. Very well. Let us stipulate that an acceptable theory of gravitation will take due note of the quantum nature of the fields to which it couples. Now what? Are we thus compelled to treat the gravitational field itself quantum mechanically?

There are a number of arguments urging the necessity of a full quantum gravity—i.e., a theory of gravity that treats the metric itself as a quantum field. There are, as well, a number of proposals for how we should go about producing this theory. I will not here be concerned to articulate the panoply of attempts to quantize gravitation theory (nor to elaborate the many problems attendant to that effort). I will instead present a partial catalogue and evaluation of the various reasons for constructing such a theory. These reasons group themselves naturally into three reasonably distinct classes.

---

* Delivered at the 5th International Conference on the History and Foundations of General Relativity, July 9, 1999.

There are what I will call problems of experiment, problems of theory and problems of meta-theory respectively. In the first class are included experiments either actually performed or detailed thought experiments. In the second class are those problems of a theoretical nature that appear to derail efforts to avoid quantization in the absence of experimental evidence. The last class will then contain, in particular, implicit as well as explicit philosophical motivations for quantizing gravity. It is this last class, I will argue, that is really responsible for the conviction (quite widespread in the physics community) that gravitation is necessarily a quantum mechanical phenomenon.

I will begin with an account of the typical problems of experiment that are offered as definitively settling the question of gravitation in favor of quantization. My account, perhaps, will not be exhaustive, but I believe it captures the flavor of the reasons of "experimental" physics and shows their inadequacy. I follow this account with some remarks about the theoretical difficulties of producing a realistic non-quantized gravitation theory that takes due notice of the quantum mechanical nature of the matter producing the gravitational field. These difficulties are not trivial, and, more to the point, I cannot resolve them. But they are no worse, at least on their face, than many of the difficulties facing those who would construct a fully quantized theory of gravity.

If I am right about these first two points, then meta-theoretical commitments of some kind are at the root of efforts to quantize the gravitational field. Focusing on just one type of commitment—theoretical unification—I argue that (as is often the case with very general principles) it fails to entail the conclusion it is used to justify. I conclude from this that the real justification for quantizing gravity has yet to be articulated.

The theory on which I will focus, in what follows, is the semiclassical theory of gravitation. Of the possible approaches to avoiding the quantization of gravity, this theory has been the most studied (albeit only as an approximation technique). Moreover it instantiates Rosenfeld's (1963) suggestion that a realistic theory of gravity could be one where the (classical) Einstein tensor is proportional to the expectation value of the (quantum) stress-energy operator.

In this theory, one constructs a quantum field theory on a curved spacetime and then allows the stress energy tensor to couple to the Einstein tensor via the semiclassical Einstein equation:

$$G_{\mu\nu} = k\langle T_{\mu\nu}\rangle.$$

For details and a particular construction of $\langle T_{\mu\nu}\rangle$ see (Wald 1999).

There are other approaches to a non-quantized gravity but I will not mention them here. It should be kept in mind, however, that even if the semiclassical proposal is inadequate, the case against treating the gravitational field classically is not, thereby, decided.

Problems of experiment

In 1975 a thought experiment was conducted (if that is indeed what one does with thought experiments) by Eppley and Hannah (1977) purporting to show that the gravitational field must be quantized. They assume the validity of semiclassical gravity, and use a gravity wave to measure the position and momentum of a macroscopic body such that $\Delta p_x \Delta x < \hbar$, thus violating the Heisenberg uncertainty principle. The key idea is that a classical wave may have arbitrarily low momentum and, simultaneously, arbitrarily short wavelength. This observation already conflicts with the de Broglie formula relating momentum to wavelength $\lambda = h/p$. But the whole point to taking seriously the semiclassical theory is to avoid directly applying quantum mechanics to the gravitational field. In order to find a conflict with quantum mechanics, it is necessary to couple such a short wavelength/low momentum gravitational wave to a quantum system. The wave may then be used to localize a particle within one wavelength while introducing vanishingly small uncertainty into the particle's momentum. Eppley and Hannah's experiment does just this.

Their experiment is, however, unrealistic in a number of ways. I cannot discuss this here, and perhaps the defects could be remedied in a different version of the experiment. Even so, their case that gravity must be quantized would still not be made. And this for two reasons: it may be that the uncertainty relations *can* be violated. They haven't really been tested in this way. Second, there are empirically adequate interpretations of quantum mechanics for which these relations are epistemological and not a fundamental feature of the world. Thus the thought experiment cannot be considered definitive.[1]

At the second Oxford symposium on quantum gravity another empirical problem for semiclassical gravity was articulated. Professor Kibble (1981) there argued for the viability of semiclassical gravity. He proposed a thought experiment that, strangely, now has come to be interpreted as having a significance diametrically opposed to the one he offered for it. His experiment consisted of a Stern–Gerlach magnet that first separated the spin up and spin down components of a particle's wave function and then passed these components by particle detectors. Detection of the particle released a heavy mass to one side or the other of the device according to the component of spin possessed by the detected particle. The whole device was included in a black box to prevent outside observation. Kibble pointed out that the standard story of quantum measurement (assuming the Einstein tensor proportional to the expectation value of the stress-energy tensor) implies that, on measurement of the spin of the particle, there would be a physically unrealistic jump in the gravitational field. The lesson he drew from his thought experiment is that the semiclassical theory of gravity would require a new theory of quantum measurement. But, he continued, we already knew that quantum measurement theory is a mess.

Page and Geilker (1981), on the other hand, regard Kibble's result as damning but not definitive evidence against a semiclassical gravitation theory. For them, its only defect is one it shared with Eppley and Hannah's experiment—it hadn't been performed. They propose to remedy this lack of experimental evidence against semiclassical gravity by performing a concrete test of the theory. Their test is essentially a classical test of the gravitational response of a torsion balance to the presence of macroscopic masses. The quantum feature is entirely captured by the method of choosing the locations of these masses. The choice is determined by what amounts to a quantum random number generator; depending on the value of some quantum variable, the masses will be sent either to the left or the right of the balance. Page and Geilker find, not surprisingly, that the balance responds only to the presence of mass and not the expectation value of where the mass *will* go.

Why should anyone have expected a different result? How does this count against the semiclassical theory? Apparently anyone using Everett's relative-state formulation of quantum mechanics would expect the torsion balance to remain fixed. On this formulation, the wave-function never collapses. Instead it "branches" out into new worlds with the distribution of worlds governed by the standard measurement probabilities of quantum mechanics. Since under this interpretation the wave function never collapses, Page and Geilker claim that the semiclassical Einstein tensor responds to the presence of matter in all branches of the universe. For example, taking $G_{ab} = k\langle\psi|T_{AB}|\psi\rangle$ as the semiclassical Einstein equation and $|\psi\rangle = c_1|\phi_1\rangle + c_2|\phi_2\rangle$ then, even if measurement shows that, after some interaction, $|\psi\rangle$ in our branch appears to have collapsed to $|\phi_2\rangle$, we still have

$$G_{ab} = k(c_1^* c_1 \langle\phi_1|T_{ab}|\phi_1\rangle + c_2^* c_2 \langle\phi_2|T_{ab}|\phi_2\rangle).$$

Page and Geilker perform a quantum experiment that they plausibly assume affects only a small subspace of the total wave-function of the universe and so casts $|\psi\rangle$ into a state like $(\frac{1}{\sqrt{2}}|\phi_1\rangle + \frac{1}{\sqrt{2}}|\phi_2\rangle) \otimes |\psi_{everythingelse}\rangle$. They set their masses according to the result of their experiment and assume, again plausibly, that Page and Geilker in the other branch do the same. Since their experiment shows that the balance responds only to the matter in our own branch of the multi-verse, they conclude that this version of semiclassical gravity, with their particular interpretation of quantum mechanics, is empirically inadequate.

In 1981, when their experiment was performed, they offered an argument that was supposed to show that only an Everett style interpretation is compatible with semiclassical gravity. Their argument assumes first that the only choices for an interpretation of quantum mechanics are (instantaneous Copenhagen style) collapse and Everett. They then claim that semiclassical gravity is incompatible with collapse. To this end they consider a superposition state $|\psi\rangle = \sum_i c_i|\phi_i\rangle$ and, in the Heisenberg picture, calculate the covariant derivative:

$$\langle\psi|T_{ab}|\psi\rangle_{;b} = \sum_{ij}(c_i^* c_j)_{;b}\langle\phi_i|T_{ab}|\phi_j\rangle \neq 0 \equiv G_{ab;b}.$$

For example, if $|\psi\rangle$ is a superposition of eigenstates of T, the expectation value for the energy may change during measurement. That is to say, if $|\psi\rangle = c_1|\phi_1\rangle + c_2|\phi_2\rangle$ where the $|\phi\rangle$ are eigenstates of T and our experiment is a measurement of T, then after the (instantaneous) measurement the $c_i s$ have changed discontinuously and it would seem miraculous if the total change had derivative 0. Then the semiclassical Einstein equation becomes inconsistent. $G_{;b}$ is identically 0 but $T_{;b}$ need not be. This argument is not entirely convincing since I can see no reason to suppose the wave function of the universe to ever be in other than an eigenstate of the stress-energy tensor (in much the same way that for conservative systems, the total state is an eigenstate of the hamiltonian). But I won't make an argument to that effect here. I will point out that Wald (1994, 78–89) has constructed a prescription for measurement in semiclassical gravity which can be given a collapse interpretation and which also satisfies

$\langle T_{ab}\rangle_{;b} = 0$. So already there is trouble with Page and Geilker's interpretation of the significance of their experiment. I won't pursue this here, but instead I will question another of Page and Geilker's assumptions.

Because they assume that the only possible interpretations of quantum mechanics are Cøpenhagen and Everett's relative state formulation, they conclude that if semi-classical gravity fails for Cøpenhagen and the relative state formulation, it fails for quantum mechanics generally. Once again, what we are really up against is the quantum measurement problem. Because they use the results of a quantum mechanical measurement to set their device (a measurement well separated from their device), by the time it is set the quantum decision is already made. So *only* a no-collapse interpretation can be used to draw conclusions about their experiment. For only on that view does the expectation value of the wave function continue to reflect the entire state of the "multi-verse." But there are many other interpretations of quantum mechanics for which the expectation value is updated as our knowledge of the wave function is updated. And for other no-collapse models the quantum state is exhaustive in the way it is classically—all values of all observables are definite all the time. This is true of the quantum logic interpretation for example. So for someone using a quantum logic interpretation of quantum mechanics, it would not make a great deal of sense to equate the Einstein tensor with the expectation value of the stress-tensor. In such a case we would not take the expectation value seriously, of course, so we would have to modify the interpretation of semiclassical gravity slightly. The natural seeming approach in that interpretation would be to equate the Einstein tensor directly to the stress-tensor in the way it is done in classical general relativity. Naturally how one would accomplish this is not obvious, but I only intend here to make it clear that, even if the semiclassical approach as outlined does fail (and at least experimentally (thought and otherwise) we don't have good evidence that it does), this failure does not constitute strong evidence that gravity is itself quantum-mechanical.

I won't go any further into this here, but I do wish to emphasize that only a peculiar[2] reading of quantum mechanics is incompatible with Page and Geilker's result. It is certainly interesting that some versions of some interpretations are incompatible with semiclassical gravity. But it is *only* interesting. Other no collapse models—such as the de Broglie–Bohm theory, and modal interpretations—as well as continuous-collapse models are apparently unaffected by these results.

Here we have an interesting footnote to debates about the proper interpretations of quantum mechanics. I have no solution to the quantum measurement problem. Nor do I find wholly satisfying any of the extant proposals for solving it. Yet I cannot worry overmuch about a proposal for a gravitation theory merely because it fails to solve the problem, or rather because it undermines one or two of the proposed solutions.

As far as I know, this is the extent of experimental evidence for quantum gravity. I think it falls somewhat short of showing the inadequacy of any semiclassical gravitation theory, including the standard, naive $G_{ab} = \langle T_{ab}\rangle$ prescription I've been considering. What is most interesting though is precisely the paucity of the evidence. There are not many experiments and those there are have not been looked at very carefully. That the problems with the experiments (and their interpretation) are so obvious and yet entirely unremarked shows, I think, that the experiments hold very little interest for researchers either in quantum mechanics, general relativity or quantum gravity. What this indicates is that, for most, the issue is not to be decided on the basis of experimental investigation. Given the deep and persistent scrutiny applied to experiments that merely confirm predictions of well established theories (the discovery of the top quark for instance), this lack of attention indicates that the conviction that gravity is quantized derives from another source.

Let me now turn, very briefly, to theoretical problems.

17.2 Problems of Theory

Mathematical physicists have identified a number of problems with the formulation of a viable semiclassical gravitation theory. These are outlined in a number of places. The following list is compiled from (Wald 1994) and (Butterfield and Isham 2001).

- The expectation value $\langle T_{\mu\nu} \rangle$ needs to be regularised to avoid divergences. Wald has done this, but there remains an ambiguity in its definition. Since his regularisation procedure is not scale invariant, there is a problem determining two conserved local curvature terms. The presence of a natural length scale for the theory would resolve this ambiguity, but it is not clear how to determine this scale.
- Some solutions of the semiclassical Einstein equations are unstable. Small changes in initial conditions produce dramatically different solutions. Some solutions have runaway behavior. Thus we need a way to distinguish physically acceptable solutions from those that are not.
- There is trouble with choosing the quantum state. "In addition," observe Butterfield and Isham, "if $|\psi_1\rangle$ and $|\psi_2\rangle$ are associated with a pair of solutions γ_1 and γ_2 to $[G_{\mu\nu} = k\langle T_{\mu\nu}\rangle]$, there is no obvious connection between γ_1 and γ_2 and any solution associated with a linear combination of $|\psi_1\rangle$ and $|\psi_2\rangle$. Thus the quantum sector of the theory has curious non-linear features, and these generate many new problems of both a technical and a conceptual nature."

These are serious problems and not mere chimeras to be banished in the bright light of philosophical reflection. But neither are they so profound to have alone derailed a dedicated research program. In the last few decades, the quantum gravity community has faced extraordinary challenges—many of which resulted in unqualified defeats for the quantum gravity program. For example the non-renormalizability of quantum gravity cast serious doubt on the proposition that the tools developed for quantum field theory could be of any use in quantizing gravity. The sheer number of programs that have flourished and then, in turn, withered in the field of quantum gravity indicates the diversity and severity of the problems to be overcome before a full theory of quantum gravity may be harvested. These problems, any one of which, perhaps, is as severe as all those facing a non-quantized gravity program, have served rather to energize than to daunt the quantum gravity community. So since neither empirical evidence nor theoretical issues suffice to make the case, why then is the conviction that gravity must be quantized so pervasive?

17.3 Problems of Metatheory

While a dedicated research program could have withstood the various conundrums outlined above, the truth of the matter is that no real research program ever sprang up. Despite Leon Rosenfeld's urging that physicists take seriously the possibility of coupling classical to quantum field, few ever did. Indeed, I know of no-one, since Wald's axiomatization of QFT in CST, who has seriously proposed trying to treat such a theory as fundamental. Why not? It is here that we encounter meta-theoretical positions.

Some, no doubt, are convinced by the arguments mentioned above. For example, Wald (1999), when asked what is wrong with the semiclassical Einstein equation, repeated essentially Page and Geilker's many-worlds objection. But others in the quantum gravity community seem motivated by more abstract concerns. Hawking, Salam, Davies and many others have advocated quantum gravitation as an essential part of a unified physics. Others advocate a unificationist position without articulating it explicitly. For example, Carlo Rovelli (2001) maintains that "we have learned from GR that spacetime is a dynamical field among the others, obeying dynamical equations, and having independent degrees of freedom. A gravitational wave is extremely similar to an electromagnetic wave. We have learned from QM that every dynamical object has quantum properties, which can be captured by appropriately formulating its dynamical theory within the general scheme of QM.

"*Therefore*, spacetime itself must exhibit quantum properties. Its properties, including the metrical properties it defines, must be represented in quantum mechanical terms. Notice that the strength of this "therefore" derives from the confidence we have in the two theories, QM and GR." It seems clear that Rovelli is using some kind of thesis about the unity of nature to extend our evidence of quantized fields to cover the case of the gravitational field. While not a complete sampling, I will take it that an important meta-theoretical impetus for quantizing gravity follows from notions of unification.

One hears a great deal about the unity of science but is rarely sure what is meant by the phrase. Oppenheim and Putnam (1958) faced, some years ago, the same quandary. Wishing to elevate the unity of science to a provisional regulative principle for the theory of science, they found it necessary first to specify its connotation. They began by enumerating three concepts of unity: 1. "Unity of language" where all the terms of science may be defined using those of one discipline; 2. "Unity of Laws" where all the laws of science can be reduced to those of one discipline; 3. "Unity of Science in the strongest sense" where the internal structure of the distinguished discipline is unified. Their treatment of the subject is not entirely relevant to my present concern. Theirs was a vision of a hierarchical structure whose various levels could be seen as reducible to the level below until, at the base, would be found a single discipline capable of supporting the entire edifice.

Considerable effort, by philosophers of science, has been expended, depending on what view of science is being promulgated, trying either to undercut or to bolster this conception. But one part of Oppenheim and Putnam's definition plays no part in their analysis. This is the idea that unity means that the single sciences must be internally unified. Clearly though it is this latter conception of unification that is at issue in discussing the unification of GR and the Standard Model. Here reductionist issues do not, at least at first blush, come into play. For here there is no suggestion of a deeper, more fundamental level of description. That is to say, we are not, for example, attempting to reduce a discipline like chemistry with its own laws to physics and in the process to show that its laws can be derived from those of physics, or that there are no new fundamental entities or processes, beyond those of physics, involved in chemical reactions. Rather, we are attempting to evaluate two distinct approaches to constructing a fundamental theory of physics. Already we are at the level of fundamental physical interactions and desire a comprehensive and, yes, unified account of the interactions. We wish to know if unification can decide the issue for us.

To find out, the first step is to answer the question "What is the nature of this unity?" Is it a unity of ontology, of methodology, of predictive content or of something else? A very general division can be made among theses of unification. On the one hand there are theses concerning nature itself—that it is unified in some way. On the other hand are theses about how to do science—the logical form theories must take, rules of thumb for constructing new theories, etc. This division corresponds to Morrison's (1994) and, loosely, to Hacking's (1996) (he introduces a further division into theses of how to do science). The idea expressed by both Hacking and Morrison is that there is no necessary connection between the unity of nature and the unity of scientific method. Morrison uses the example of the electro-weak theory. In constructing the theory, physicists were guided by analogy with electromagnetism. Then later they found that electromagnetism and the weak force could be subsumed under a single theoretical framework and thus unified. Morrison argues that this unification is not complete in the sense of a unified ontology. For example, electromagnetic interactions are mediated by a massless force carrier while the force carrier in weak interactions is massive. This distinction comes from the particular way in which the electroweak symmetry is broken, but nevertheless, it introduces a sharp division in the ontology of the two theories. So here we have a unity of theoretical structure but no unity of ontology.

Ian Hacking has spoken out against the idea of a unified science. He advocates an increasingly popular pluralism in the sciences and, in particular, rejects the idea that some kind of universal method characterizes scientific activity. While remaining neutral concerning his conclusions, I will adopt certain of Hacking's accounts of unity which, I think, illuminate the issue of unification and quantum gravity. Hacking provides a tripartite taxonomy of unification "theses": metaphysical, concerning what there is; "practical precepts"; and theses of scientific reasoning—e.g., logical and methodological imperatives. Of these, only the metaphysical will interest me here. For I am not so much concerned with the implementation of unificationist ideals as with the ideals themselves.

Under the heading of metaphysical theses, Hacking includes three distinct notions—interconnection, structure and taxonomy. The first of these implies that, at root, all phenomena are related in some way, that no class of phenomena can be fully characterized in isolation. His example is Faraday's conviction that light must be affected by magnetic fields. (Lest this example be misunderstood, let me make clear that Faraday's claim here is not that light and magnetism are aspects of the same electromagnetic field, but rather that *all* phenomena are to some degree mutually conditioned.) As Hacking characterizes it, such a thesis is fully compatible with a non-quantized gravity along the lines of QFT in curved spacetime. Clearly such a theory allows for, indeed demands, fundamental interaction between gravitational and quantum fields. If the unity of physics is to be characterized in this way, such unity provides no clear motivation for quantizing the gravitational field. For example, the thesis does not require that all fields share essential features but rather that they have domains of overlap and interaction with each other. In QFT in curved spacetime, we have that.

I will return to the structural thesis which, for Hacking, has the best shot at requiring a quantized gravitational field and first address the taxonomic thesis: "there is one fundamental, ultimate, right system of classifying everything: nature breaks into what have been called 'natural kinds'." By itself this thesis is entirely consistent with a classical gravitational field. It is already commonly supposed that there is something unique about this field that makes it stand out on its own. To claim that all fields, objects and what have you may be uniquely specified according to some overarching taxonomical classification adds little of relevance to the project of finding out what this unique something might be.

On the other hand, a denial of the taxonomical thesis might prompt one to wonder if the semiclassical approach is coherent at all. For example, if the gravitational field cannot be notionally separated from the electromagnetic field, then it makes no sense to quantize one and not the other. I will not address this issue fully, but will attempt to deflect it as follows: if the thesis is rejected we may not *appeal* to taxonomic unification in constructing our theories. But on the other hand, we may still find, as a technical matter, that we are able to construct a theory that does in fact distinguish between the gravitational fields and others. My opinion is that, even among those who deny the taxonomic thesis about *things*, few will maintain the much stronger view that *no* separations can be made between the types of entities, fields, processes etc. that populate the world. It is this stronger claim that is required to undercut the semiclassical approach to gravitation.

Then of Hacking's metaphysical theses, only the structural remains. His characterization of this thesis is somewhat vague. He appeals to Wittgenstein's notion that "there is a unique fundamental structure to the truths about the world." The idea seems to be that we can discover all the truths about the world if we have access to the core truths—these presumably include those of logic, mathematics and some central or fundamental physical principles. As far as I am aware, the primary purpose for the structural thesis is to support ideas about the transitivity of scientific confirmation. For example, Michael Friedman (1983) uses the image of a unified structure of scientific truths to illustrate how such diverse phenomena as gas behavior and chemical bonding can be explicated by appeal to the molecular hypothesis. On this basis, he argues that unified theoretical structures can be better confirmed than can dis-unified structures. This is because confirmation can come from a wide variety of phenomena. Note though that this use of the structural thesis tells us very little about the nature of the scientific truths involved. The postulation of a unified structure of truths may be useful for many different purposes, but not for specifying the *content* of a good theory.

Hacking alludes to an extreme version of the structural thesis. This version affirms the existence of a single "master law" that is sufficiently rich to allow for the derivation of all other laws from it alone. It is only this version, he says, that would fall should there be "no connection between gravitational phenomena and electromagnetic phenomena." Despite the dubiousness of the existence of such a law, I still do not see any fundamental problem its existence would pose for QFT in curved spacetime. The claim would have to be that some one physical process is at root the sole process operative in the world, and that all other processes are successive concatenations and permutations of this one process. So here's a master law: the quantum fields of the standard model propagate in a curved spacetime and the "back reaction" of these fields on the metric is governed by the semiclassical Einstein equation. This is not very impressive as master laws go, but it seems to satisfy the principle in question.

Finally I want to correct a significant defect in Hacking's account of the metaphysical theses of scientific unity—he does not mention the most obvious of the claims a unificationist might make. This is the claim that there is, at root, one and only one kind of matter.

If the claim is meant as an expression of concern about unduly inflating the number of entities the theory identifies, then there seems no reason to decide in favor of full quantization over the semiclassical picture. There are no extra things associated with semiclassical gravity that are absent in a full quantum gravity. Indeed the graviton, or whatever is presumed to carry the gravitational force, is absent in semiclassical gravity. Here the metric couples directly to the stress energy of the quantized matter field. There is, to be sure, the metric field on spacetime, but it would, in any case, be present in full quantum gravity. The ontological structure of the theory suffers no enlargement in the semiclassical case. A unification strategy based on parsimony of ontology thus affords no advantage to quantizing the Einstein tensor.

One could make the more radical ontological objection that it is precisely the presence in the theory of both classical and quantum fields that decides the case for quantization. Surely this duality is contrary to the very idea of a unified quantum description of nature. But I take this to be nothing other than the point at issue. Does unification, as a guiding principle rule out the co-existence of classical with quantum fields? To answer this question by identifying unity with a thorough-going quantum mechanical description is to not answer at all. One could, presumably level the same sort of objection against the claims of unification of the electro-weak theory. One might adopt a position according to which unification demands electrically neutral force carriers. The suggestion might be that, for a unified ontology, the nature of the fields must be consistent and thus that a charged force carrier violates ontological unity. Again, it might be objected that quantization is just not like properties such as charge and mass but is instead an essential feature of all fields. This claim may be correct. But it can hardly be taken as a principled objection to its own denial. It is a very specific claim—all fields are, of necessity quantized. Such a claim cannot be regarded as simply following from the very idea of unification.

If we adopt the unity of physics as a legitimate consideration in constructing our theories, then, for this principle to do any work, it must be sharpened. As it stands the ideal of unification tells us very little about the nature of the world. It is thus incapable of determining which theories can best describe the world.

What is the point to all this? If very general principles, like the unification of physics, are inadequate as motivations for quantizing gravity, perhaps we should seek elsewhere. For better or worse, the semiclassical project is dead. It appears that diagnoses of its demise are as varied as the programs that have replaced it. In my opinion, there is much to be learned from this observation alone. There is no single motivating principle driving the search for quantum gravity. Instead particular programs may be seen as individual responses: not to a common problem but rather to a common conviction arising from a number of different problems. To know how well a given program has succeeded we must, in part, understand the problems it is meant to solve, and the various approaches have their own sets of problems. Why quantize gravity? That is a question that ought to be reserved for particular programs of quantization, and one whose answer will, ultimately, shed light on the methods and success of these programs.

It is, I believe, worthwhile to observe that no single explanation exists for the conviction that gravity must be quantized. As a corollary, there is no consensus about what is expected from a quantum theory of gravity. To become aware of this point is to recognize the possibility of classifying programs in quantum gravity according to their motivations. In the absence of ready data for evaluating the success of quantization programs, some other criterion must be made available. Quantum gravity is sometimes portrayed as a panacea for the troubles afflicting a world described very well by GR in one regime and in another by QFT. What I wish to point out is that some of these problems may not exist as they are commonly understood. More importantly, I wish to observe that sharpening our ideas about what is wrong with non-quantized gravity will make it clear which of these problems can be expected to submit to solution via quantization and which will not. Such clarity is essential in judging the success of various quantization programs.

My thesis is simple. Standard arguments from physics as well as general (but not necessarily uncontroversial) arguments from the theory of science do not compellingly indicate that gravitation theory should be quantized. I consider this to be an important foundational issue in physics in its own right. But I suggest that producing better arguments favoring quantization, may, as an added bonus, result in new insights into how best to quantize gravity.

References

Butterfield, J. and C. J. Isham, C. J. (2001). Spacetime and the philosophical challenge of quantum gravity. In *Physics Meets Philosophy at the Planck Scale*. Craig Callender and Nick Huggett, eds. Cambridge University Press, Cambridge. Preprint cited here, gr-qc/9903072.

Callender, Craig and Huggett, Nick, eds. (2001), *Physics Meets Philosophy at the Planck Scale*, Cambridge University Press, Cambridge, pp. 1–30.

Eppley, K. and Hannah, E. (1977). The Necessity of Quantizing the Gravitational Field. *Foundations of Physics* **7**, 51–65.

Friedman, M. (1983). *Foundations of Spacetime Theories*. Princeton University Press, Princeton.

Hacking, I. (1996). The Disunities of Science. In: *The Disunity of Science*. P. Galison and D. Stump, eds. Stanford University Press, Stanford.

Kibble, T.W.B. (1981). Is a semiclassical Theory of Gravity Viable? In: *Quantum Gravity 2, A Second Oxford Symposium*. C.J. Isham, R. Penrose and D.W. Sciama, eds. Oxford: Clarendon, 63–80.

Mattingly, James (in preparation). Why Eppley and Hannah's Experiment Isn't.

Morrison, M. (1994). Unified Theories and Disparate Things. *PSA* **2**, 365–373.

Oppenheim, P. and H. Putnam, H. (1958). Unity of Science as a Working Hypothesis. In: *Minnesota Studies in the Philosophy of Science*. Vol.II, *Concepts, Theories and the Mind-body Problem*. H. Feigl, M. Scriven and G. Maxwell eds. University of Minnesota Press, Minneapolis.

Page, D.N. and Geilker, C.D. (1981). Indirect evidence for quantum gravity. *Physical Review Letters* **47**, 979–982.

Rosenfeld, L. (1963). On quantization of fields. *Nuclear Physics* **40**, 353–356.

Rovelli, C. (2001). Quantum spacetime: what do we know? In: *Physics Meets Philosophy at the Planck Scale*. Craig Callender and Nick Huggett eds. Cambridge University Press, Cambridge, 2001. Preprint cited here http://xxx.lanl.gov/gr-qc/9903045.

Wald, R.M. (1994). *Quantum Field Theory in Curved Spacetime and Black Hole Thermodynamics*. Chicago University Press, Chicago.

Wald, R.M. (1999). Private Communication.

Notes

[1] Since the presentation of this paper at HFGR5, I have shown that, for example, their measurement device exits only within its own black hole. For a presentation of this and other problems see (Mattingly in preparation) as well as (Calendar and Hugget 2001, 6–12). Both of these present arguments that are much more careful than the hasty remarks above.

[2] The raw "many-worlds" understanding of Everett's relative state formulation of quantum mechanics is rarely taken seriously. Most advocates of that formulation now combine it with some version of decoherence or consistent-histories for which the claim that the expectation value after an experiment is the same as it was before the experiment is not at all obvious. See, e.g., (Omnès 1999) for a discussion of how a decoherence reading of the many worlds interpretation prevents interference in a "real" measurement (Chapter 19), and also, in general, how to understand this interpretation without requiring the kind of overlap envisioned by Page and Geilker.

18

Einstein in the Daily Press:
A Glimpse into the Gehrcke Papers

Milena Wazeck

Max-Planck-Institut für Wissenschaftsgeschichte, Wilhelmstraße 44, 10117 Berlin,
Germany; wazeck@mpiwg-berlin.mpg.de

18.1 Introduction

The Max Planck Institute for the History of Science has recently acquired what has
been preserved of the Ernst Gehrcke papers.[1] All that is known about the history of
these papers is that the rest of them were lost in World War II. The following will
provide an overview of this material and a glimpse into the papers, in particular into
the Gehrcke newspaper article collection. The focus will be on Einstein's opponents in
the daily press during the run-up to the centennial of the German Society for Natural
Scientists and Physicians in the summer of 1922.

Ernst Gehrcke is known as a fervent critic of Einstein and a leading figure among
Einstein's German opponents. In particular his name is linked to a meeting at the
Berlin Philharmonic Hall in August 1920, which was organized by the anti-Semitic
agitator Paul Weyland and set up chiefly to oppose Einstein.[2]

From 1902 until 1946, Gehrcke was employed at the Physikalisch-Technische
Reichsanstalt, and became director of the department of optics in 1926. Although
Gehrcke, an experimentalist and specialist in optics, is not one of the well-known
physicists of the time, his work is recognized through the Lummer–Gehrcke plate, the
cathode-ray oscilloscope and the multiplex interference spectroscope.

Gehrcke's interests outside physics were broad, ranging from patent law and the
Paleolithic age to climatic research, which during the 1930s became an increasingly
important part of his work. He developed an artificial healing climate, which was applied as therapy for tuberculosis and other respiratory diseases in the Gehrcke climate institutes, which he founded. In fact, the majority of the papers contain material
concerning Gehrcke's medical interests, for example, correspondence with patients or
medical magazines. Furthermore, the papers contain:

- numerous offprints and booklets presenting unorthodox theories of space, time and
  gravitation, some explicitly opposing the theory of relativity,
- correspondence with the physicists Philipp Lenard, Stjepan Mohorovičic, Ludwig
  Glaser, Hermann Fricke, Johannes Stark, Otto Lummer and the philosophers Oskar
  Kraus, Melchior Palagyi, Leonore Frobenius-Kühn, and others,

- some drafts and manuscripts by Gehrcke, for example, "Über das Uhrenparadoxon in der Relativitätstheorie" (Gehrcke 1921, 428) and "Die erkenntnistheoretischen Grundlagen der verschiedenen physikalischen Relativitätstheorien" (Gehrcke 1914, 481–487), and
- all the parts rescued from the Gehrcke newspaper article collection.

18.2 The Newspaper Article Collection

In 1924 Gehrcke's booklet, *Die Massensuggestion der Relativitätstheorie*, appeared. With this he aimed to reveal the theory of relativity as a "suggestion to the masses" introduced by propaganda in the daily newspapers. The newspaper article collection is the material upon which Gehrcke based his booklet (Gehrcke 1924, 1).

To acquire the articles, Gehrcke subscribed to clipping services that sent him all articles containing the keywords "Einstein" or "relativity." Although these clipping services were very representative, they did not cover every single newspaper. Other articles from the Gehrcke collection were sent from friends. Since most research on the reception of relativity in the press has so far focused on articles appearing in the major newspapers,[3] the Gehrcke collection emerges as an unusual source due to its extraordinary richness of articles, which also came from small and regional newspapers.

In the following a glimpse is given at parts of this collection. The article collection is made up of twenty-one folders, eight of which were lost during the war. The collection contains altogether about 3000 articles—from the over 5000 constituting the original collection—(Gehrcke 1924, 1). Most of them were mounted and pasted, some were loose, and the majority published in the years 1921–1923. Gehrcke organized the folders more or less thematically in preparation for his booklet, in which he focuses on articles covering specific events, such as Einstein's trips to France, Italy, England, America, and Japan, as well as his various lectures.

18.2.1 The French folder

What may be called the French Folder is the most extensive, with over 650 articles covering Einstein's trip to France in March 1922, and reflects the overwhelming reception of this trip among the general public. The press (mostly French and German) covered Einstein's schedule in great detail; his arrival, his lectures at the Collège de France, his visit to World War I battlefields, and his cancellation of the meeting at the Académie des Sciences due to a planned boycott by its members. In addition, the daily newspapers provided popular accounts of the theory of relativity, the positions of French scientists on Einstein's theory, as well as anti-German or anti-Semitic sentiments.[4]

Einstein's late arrival, for instance, led to various speculations and anecdotes reported in the press. Under headings such as "The False Einstein"[5] and "The Fuss over Einstein in Paris"[6] it was announced:

Fig. 18.1. Headlines from the French Folder. *La France*, March 24, 1922; *Presse*, April 10, 1922; *Telegramme*, April 9, 1922; *Chicago Tribune*, April 1, 1922.

The arrival of Einstein, the first German scholar for whom the Collège de France has given an honorable reception, attracted numerous journalists, photographers and spectators to the Gare du Nord; but the gentleman who stepped off the train was not Einstein but a Polish minister received by members of the embassy. Neither the public nor the photographers recognized the mistake in time. Thus the Polish minister was admired and photographed with an interest he had not expected. A woman from the crowd shook her head and said to me: And all this for a German! It was indeed a surprise to see the alleged German being received by the Polish military attaché in uniform. In fact, Einstein had been in Brussels.[7]

Even a purely ironic article such as this is referring to the anti-German sentiments omnipresent at that time in French society. In fact, Einstein was the first German scholar to be officially received in France after World War I when anti-German attitudes were still high. The question of Einstein's nationality (Swiss or German) was extensively discussed by the newspapers after numerous French newspapers presented Einstein to the public as a Swiss mathematician, evidently to avoid anti-German sentiments:

The Société Française de Physique has just invited the celebrated mathematician Einstein to give a series of lectures on the special and general theory of relativity at the Collège de France. Mr. Einstein will arrive in Paris on March 28th. He will give six lectures, one of which will take place at the Société de Physique and one at the Société de Philosophie. He will remain in Paris for ten days. At the Académie des Sciences, Mr. Painlevé will comment on the theories of Einstein in the presence of the Swiss mathematician.[8]

This announcement immediately provoked the German papers to react, regardless of their political affiliation—a point to which I shall return later. The social democratic newspaper *Vorwärts* comments:

The Temps calls Einstein a Swiss mathematician, of course, a kraut would not be allowed to appear in Paris.[9]

And the national conservative *Leipziger Neueste Nachrichten:*

> The French hatred of the Germans has now come to a point where it can only
> be seen as comical. The derogation of everything German can, in the end
> when faced with the truth, find no other way out than to revert to falsification.
> This is no longer unspeakably cowardly, but just ridiculous. There is no future
> for this nation.[10]

The collection also includes folders containing articles on Einstein's trip to America (90 articles), to England in June 1921 (174 articles), to Italy in October 1921 (68 articles), and to Japan in November 1922 (98 articles).[11] As nationalism is a major topic in the reactions to these trips, the articles from the Gehrcke collection provide a hitherto unexploited source for insights into nationalism in science.

18.2.2 The movie folder

In 1922, the Colonna Movie Company in cooperation with a group of scientists[12] produced a movie on the theory of relativity for the general public. It is unclear whether this has been preserved. In any case, one can get an impression of the film's content and impact by the over 70 articles in Gehrcke's folder on the "Einstein movie."

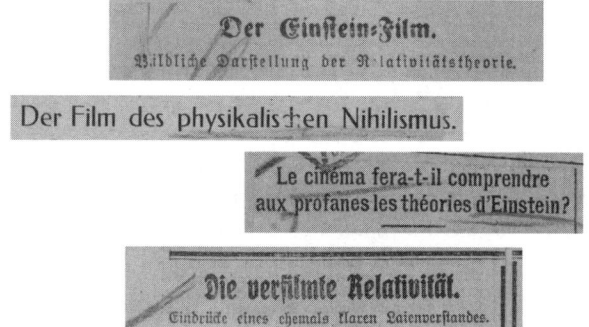

Fig. 18.2. Headlines from the Movie Folder. *Vossische Zeitung*, April 6, 1922; *Kino-Rat* No. 9/10, 1922; *Avenir* May 4, 1922; *Berliner Lokal-Anzeiger*, May 8, 1922.

In particular, this early approach to the "Public Understanding of Science" with the help of modern media (with animation and special effects) provoked many satirical reactions such as "Relativity filmed: Impressions of a previously clear layman's mind"[13] (Aros 1922):

> The accurate clock . . . plays an important role in the explanation of the theory
> of relativity. We are shown that a clock on the street indicates a completely
> different time than a clock carried by a man riding the subway . . . . And if I
> mention finally that the same train can, at the same time, be twelve, eighteen

and then finally even twenty-four meters long, and that all these measurements are correct, I will probably, like the film, have given the layman an illustrative description of the most famous of all theories.[14]

This article is somewhat unclear about whether the satire applies to the film medium or to the theory of relativity, while other statements about the film are clearly primarily intended to go against Einstein's theory. Polemics such as "The Film of Physical Nihilism"[15] (*Kino-Rat* 1922) equated the theory of relativity with ethical relativism:

> Einstein creates a universe using the imperfection of our sensory perception. He preaches to us: All your perceptions are relative, therefore you must construct a relative universe following my recipe. That is nothing but the most unproductive scientific nihilism and in accordance with the political past of the professor, who belongs to political parties, which intend to relativize the national sense of honor... All Einsteinians with their comprehension-simulating bolshevik-zionist clique cannot deny the fact that time, space and matter exist infinitely and that, from a given center, one can indeed develop an absolute world-view.[16]

This fabrication of a close connection between Einstein's "dubious character" and the "relativism" of the theory of relativity is the basic structure of argumentation in the anti-Semitic attacks against Einstein, which are already rife in the early 1920s.[17]

18.2.3 The Eclipse Folder

The Eclipse Folder is of particular interest with regard to the public discussions of the three experimental tests for general relativity: the precession of the perihelion of Mercury, the gravitational red shift, and the gravitational bending of light near a massive body.[18] At that time, the latter was only observable during a total eclipse of the sun. The test was carried out by the British astronomer Arthur Eddington during the eclipse of May 29, 1919, and his results confirmed general relativity's prediction.[19]

Fig. 18.3. Headlines from the Eclipse Folder. *Berliner Morgenzeitung*, July 20, 1922; Volksrecht, April 23, 1923; *Der Tag*, April 8, 1923; *Deutsche Zeitung*, April 27, 1923.

The announcement in the media of the results from the Eddington expedition triggered a public Einstein controversy. Beside the celebration of the "New Giant in World History,"[20] there were immediate doubts in the press about the accuracy of measurement and the significance of the results. These doubts did not diminish for many years and were voiced in particular by scientists who aimed to gain public support for their dispute with relativity.

The eclipse in September 1922 and the Dutch–German Solar Eclipse expedition to Christmas Island are covered by 250 articles in the Gehrcke collection. Their tone ranges from enthusiastic to polemic, from *Einstein's Triumph (Vorwärts 1923)* to *Einstein's Fantasies* (Riem 1923b).

The science popularizer Rudolf Lämmel wrote in the Swiss social democratic newspaper *Volksrecht*:

> One of the most controversial theories of the Einstein school can be regarded as being finally confirmed by this result. ... The meaning of this confirmed result reaches far beyond the theory of general relativity and interferes deeply with our traditional physical knowledge. It is not merely the confirmation of the deflection of a light ray in a gravity field, but also the unchallengeable fact that a ray of light is of a material nature. The acceptance of the hypothetical world ether is no longer required for explaining the physical characteristics of phenomena such as light and electromagnetism .... space is empty, and the only visible thing arriving from the infinite depths of the universe to our planet is light ... [21] (Lämmel 1922).

Numerous newspapers paraphrase the explanations given by Astronomer Royal Sir Frank Dyson at a press conference on the meaning of the eclipse observation. Here the difficulties in communicating complex scientific ideas in a way that is understandable to the general public become evident. His attempt to summarize the gist of relativity culminates in the statement of a rather meaningless "general theorem." The articles, closely paraphrasing Dyson, all conclude that:

> Even if Einstein's whole theory cannot be expressed by a simple formula, the general theorem is accepted as being valid, that the characteristics of space, which until now were considered as absolute, are related to special circumstances and thus depend on special circumstances.[22]

More than 20 articles in the Eclipse Folder are explicitly anti-Einstein. Among these 20 articles were, for instance, several published by the Potsdam astronomer and Einstein opponent Johannes Riem (Riem 1922, 1922a, 1923, 1923a, 1923b, 1923c). In this series of articles Riem defends Johann Georg Soldner's formula for the gravitational bending of light on the base of classical physics (Lenard 1921) and accuses Einstein of plagiarism.[23] Riem emphasizes that:

> As shown by Soldner's activities, this effect [the deflection of light] has nothing to do with the theory of relativity. He [Soldner] indicated a physical cause, while the theory of relativity is nothing but a scientifically implausible and also philosophically impossible speculation developed on an extremely dubious basis.[24]

Riem's articles all appear in right-wing papers, while Einstein's Triumph was celebrated in the liberal papers. We shall look at this connection between political affiliation and attitude for or against relativity in more detail in the following section.

18.2.4 The Leipzig folder

The Leipzig Folder contains articles related to the centennial celebration of the German Society of Natural Scientists and Physicians in Leipzig in September 1922. Following the first wave of polemics in August 1920, the run-up to this celebration prompted the second anti-Einstein wave. As these more than 100 articles in the Leipzig folder provide a valuable source illustrating the course this anti-Einstein wave took, we shall look at this in somewhat closer detail. .

On August 5th, an article with an apparently harmless headline appeared in the *Leipziger Neueste Nachrichten*: "Is Professor Einstein coming to Leipzig?" But the message was anything but harmless. After foreign minister Walter Rathenau was murdered by right-wing extremists on June 24, Einstein was warned that he would be one of the next victims. He therefore decided to withdraw from public life for some time and cancelled his plans to give the plenary lecture on the theory of relativity at the centennial celebration.[25] These events had great resonance in the press; for example, there are nearly 60 articles in the Gehrcke collection referring to Einstein's murder threat.

Max Planck, the chairman of the German Society of Natural Scientists and Physicians, was shocked that a gang of murderers could dictate the itinerary of a scientific society and expressed this in a letter to Max von Laue on July 9.[26] On the other hand, he also saw a positive effect, as he wrote in a letter to Wilhelm Wien on the same day:

> Taken purely objectively, this switch [von Laue speaking instead of Einstein] perhaps even has the advantage that those who still believe that the principle of relativity is at bottom Jewish advertising for Einstein will be set right.[27]

Apparently Planck was convinced that it made sense to separate the theory from Einstein's person. The dangers of such a separation became more obviously apparent much later after the Nazis' rise to power when Planck felt compelled to consent to Einstein's exclusion from the Prussian Academy.[28]

The articles in the Leipzig folder are all centered on the three topics Planck addressed in his letter:

- the murder threat,
- Einstein's opponents who believed there was nothing but propaganda behind his theory,
- the centennial and its highlight: the plenary lectures by Max von Laue and Moritz Schlick on the theory of relativity.

The democratic and the right-wing press

In the following, some reactions to the three topics mentioned above will be shown from the democratic and the right-wing press in Germany. Of course, the journalistic landscape was much more differentiated than is suggested by these two camps, but it is nevertheless possible to make a rough distinction between newspapers generally supporting and newspapers generally rejecting the democratic system of the Weimar republic as such.[29] This division between the democratic and right wing press corresponds to the division between general support and rejection of the theory of relativity in the press. Thus the "democratic press" includes, for instance, the well-known liberal papers *Vossische Zeitung* and *Berliner Tageblatt* as well as the semi-official *Deutsche Allgemeine Zeitung*.[30]

The right-wing press is often affiliated with the right-wing parties of the opposition, namely the *Deutschnationale Volkspartei* (German National Peoples' Party) via its member Alfred Hugenberg, the press and movie-industry tycoon. Among the more familiar Hugenberg papers are the *Rheinisch-Westfälische Zeitung* and the nationalist newspaper *Der Tag*. The *Deutsche Zeitung* and *Die Wahrheit* are particularly well known as coming from the anti-Semitic camp. We shall begin in chronological order with the reactions to the murder threat as the first event.

18.2.5 The murder threat

After the murder threat was made known, there was an immediate and clear expression of solidarity with Einstein on the democratic side. Thus the *Berliner Tageblatt* speaks of a: "moral degeneration which prevails in broad circles of right-wing radicalism."[31] And the *Dresdener Volkzeitung* comments:

> It is a disgrace for all of Germany that a world-famous scholar can be put on the list for assassination and chased out of the country by unthinking, reactionary scoundrels ... .[32]

The *Nationalzeitung*, a national liberal newspaper, was skeptical about the reliability of the source:

> Only by hearsay has it been mentioned that Professor Einstein also belongs to the various prominent republicans against whom assassinations have been planned.[33]

At first sight, the *Nationalzeitung* seems to have been correct in speaking about "*only* hearsay": According to the statements in the trial against members of the murder gang "Organisation Consul," various names and lists circulated among the extremists, but never Einstein's.[34] The great resonance of this "hearsay" on the murder threat in the press shows that there was no doubt at all among the public that there *could* be a murder threat to Einstein. Characteristically, only the *right-wing press* voiced such doubts. *Die Wahrheit* comments:

> Einstein should not have taken such nonsense seriously; then the intended "honor" [the plenary lecture] in the grand manner would not have eluded him; for it is not believable that such crazy people who toy with murderous intentions actually exist.[35]

In view of the contemporary context, characterized by more than 350 cases of political murder motivated by right-wing radicalism from 1919 to 1922,[36] this seemingly innocuous comment can actually be understood as an attempt to downplay the real danger to Einstein's life at this time.

Another view from the right: The *Rheinisch-Westfälische Zeitung* under the heading "The Fugitive Relativity"[37] reports that:

> ... the flight he [Einstein] staged is to be interpreted as advertising, intended to make his by now considerably faded star shine in new glory, and is hardly the gist of the matter in this affair.[38]

For the right-wing press the situation was clear: Einstein's escape was not to be taken seriously and was—ultimately—nothing but propaganda.

Following the course of events, we shall now discuss the other two topics Planck mentioned: Einstein's opponents and the centennial celebration.

18.2.6 The "declaration of protest" of Einstein's opponents in the run-up to the centennial celebration

The criticism of the theory of relativity in the 1920s is interwoven with personal attacks on Einstein—the pacifist, the democrat, the internationalist, the Jew.[39] This combination is at the core of the "joint effort" by his opponents in September 1922, the "declaration of protest" in the run-up to the centennial celebration.

Among the nineteen signatories of the declaration,[40] all of them doctors or professors of physics, mathematics or philosophy, are the physicists Philipp Lenard, Ernst Gehrcke, Hermann Fricke and Ludwig Glaser. Five of the signatories[41] would later contribute to the pamphlet *100 Autoren gegen Einstein*, published in 1931 (Ruckhaber et al. 1931; Gönner 1993b). The declaration was labeled a scientific statement, but intended and understood as a political statement. Not surprisingly, it was supported by the right-wing press, as will be shown. The declaration rests upon the shared assumption of what I will call the "oppression theory," according to which criticism of the theory of relativity is oppressed by the organized use of propaganda in the scientific and public spheres. The declaration reads as follows:

> [The undersigned] deplore most deeply the deceiving of public opinion, which extols the theory of relativity as the solution to the riddle of the universe, and which keeps people in the dark about the fact that many scholars ... reject the theory of relativity ... as fundamentally misguided and logically untenable fiction. The undersigned regard it as being irreconcilable with the seriousness and dignity of German science, when a theory disputable in the highest degree is conveyed to the layman so prematurely and in such a charlatan manner, and when the Society of German Natural Scientists and Physicians is used to support such efforts.[42]

The intention of the declaration is outlined once more by one of the initiators, physicist and patent clerk Hermann Fricke, in the right-wing newspaper *Der Tag* on September 28th:

It appears as if any resistance against the theory is to be vigorously suppressed from the start ... .[43]

Evidently, Fricke felt that the Einstein opponents were being oppressed. *Die Wahrheit* prints the pamphlet, introducing it as a "protest of German and foreign scholars against the propaganda for the benefit of Professor Einstein."[44]

J.E.G. Hirzel[45] in the *Luzerner Neueste Nachrichten*, also openly argues anti-Semitically in his "explanation" of why the overwhelming majority of the democratic press just ignores the pamphlet:

> ... the major press in Germany is almost exclusively in the hands of Einstein's fellow countrymen and is not able to find any fault with him. He is their protégé and darling. A public discussion is prevented in the exclusive interest of Einsteinianism.[46]

Twenty-two articles in the Gehrcke Collection refer directly to the declaration. Most of them are published by right-wing newspapers such as *Deutsche Zeitung, Neue Preussische (Kreuz-) Zeitung* and the *Rheinisch-Westfälische Zeitung*, which reprint the declaration without comment or with sympathizing comments. Only very few are published by democratic newspapers.

For our purposes, it is useful to distinguish here between conservative-democratic newspapers and liberal-democratic newspapers. This is because the conservative newspapers (*Frankfurter Nachrichten and Düsseldorfer Nachrichten*) reprint or paraphrase the declaration without comment while the liberal newspapers (*Berliner Tageblatt, Leipziger Volkszeitung, Frankfurter Zeitung*) express criticism. The Leipziger Volkszeitung, for example, ridicules the pamphlet as a document "characterizing these luminaries of the university" (*Leipziger Volkszeitung*) and the *Berliner Tageblatt* sees the declaration in the tradition of the "spirit" of the first anti-Einstein wave:

> The embittered opponents of Einstein ... thus regard it "to be irreconcilable with the seriousness and dignity of German science" when an "unproven hypothesis" is put to a forum of mathematicians for discussion, but they apparently find it thoroughly dignified of German science to present this "unproven hypothesis" with all its difficult scientific issues to the layman in the Berlin Philharmonic Hall.[47]

Here, the *Berliner Tageblatt* unmasks in one sentence the dubious argumentation of Einstein's opponents.

18.3 Summary

An examination of the sample of newspapers that Gehrcke obtained from his clipping services clearly disproves the oppression theory by Einstein's opponents who claimed that the published opinion oppressed critical views on his theory.

On the contrary, of the more than one hundred articles in the Gehrcke collection covering the centennial of the German Society of Natural Scientists and Physicians in

1922, the majority of these articles published after the publication of the declaration referred to it, some were even sympathetic towards it. While several articles in the Leipzig Folder were written by Einstein's opponents,[48] not a single one was written by one of his followers.

At least to some extent the articles in the collection may be considered as a representative sample as they came from clipping services. Also, as far as is known, Gehrcke did not select articles as his collection contains newspapers from the radical right as well as the *Rote Fahne*, the communist party newspaper. Thus the Gehrcke material can be used for research on a broad array of questions, in particular concerning the way in which a scientific theory can enter the public sphere either by triggering political debates or by becoming a topos of everyday thinking as when one newspaper writes: "Only a few understand the theory of relativity, but nobody understands the new tariff law" (Heldt 1921).

References

Aros (1922). Die verfilmte Relativität. Eindrücke eines ehemals klaren Laienverstandes. *Berliner-Lokal-Anzeiger*, May 8.

Biezunski, Michel (1991). *Einstein à Paris. Le temps n'est plus ... .* Presses Universitaires de Vincennes, Paris.

Crelinsten, Jeffrey (1980). Einstein, Relativity and the Press. In *Physics Teacher* **18**: 115–122.

— (1980a). Physicists Receive Relativity. In *Physics Teacher* **18**: 187–193.

Earman, John and Glymour, Clark (1980). Relativity and Eclipses: The British Eclipse Expeditions of 1919 and Their Predecessors. In *Historical Studies in the Physical Sciences* **11**: 49–85.

Elton, Lewis (1986). Einstein, General Relativity, and the German Press. 1919–1920. In *Isis* **77**: 95–103.

Fricke, Hermann (1922). *Der Tag*, September 28.

Gehrcke, Ernst (1914). Die erkenntnistheoretischen Grundlagen der verschiedenen physikalischen Relativitätstheorien. In *Kant-Studien*, Bd. XIX. de Gruyter, Berlin, 481–487.

— (1921). Über das Uhrenparadoxon in der Relativitätstheorie. In *Die Naturwissenschaften*, Bd. IX. Springer, Heidelberg, 482.

— (1924). Die Massensuggestion der Relativitätstheorie. Meusser, Berlin.

Gönner, Hubert (1993a). The Reaction to Relativity Theory I: The Anti-Einstein Campaign in Germany in 1920. In *Science in Context* **6**: 107–133.

— (1993b). The Reaction to Relativity III: A Hundred Authors Against Einstein. In *Einstein Studies*, vol. 5, Birkhäuser Boston, 248–273.

Grundmann, Siegfried (1967). Das moralische Antlitz der Anti-Einstein-Liga. In *Wissenschaftliche Zeitschrift der TU Dresden* No. 5.

— (1998). *Einsteins Akte. Einsteins Jahre in Deutschland aus Sicht der deutschen Politik.* Springer, Berlin.

Gumbel, Emil (1922). *Vier Jahre politischer Mord.* Verlag der Neuen Gesellschaft, Berlin-Fichtenau.

Heilbron, John Lewis (1988). *Max Planck. Ein Leben für die Wissenschaft 1858–1947.* Hirzel, Stuttgart.

Heldt, A (1921). Wirtschaftliche und politische Gegenwartsfragen in Amerika. Esslinger Zeitung, March 9.

Hentschel, Klaus (1992). *Der Einstein-Turm. Erwin F. Freundlich und die Relativitätstheorie. Ansätze zu einer dichten Beschreibung von institutionellen, biographischen und theoriegeschichtlichen Aspekten.* Spektrum, Heidelberg (u.a.).

Hermann, Armin (1994). *Albert Einstein. Der Weltweise und sein Jahrhundert.* Piper, Munich (u.a.).

Hirzel, J.E.G. (1922). *Luzerner Neueste Nachrichten*, October 28.

Jaki, Stanley L. (1978). Johann Georg von Soldner and the Gravitational Bending of Light, with an English Translation of His Essay on It Published in 1801. In *Foundations of Physics* **8**: 927–950.

Kirsten, Christa and Treder, Hans-Jürgen (1979). *Albert Einstein in Berlin 1913–1933*, vol. 1. Akademie-Verlag, Berlin.

Kino-Rat Heft 9 / 10 (1922). Der Film des physikalischen Nihilismus.

Kleinert, Andreas (1993). Paul Weyland, der Berliner Einstein-Töter. In *Naturwissenschaft und Technik in der Geschichte. 25 Jahre Lehrstuhl für Geschichte der Naturwissenschaft und Technik am Historischen Institut der Universität Stuttgart*, ed. H. Albrecht. Verl. für Geschichte der Naturwiss. und der Technik: Stuttgart, 198–232.

Kühn-Frobenius, Leonore (1922a). Die Relativitätstheorie auf dem Naturforschertag. (Part 1) *Rheinisch-Westfälische Zeitung*, September 19.

— (1922b). Die Relativitätstheorie auf dem Naturforschertag. (Part 2). *Rheinisch-Westfälische Zeitung*, September 20.

— (1922c). Die Hundertjahrfeier Deutscher Naturforscher und Aerzte in Leipzig. (Part 1) *Frankfurter Nachrichten und Intelligenz-Blatt*, September 20.

— (1922d). Hundertjahrfeier Deutscher Naturforscher und Aerzte in Leipzig. (Part 2) *Frankfurter Nachrichten und Intelligenz-Blatt*, September 22.

Lämmel, Rudolf (1923). Einsteins Triumph. Die neue Bestätigung der Relativitätstheorie, *Volksrecht*, April 23.

Lenard, Philipp (1921). Über die Ablenkung eines Lichtstrahls von seiner gradlinigen Bewegung durch die Attraktion eines Weltkörpers, an welchen er nahe vorbeigeht, von Johann Georg Soldner 1801. Mit einer Vorbemerkung von P. Lenard. In *Annalen der Physik* **65**: 593–604.

— (1922a). Ueber Aether und Uräther. Ein Mahnwort an deutsche Naturforscher. *Deutsche Zeitung*, September 15.

— (1922b). Einstein und die deutschen Naturforscher. *Süddeutsche Zeitung*, September 15.

Mendelssohn, Peter de. (1982). *Zeitungsstadt Berlin.* Ullstein, Frankfurt, M. Berlin, Vienna.

— Moores, Kaaren M. (1997). *Presse und Meinungsklima in der Weimarer Republik.* Univ. Diss, Mainz.

Pais, Abraham (1994). *Einstein Lived Here*. Clarendon Press, Oxford.

Renn, Jürgen et al. (1999). Albert Einstein: alte und neue Kontexte in Berlin. In *Die Königlich Preußische Akademie der Wissenschaften zu Berlin im Kaiserreich*. J. Kocka, ed. Akademie Verlag, Berlin, 333–354.

Riem, Johannes (1922). Nachdenkliches zur Sonnenfinsternis im September. *Rheinisch Westfälische Zeitung*, July 29.

— (1922a). Von der deutschen Sonnenfinsternis-Expedition. *Deutsche Tageszeitung*, July 27.

— (1923). Die Bestätigung des Soldner-Effekts. *Rheinische-Westfälische Zeitung*, April 23.

— (1923a). Die Bestätigung des Soldner-Effekts. *Weser-Zeitung* (Bremen), April 5.

— (1923b). Die Einsteinschen Phantasien. Sonnenfinsternisbeobachtung und Einsteineffekt. *Deutsche Zeitung*, April 27.

— (1923c). Zu Einsteins Relativitätstheorie. *Mannheimer General-Anzeiger*, April 5.

Rowe, David. (2002) Einsteins Encounters with German Anti-Relativists. In *The Collected Papers of Albert Einstein, vol. 7, The Berlin Years: Writings 1918–1921*. Princeton University Press, Princeton, 101–113.

Ruckhaber, Erich, Israel, Hans, and Weinmann, Rudolf (eds.) (1931). *100 Autoren gegen Einstein*. Voigtländer: Leipzig.

Sabrow, Martin (1994). *Der Rathenaumord. Rekonstruktion einer Verschwörung gegen die Weimarer Republik*. Oldenbourg, Munich.

Schwerdt (1922). Die Relativitätstheorie im Film. In *Tägliche Rundschau*, March 22.

Seelig, Carl (1954). *Albert Einstein. Eine dokumentarische Biographie*. Europa-Verl, Zürich (u.a.).

Stöber, Rudolf (2002). *Deutsche Pressegeschichte. Einführung, Systematik, Glossar*. UVK-Medien, Konstanz.

Vorwärts (1923). Einsteins Triumph, April 17.

Notes

[1] These papers will be digitized and made accessible within the framework of a research project at the Max Planck Institute for the History of Science. As part of this project, ...y dissertation deals with amateur scientists' opposition to the theory of relativity in the early 20th century.

[2] See (Gönner 1993, 107–133; Rowe 2002) for further references. For Weyland see (Kleinert 1993, 198–232). The typescript of Gehrcke's talk with handwritten corrections is preserved in the Gehrcke papers.

[3] Pais, for example, focused on the *New York Times*. See (Pais 1994); Crelinsten concentrated on the *Times* (London) and the *New York Times*. See (Crelinsten 1980, 115–122; 1980a, 187–193); Elton focused on leading German newspapers, particularly on the *Vossische Zeitung*. See (Elton 1919–1920, 95–102).

[4]For the reaction in France, see in particular (Biezunski 1991).

[5]"Der falsche Einstein," *12 Uhr Mittagszeitung*, March 31, 1922.

[6]"Das Einsteintheater in Paris," *Allgemeine Zeitung für Mitteldeutschland*, April 2, 1922.

[7]"Die Ankunft Einsteins als des ersten deutschen Gelehrten, dem das Collège de France einen ehrenvollen Empfang bereitet, hatte zahlreiche Journalisten, Photographen und Zuschauer nach der Gare du Nord gebracht; aber der Herr, der dem Zuge entstieg, war nicht Einstein, sondern ein polnischer Minister, der von Mitgliedern der Gesandtschaft empfangen wurde. Weder das Publikum noch die Photographen erkannten den Irrtum rechtzeitig. So wurde der polnische Minister mit einem Interesse bestaunt und photographiert, das er nicht erwartet hatte. Eine Frau aus dem Volke meinte zu mir kopfschüttelnd: Und das alles für einen Deutschen! Es war in der Tat eine Ueberraschung, den vermeintlichen Deutschen vom polnischen Militärattaché in Uniform empfangen zu sehen. In Wirklichkeit hatte sich Einstein in Brüssel aufgehalten." "Der falsche Einstein," *12 Uhr Mittagszeitung*, March 31, 1922.

[8]"La société française de physique vient d'inviter le célèbre mathématicien Einstein à venir faire, au Collège de France, une série de conferences sur les théories de la rélativité simple et généralisée. M. Einstein arrivera Paris le 28 mars. Il donnera six conférences, dont une la Société de Physique et une la Sociéte de Philosophie. Il restera à Paris une dizaine de jour. M. Painlevé fera un commentaire, à l'Académie des Sciences, sur les théories d'Einstein, en présence du savant mathématicien suisse." *Le Temps, Ère Nouvelle, Victoire, Èclair, Petit Parisien, Rappel*, March 21, 1922.

[9]"Der "Temps" nennt Einstein einen Schweizer Mathematiker, natürlich, ein Boche dürfte in Paris nicht auftreten." *Vorwärts*, March 31, 1922

[10]"Hiermit ist der Deutschenhaß der Franzosen an dem Punkt angelangt, wo er nur noch komisch wirkt. Die Verkleinerung alles dessen, was aus Deutschland kommt, die sich schließlich vor der Wahrheit nicht anders helfen kann als dadurch, daß sie fälscht. Das ist nicht mehr unsäglich erbärmlich, das ist nur noch lächerlich. Dieser Nation kann nicht die Zukunft gehören." *Leipziger Neueste Nachrichten*, March 22, 1922.

[11]The small Swedish Folder is unusual in so far as it mainly contains material from the 1930s and 1950s, including articles on the occasions of Einstein's 75th birthday (on March 14, 1954) and his death (on April 18, 1955), some caricatures and cartoons.

[12]Georg Nicolai, Rudolf Lämmel, Otto Buek, Otto Fanta

[13]"Die verfilmte Relativität. Eindrücke eines ehemals klaren Laienverstandes." (Aros 1922).

[14]"Eine große Rolle bei der Erklärung der Relativitätstheorie spielt ... die richtiggehende Uhr. Es wird uns gezeigt, daß eine Uhr, die auf der Straße geht, ganz andere Zeiten anzeigt, wie eine Uhr, die ein Mann bei sich hat, wenn er mit der Untergrundbahn zu fahren hat. ... Wenn ich schließlich noch erwähne, daß ein und derselbe Eisenbahnzug einmal zwölf, dann wieder achtzehn und

schließlich sogar vierundzwanzig Meter lang sein kann, und daß all diese Maße richtig sind, werde ich wohl ebenso wie der Film dem Laien eine anschauliche Darstellung der berühmtesten aller Theorien gegeben haben." (Aros 1922).

[15]"Der Film des physikalischen Nihilismus."

[16]"Einstein baut ein Weltgebäude auf den Unvollkommenheiten unserer Sinneswahrnehmungen auf. Er predigt uns: Alle deine Wahrnehmungen sind relativ, folglich mußt du dir ein relatives Weltall nach meinem Rezept zurechtzimmern. Das ist nichts anderes als wissenschaftlicher Nihilismus unfruchtbarster Art im Einklang mit der politischen Vergangenheit des Professors, der Parteien angehört, die die Relativität des nationalen Ehrgefühls auf ihre Fahne geschrieben haben. ... Alle Einsteinler mitsamt ihrem Verständnis heuchelnden bolsche-zionistischen Klüngel können die Tatsache nicht aus der Welt schaffen, daß Zeit, Raum und Materie unendlich bestehen und daß man von einem gegebenen Mittelpunkt aus sehr wohl ein absolutes Weltbild konstruieren kann." (*Kinorat* 1922).

[17]See (Rowe 2002; Gönner 1993a; Grundmann 1967; Grundmann 1998, esp. p. 142ff.).

[18]The Einstein Tower is represented in the Gehrcke Collection in a folder with 31 articles, mainly featuring well-known photographs. For a historical discussion of the role of the Einstein Tower in the experimental verification of general relativity, see (Hentschel 1992).

[19]For details and a modern assessment see (Earman and Glymour 1980).

[20]A headline from the *Berliner Illustrierte Zeitung*, December 14, 1919.

[21]"Eine der umstrittensten Theorien der Einsteinschen Lehre darf durch dieses Ergebnis wohl endgültig als bestätigt gelten. ... Die Bedeutung des nunmehr gefundenen Ergebnisses geht weit hinaus über die allgemeine Relativitätstheorie und greift aufs Tiefste in unsere bisherige physikalische Erkenntnis ein. Es handelt sich ja nicht allein um den bloßen Nachweis der Ablenkung des Lichtstrahls in einem Schwerefeld, sondern um die nun nicht mehr zu bestreitende Erkenntnis, daß der Lichtstrahl materieller Natur ist. Es bedarf fortan nicht mehr der Annahme des hypothetischen Weltäthers, um die physikalischen Erscheinungen des Lichtes und der Elektrizität zu erklären ... der Raum ist leer, und das einzige, was aus den unendlichen Tiefen des Universums wahrnehmbar bis zu unserem Planeten gelangt, das Licht ... " (Lämmel 1922).

[22]"Wenn auch die ganze Theorie Einsteins sich durch eine einfache Formel nicht ausdrücken lasse, so werde der allgemeine Lehrsatz als gültig angenommen, daß die Eigenschaften des Raumes, die bisher als absolut gegolten haben, in einem Verhältnis zu besonderen Umständen stehen, daß sie also sich nach besonderen Umständen richten." "Die Bestätigung der Relativitätstheorie," *Berliner Börsen-Courier*, April 16, 1923; "Der Triumph der Einsteinschen Theorie," *Vossische Zeitung*, April 10, 1923; "Einsteins Relativitätstheorie endgültig bestätigt?," *Neue Preussische Zeitung* (Kreuz), April 16, 1923; "Bestätigung der Einsteinschen Relativitätstheorie," *Berliner*

Börsen-Zeitung, April 17, 1923, "Die Bestätigung der Einstein-Theorie," *Berliner Tageblatt*, April 18, 1923, "Bestätigung der Einstein-Theorie," *Vorwärts* April 16, 1923; "Einstein hat Geltung," *Dresdener Anzeiger*, April 17, 1923, "Die Einstein-Theorie bestätigt," *Elbinger Zeitung*, April 16, 1923; "Bestätigung der Einstein-schen Theorie," *Königsberger Hartungsche Zeitung*, April 16, 1923; "Einsteins Relativitätstheorie. Neue Mitteilungen," *Ostsee-Zeitung*, April 16, 1923; "Die Bestätigung der Einstein-Theorie," *Neuer Görlitzer Anzeiger*, April 17, 1923; "Einsteins Lob bei den Engländern und Amerikanern," *Generalanzeiger für Stettin*, April 17, 1923; *Schlesische Zeitung*, April 23, 1923; "Die Bestätigung der Relativitätstheorie," *Bote aus dem Riesengebirge*, April 17, 1923; "Bestätigung der Einstein-Theorie," *Lübecker Generalanzeiger*, April 22, 1923; "Englische Bestätigung der Einsteinschen Relativitätstheorie," *Der Gesellige*, April 23, 1923; "Die Bestätigung des "Einstein–Effektes"," *Dresdener Neueste Nachrichten*, April 20, 1923; "Der Triumph der Einsteinschen Theorie," *Pester Lloyd*, April 20, 1923.

[23] Philipp Lenard reprinted Soldners work in 1921 (Lenard 1921). See also (Jaki 1978).

[24] "Mit der Relativitätstheorie hat die Sache (der Lichtablenkung) nichts zu tun, wie der Vorgang Soldners zeigt. Dieser hat einen physikalischen Grund angegeben, während die Relativitätstheorie nicht ist als eine auf höchst zweifelhafter Grundlage aufgebaute naturwissenschaftlich unmögliche und philosophisch ebenso unmögliche Spekulation." (Riem 1923).

[25] "... ich bin nämlich von seiten durchaus ernst zu nehmender Menschen - von mehreren unabhängig davor gewarnt worden, mich in der nächsten Zeit in Berlin aufzuhalten und insbesondere davor, irgendwie in Deutschland öffentlich aufzutreten. Denn ich soll zu der Gruppe derjenigen Personen gehören, gegen die von völkischer Seite Attentate geplant sind." Einstein to Planck, July 6 1922, quoted in (Seelig 1954, 213f.).

[26] "So weit haben es die Lumpen wirklich gebracht, daß sie eine Veranstaltung der deutschen Wissenschaft von historischer Bedeutung zu durchkreuzen vermögen." Planck to Laue, July 9, 1922, quoted in (Hermann 1994, 281).

[27] "Rein sachlich genommen hat dieser Wechsel vielleicht sogar den Vorteil, daß diejenigen, welche immer noch glauben, daß das Relativitätsprinzip im Grunde eine jüdische Reklame für Einstein ist, eines besseren belehrt werden." Planck to Wien, July 9, 1922, quoted in (Heilbron 1988, 127).

[28] "Einstein [hat] selber durch sein politisches Verhalten sein Verbleiben in der Akademie unmöglich gemacht ..." (Kirsten and Treder 1979, 267). See also (Renn et al. 1999).

[29] See (Mendelssohn 1982, 371f.) and (Moores 1997, 76ff.).

[30] The following categorization of newspapers is based on (Mendelssohn 1982) and (Stöber 2002, esp. 202–237).

[31]" ... moralischen Verwilderung, die ... in weiten Kreisen des Rechtsradikalismus eingerissen ist." *Berliner Tageblatt*, August 5, 1922.

[32]"Es ist eine Schande für ganz Deutschland, dass ein weltberühmter Gelehrter von einem ungeistigen reaktionären Halunkentum auf die Attentatliste gesetzt und außer Landes gehetzt werden kann ... " *Dresdener Volkszeitung*, August 5, 1922.

[33]"Es sei lediglich gerüchteweise erwähnt worden, dass zu den verschiedenen prominenten Republikanern, gegen die Attentate geplant seien, auch Professor Einstein gehöre." *Nationalzeitung*, August 6, 1922.

[34]The name Einstein never appears in the comprehensive study of Martin Sabrow. See (Sabrow 1994).

[35]"Einstein hätte solchen Blödsinn nicht ernst nehmen sollen, dann wäre er der ihm zugedachten "Ehrung" [der Festvortrag] in großem Stil nicht entgangen; denn, dass es wirklich verrückte Menschen geben sollte, die sich mit dergleichen Mordabsichten tragen, ist nicht glaubhaft." *Die Wahrheit*, September 23, 1922.

[36]See (Gumbel 1922, 78).

[37]"Die flüchtige Relativität," *Rheinisch-Westfälische Zeitung*, August 5, 1922.

[38]" ... die von ihm [Einstein] in Szene gesetzte Flucht als Reklame auszulegen ist, die seinen schon merklich verblassten Stern in neuem Glanze erstrahlen lassen soll, dürfte wohl des Pudels Kern in dieser Affäre bedeuten." *Rheinisch-Westfälische Zeitung*, August 5, 1922.

[39]See, for instance, (Gönner 1993a; Grundmann 1967; Grundmann 1998; Rowe 2002).

[40]Johannes Riem, M. Wolff, A. Krauße, Josef Kremer, Ernst Gehrcke, Rudolf Orthner, Stjepan Mohorovicic, Hermann Fricke, Philipp Lenard, Melchior Palagyi, E. Hartwig, Leonore Kühn-Frobenius, Ludwig Glaser, Karl Strehl, R. Geißler, Karl Vogtherr, Sten Lothigius, Vincenz Nachreiner, and Friedrich Lipsius.

[41]Karl Strehl, R. Geißler, Karl Vogtherr, Sten Lothigius, and Vincenz Nachreiner.

[42]"[Die Unterzeichneten] beklagen aufs tiefste die Irreführung der öffentlichen Meinung, welcher die Relativitätstheorie als Lösung des Welträtsels angepriesen wird, und welche man über die Tatsache im Unklaren halt, dass viele ... Gelehrte ... die Relativitätstheorie ... als eine im Grunde verfehlte und logisch unhaltbare Fiktion ablehnen. Die Unterzeichneten betrachten es als unvereinbar mit Ernst und Würde der deutschen Wissenschaft, wenn eine im höchsten Maße anfechtbare Theorie voreilig und marktschreierisch in die Laienwelt getragen wird, und wenn die Gesellschaft Deutscher Naturforscher und Ärzte benutzt wird, um solche Bestrebungen unterstützen." *Die Wahrheit*, September 23, 1922.

[43]"Es hat den Anschein, als ob jeder Widerstand gegen die Theorie von vornherein gewaltsam unterdrückt werden sollte ... ." (Fricke 1922).

[44]"Ein Protest deutscher und auswärtiger Gelehrter gegen die Stimmungsmache zugunsten des Professors Einstein geht uns mit der Bitte um Veröffentlichung zu." "Einstein." In *Die Wahrheit*, Sepember 23, 1922.

[45]Hirzel is a pseudonym of the Swiss amateur scientist and Einstein opponent Johann Heinrich Ziegler. Zieglers quarrel with relativity against the background of his (amateur scientific) theory about the world will form part of my dissertation.

[46]" ... die große Presse ist in Deutschland fast ausschließlich in den Händen der Volksgenossen Einsteins und lässt diesem nichts anhaben. Er ist ihr Schützling und Schoßkind. Man verhindert eine öffentliche Discussion im ausschließlichen Interesse des Einsteinianismus" (Hirzel 1922).

[47]"Die verbissenen Gegner Einsteins ... betrachten es also "als unvereinbar mit dem Ernst und der Würde deutscher Wissenschaft," wenn eine "unbewiesene Hypothese" vor einem Forum exakter Wissenschaftler zur Diskussion gestellt wird, aber sie halten es anscheinend durchaus für würdig, der deutschen Wissenschaft diese "unbewiesene Hypothese" mit all ihren schwierigen wissenschaftlichen Fragen in der Berliner Philharmonie dem Laienurteil zur Erledigung vorzusetzen."

[48]See, for example, (Kühn–Frobenius 1922a, 1922b; Hirzel 1922; Fricke 1922; Lenard 1922a, 1922b).

19

Syracuse: 1949–1952

Joshua Goldberg

Department of Physics, Syracuse University, Syracuse, NY 13244-1130, U.S.A.;
goldberg@physics.syr.edu

19.1 Introduction

The year 1949 saw the publication of two papers by Peter Bergmann: *Non-Linear Field Theories* in the Physical Review [1] and, together with Johanna Brunnings, in the Reviews of Modern Physics [2], with the ambitious title *Non-Linear Field Theories II: Canonical Equations and Quantization*. These papers lay the foundation for the research of the Syracuse group working to quantize Einstein's theory of general relativity. A year later saw the publication by Paul Dirac of two papers in the Canadian Journal of Mathematics [3,4] which were based on a series of lectures he gave in Vancouver in August and September of 1948. These papers, while not concerned with general relativity, dealt with the problem of constructing a canonical formalism for a theory with constraints among the momenta and configuration space variables. Later that year, Pirani and Schild published *On the Quantization of Einstein's Gravitational Field Equations* [5]. This was their construction of the Hamiltonian for general relativity based on the ideas put forward by Dirac. This amazing flurry of work was all independent except for the stimulation by Dirac of Pirani and Schild. Dirac's papers do not mention general relativity. According to Felix Pirani [6], he and Alfred Schild attended the Vancouver lectures. Alfred immediately saw the connection with general relativity and suggested that Felix work with him on the problem for his doctoral dissertation.

Although the work in Syracuse began without knowledge of the Dirac lectures or that Pirani and Schild existed, I feel that it is important for me to comment and to compare the different basic ideas which led to different constructions of the Hamiltonian. Dirac's first work in general relativity was published in 1958 [7,8]. It is very closely related to ideas in the Vancouver lectures, but very different from the other constructions. I will take the lecturer's prerogative of including material which is outside my defined limits because the ideas belong inside. Therefore, I shall begin with a comparison of Bergmann and Dirac approaches with emphasis on what theoretical ideas motivated them. Then I will discuss the work at Syracuse in the time frame of my title.

19.2 Fundamental Motivation

Peter Bergmann began his research with the intent of bringing together Einstein's theory of gravitation with the quantum theory of fields. General relativity is a non-linear field theory in which the underlying space-time geometry is not specified. The field equations, which determine that space-time geometry, are covariant with respect to continuous coordinate transformations which are piecewise four times differentiable. Quantum field theory, as we know it, is defined on a flat given space-time background, generally Minkowski space. In this background, one can construct a Hamiltonian and the associated Poisson brackets which can be carried over to the commutation relations of the operators of the quantum field theory. For general relativity, the field is the metric. It most certainly is not flat. The first problem for a quantum theory of gravity, then, is whether one can construct a Hamiltonian without having a background space-time. Then, having done so, can one find the Poisson bracket structure for observables and the appropriate Hilbert space that can lead to a generalized quantum field theory for the gravitational field. This is the task which Peter set for himself upon arriving at Syracuse University in 1947.

In the 1949 paper he states:

> The purpose of the present program is to analyze each of the two theories for its essential and, presumably relatively permanent contributions to our present knowledge and, thus, to construct what might be called skeletonized theories. An attempt will be made to see whether such a covariant theory is at all susceptible to quantization and whether the result will be an improved theory.

However, his particular approach to the problem came from his work as an assistant to Einstein from 1936 to 1940. Therefore, he goes on to say,

> $\cdots$ the theory of relativity contains two great permanent achievements: (a) it is the only theory of gravitation which explains reasonably the equality of inertial and gravitational mass (the so-called principle of equivalence); (b) it is the only classical field theory in which the equations of motion of particles in the field are contained in the field equations, instead of being logical juxtapositions.

The latter statement was based on the then recent determination of the equations of motion for compact sources by Einstein, Infeld, and Hoffmann [9–11]. Therefore, it was important that one treat the full non-linear theory. In a linearized, perturbative, version with a background Minkowski space, the essential character of general relativity is lost. First of all, with linearization, a flat rigid structure is substituted for the relation between matter and geometry, the crucial property of general relativity. Second, with linearization one loses, not only the self interaction, but also the interaction between source and field which leads to limitations on the motion of matter. While one may build up these interactions by successive approximation, the important structure of the unperturbed field is buried in the formalism of approximation. Furthermore, strong fields, where quantum interactions should be very important, are not small deviations from Minkowski space.

Therefore, Peter wished to study a general class of non-linear field theories which are covariant under the group of general coordinate transformations, the invariance group, and whose field equations define the motion of particles. Covariance refers to the existence of a group of transformations which leave the field equations unchanged, but which depend on a number of arbitrary functions of the space-time coordinates. As a result, a solution of the field equations can be mapped by an invariant transformation to another solution which remains the same on an initial surface, but is different in the future. Thus, the propagation of initial data is not unique. This property results in certain identities among the field equations themselves as will become evident.

The two goals, of covariance and equations of motion for particles, led him to study a general field theory whose field equations are derived from the variation of an action whose Lagrangian density is a function of generalized variables $y_A(x)$, $A = 1 \cdots N$ and their first derivatives. It is understood that the metric tensor describing the underlying geometry is included among the y_A. Arbitrary variations of the field variables in the action,

$$S = \int_V L(y_A, y_{A,\rho}) d^4 x, \qquad (19.1)$$

which vanish on the boundary of the four-dimensional domain, V, lead to the field equations

$$L^A \equiv \partial^A L (\partial^{A\rho} L)_{,\rho} = 0, \qquad (19.2)$$

$$\partial^A L \equiv \frac{\partial L}{\partial y_A} \qquad \partial^{A\rho} L \equiv \frac{\partial L}{\partial y_{A,\rho}}.$$

He assumed that the Lagrangian is covariant and the field equations invariant under the variation induced by a general coordinate transformation that he writes as

$$\bar{\delta} y_A = u_{A\mu}{}^\nu \xi^\mu{}_{,\nu} y_{A,\mu} \xi^\mu; \qquad (19.3)$$

the Lie derivative for the transformation $\delta x^\mu = \xi^\mu$; $u_{A\mu}{}^\nu$ is linear in the field variables, in general with constant coefficients. These transformations can be shown to form a group and, because they depend on arbitrary functions, the descriptors $\xi^\mu(x)$, they lead, as noted above, to differential identities, strong conservation laws, and to constraints among the momenta and field variables. The identities and constraints are a reflection of the differentiability of the mappings induced by the transformations.

Further, the EIH result shows that the field equations determine the motion of the sources of the field. As a result, before one has a solution, one cannot predict where the particle sources of the field might be located. Therefore, Peter introduced a parametric description of the coordinates, in terms of which he wished to describe the motion,

$$x^\mu = x^\mu(t, u^i), \qquad i = 1 \cdots 3. \qquad (19.4)$$

It was Peter's hope that quantization of the Einstein theory would lead to the Schrödinger equation or it's generalization for particles. The introduction of parameters led to difficulties which were overcome in two long papers [2, 12]. However, the parameters produce four additional constraints, including the Hamiltonian. As a result, they did not lead to additional degrees of freedom for particles and the idea was quickly dropped.

Rather than the concern with non-linear field theories and general covariance, Dirac's motivation came from Lorentz invariant theories with constraints. In reference [3], he set a particle-like model in an N-dimensional configuration space with the assumption that the velocities can not be solved for in terms of the coordinates and the momenta,

$$p_n = \frac{\partial L(q, \dot{q})}{\partial \dot{q}^n}, \quad n = 1 \cdots N.$$

As a result there are constraints

$$\phi_m(q, p) = 0, \quad m = 1 \cdots M < N, \tag{19.5}$$

among the coordinates and momenta. Dirac distinguishes between *strong* and *weak* equations with zero right-hand sides. The variation of a strong equation with respect to its variables remains zero whereas the variation of a weak equation does not. The product of two weak equations is a strong equation, but there are other possibilities as well. In Dirac's usage, equations involving coordinates, velocities, and momenta may be strong.

As a result of the existence of constraints, he finds that the Hamiltonian is not unique, but one can always add a linear combination of the constraints with coefficients which may depend on the velocities as well as the coordinates and momenta. Propagation of these constraints leads to further constraints $\chi_k = 0, k = 1 \cdots K$. The totality of constraints can then be divided into first class constraints, whose Poisson brackets with all the constraints vanish modulo the constraints, and second class constraints whose Poisson brackets with other constraints do not vanish modulo the constraints. Only the first class constraints contribute to the Hamiltonian and the second class constraints can be eliminated as redundant canonical degrees of freedom. Dirac's motivation is in understanding the algebraic structure of the constraints, not in the existence of a group of invariant transformations. However, he, too, introduces an arbitrary parameter so that the time can become dynamical. He shows that this leads to the Hamiltonian as a constraint so that no new degree of freedom is added to the theory.

In the second paper, Dirac introduces a field theory in Minkowski space. He is not thinking about general covariance. But, he is concerned with Hamiltonian theories with constraints and with maintaining the four-dimensional symmetry of Lorentz invariance, while at the same time introducing an arbitrary space-like surface in terms of which to define the canonical formalism. To accomplish this, he introduces a parametric description of the Minkowski space coordinates as in (19.4) above. In order to assure that the surface $t = 0$ is space-like, Dirac introduces the time-like unit normal l_μ with the properties

$$l_\mu \frac{\partial x^\mu}{\partial u^i} = 0, \quad l^\mu l_\mu = 1.$$

Then he proceeds to define quantities on the surface which are covariant under a change in parameters, which leaves the surface $t = 0$ unchanged. For tensors, these are the normal components to the hypersurface and the forms in the hypersurface. For example,

$$V_L = V^\mu l_\mu \quad \text{and} \quad V_\mu \frac{\partial x^\mu}{\partial u^i} = V_i. \tag{19.6}$$

Clearly, the decomposition of the field with respect to the time-like normal vector of the arbitrary space-like surface reflects the Lorentz invariance of the theory. This geometric decomposition of variables led to his particular treatment of the metric tensor when he later came to discuss general relativity. Peter referred to this decomposition as "D-invariance."

I do not want to discuss this further as it will take me too far from my purpose in discussing the work of the Syracuse group. But, I think it is important to see the difference between Peter Bergmann's view and goal and that of Paul Dirac at this time. Peter was thinking about quantization of a non-linear field theory which is covariant under a group of arbitrary coordinate mappings, in which the metric is part of the dynamical stucture, and in which the equations of motion for particles will be determined. Dirac is interested in theories in Minkowski space that may have constraints either imposed or intrinsic. Dirac is thinking more algebraically and Bergmann more group theoretically. Both introduce a parametric description of the coordinates, but ultimately conclude that it is useless.

It was Alfred Schild who saw that, once one introduced an arbitrary space-like surface, the Dirac formalism could be applied to general relativity. In a straightforward application of the Dirac approach with some clever mathematical manipulations, together with Felix Pirani he constructed the first explicit expression for the Hamiltonian [5]. However, they did not complete the decomposition of the metric or of the field variables for the Maxwell field with respect to l^μ, the normal. More important, they did not examine the propagation of the constraints until later [13].

In the following sections, I will sketch the results of the Bergmann group in the early period, 1949–52. First I discuss the derivation of conservation laws and equations of motion from the invariance under diffeomorphisms. Then, in Section 4, the construction of the Hamiltonian in the parameter formalism will be presented. In Section 5, the parameters will be dropped and the secondary constraints will be examined. Section 6 will examine how the results obtained in the canonical formalism appear in the Lagrangian formalism. Finally, in Section 7, preliminary steps to a quantum theory of gravity will be described.

19.3 Invariance, Conservation Laws, and Equations of Motion

If the field equations are to be invariant in form under a mapping $y_A(x) \rightarrow y_A(x) + \bar{\delta} y_A(x)$, the Lagrange density should be unchanged in form except for the addition of a total divergence. Thus, in general we have

$$\delta L = L^A \bar{\delta} y_A(x) + (\partial^{A\rho} L \bar{\delta} y_A(x))_{,\rho} = Q^\rho_{,\rho}.$$

This can be rewritten as

$$L^A \bar{\delta} y_A(x) = (Q^\rho \partial^{A\rho} L \bar{\delta} y_A(x))_{,\rho}. \qquad (19.7)$$

Thus, when the field equations are satisfied, invariant transformations give us a conservation law. If the mapping is defined by (19.3), after integrating over an arbitrary four dimensional domain with a descriptor ξ^μ which vanishes on the boundary, we find an identity for the field equations and, as as a result, a *strong* conservation law:

$$(u_{A\mu}{}^\nu L^A)_{,\nu} + L^A y_{A,\mu} \equiv 0, \qquad (19.8a)$$

$$T^\nu_{,\nu} \equiv 0, \qquad (19.8b)$$

$$T^\nu = Q^\nu \partial^{A\nu} L\, \bar{\delta} y_A u_{A\mu}{}^\nu L^A \xi^\mu.$$

The identity in (19.8b) implies the existence of a superpotential such that

$$T^\nu = U^{[\nu\mu]}_{,\mu}, \qquad (19.9)$$

the square brackets imply skew symmetry. From the definition of T^ν, we get an expression for the field equations in terms of the superpotential:

$$u_{A\mu}{}^\nu L^A \xi^\mu = -U^{[\nu\mu]}_{,\mu} + t^\nu, \qquad (19.10)$$

$$t^\nu = \partial^{A\nu} L \bar{\delta} y_A Q^\nu.$$

Note that for $\nu = 0$, the right-hand side is free of second time derivatives. Thus, this combination of the field equations corresponds to the secondary constraints in the canonical formalism.

If there are sources P^A which are not described by the fields, so that $L^A = -P^A$, the above equation becomes a relation between the sources and the field:

$$u_{\mu A}{}^\nu P^A \xi^\mu = U^{[\nu\sigma]}_{,\sigma} t^\nu. \qquad (19.11)$$

Thus, the strong conservation law becomes a conservation law for matter interacting with the field:

$$(u_{\mu A}{}^\nu P^A \xi^\mu + t^\nu)_\nu = 0.$$

Assume that there are several localized sources — even point particles — and consider a space-like surface in which a two-surface surrounds one of the local sources. In (19.10), let $\nu = 0$, integrate the resulting expression over the space-like interior of the two-surface, and obtain

$$\oint_{\partial V} U^{[0s]} n_s dS = \int_V (u_{\mu A}{}^0 P^A \xi^\mu + t^0) dV.$$

This relates the matter and field in the interior to a surface integral which involves only field variables. Next in (19.10), let $\nu = s$ and integrate the resulting expression over the two surfaces to obtain

$$\oint_{\partial V} U^{[s0]}{}_{,0} n_s dS = \oint_{\partial V} (u_{\mu A}{}^s P^A \xi^\mu + t^s) n_s dS.$$

For localized sources P^A vanishes on the surface, so the expression says that the rate of change of some parameters defining the matter variables are related to the flux of matter through the surface. Through the field equations, one can show that the conditions on the sources are independent of the surface as long as the matter is confined within the surface. The relations above depend on the choice of descriptor ξ^μ. We may loosely think that choosing a vector normal to the space-like surface leads to an energy condition, while vectors on the surface lead to conditions on momentum or angular momentum.

These are the equations of motion for matter in general relativity [14]. As applied by Einstein, Infeld, and Hoffmann in the slow motion approximation, this leads to the equations of motion for point particles. This is also the basis for the calculations by Damour and Deruelle [15], and Iyer and Will [16].

19.4 The Parameter Formalism

The Bergmann group worked with the Einstein Lagrangian which is homogeneous quadratic in the first derivatives of the metric. Therefore, in the more general discussion, the Lagrangian was assumed to have the form

$$L = \Lambda^{A\rho B\sigma} y_{A,\rho} y_{B,\sigma},$$

and when parameters are introduced (dot$= \partial/\partial t$, $|s = \partial/\partial u^s$),

$$L' = JL, \qquad J \equiv \det\left(x^\mu{}_{|s}, \dot{x}^\mu\right) \tag{19.12}$$

which is clearly homogenous of degree 1 in the derivatives with respect to t.

With respect to the mappings (19.2), the Lagrangian is assumed to be of the same functional form except for the addition of a divergence. The invariance of the Lagrangian leads to identities among the field equations. These identities show that the definition of the momenta, $\pi^A = \partial J L / \partial \dot{y}_A$, cannot be inverted to solve for $\dot{y}_A$. As a result there are constraint equations involving only y_A and π^A which I shall write as

$$\phi_\mu(\pi^A, y_A, y_{A|s}, x^\rho{}_{|s}) = 0. \tag{19.13}$$

In addition, invariance under parameter changes yield four more constraints for the momenta conjugate to the coordinates, $\lambda_\rho = \partial J L / \partial \dot{x}^\rho$,

$$\lambda_\rho J t_{,\rho} L + y_{A,\rho} \pi^A = 0.$$

These latter equations plus the homogeneity of the Lagrangian with respect to the velocities tell us that the Hamiltonian is zero:

$$H = \lambda_\rho \dot{x}^\rho + \pi^A \dot{y}_A J L = 0,$$

which, as usual, one can show is independent of the velocities. Thus, there exists a function $H(\pi^A, y_A, y_{A|s}, \lambda_\rho, x^\rho, x^\rho{}_{|s}) = 0$ which generates the field equations. However, this Hamiltonian is not unique for one can add to it a linear combination of all the constraints.

This lack of uniqueness arises because the relationship between the momenta and velocities is singular. Therefore, one cannot invert the relation defining the momenta. However, a quasi-inverse can be found. The quasi-inverse also is not unique and it has null vectors which can be paired with those for the matrix connecting the velocities to the momenta. This lack of uniqueness can be exhibited explicitly and the Hamiltonian was constructed [12]. Thus, the lack of uniqueness results in the addition of an arbitrary linear combination of the constraints to the Hamiltonian. Up to this point, the only constraints are the *primary* constraints, those which come directly from the definition of the momenta.

Once one had the Hamiltonian, one formally formed the commutation relations based on the Poisson brackets and, looking forward to the quantum theory, said that the quantum state vector must vanish when operated on by the Hamiltonian or any of the constraints. Of course, such a formal statement was not quantization, but only an indication of the direction future work to quantize the field should go.

This is all I want to say about the theory with parameters. One soon found that the parameters were an unnecessary complication and after the fall of 1950 they were abandoned. As a result, the issue of *secondary* constraints was not examined in the parameter formalism.

19.5 Secondary Constraints

The construction of the Hamiltonian without the use of parameters was carried out by Robert Penfield [17]. He followed the technical arguments of the parameter construction. It was much simpler because one did not have the coordinates as additional variables and, therefore, no coordinate constraints. It also meant that the Hamiltonian did not have to vanish although the ambiguity with respect to the primary constraints remained.

Penfield also studied the propagation of the primary constraints [18]. He found that the propagation led to additional constraints. He recognized that these secondary constraints were just those field equations which lacked second time derivatives and therefore could be expressed directly in terms of the π^A and y_A. Propagation of the secondary constraints was carried out in several different ways [14,19,20] with the result that there were no tertiary constraints with the invariance group as defined above.

In discussing the secondary constraints, the question arose as to how they are related to the number of time derivatives in the transformation group. Therefore, Anderson and Bergmann [19] studied a generalized transformation group defined by

$$\bar\delta y_A = \sum_{p=0}^{P} f_{Ai}{}^{\mu_1\cdots\mu_p} \xi^i_{,\mu_1\cdots\mu_p} y_{A,\rho}\delta x^\rho. \tag{19.14}$$

Here the index $i = 1 \cdots I$ ranges over the number of descriptors which includes the coordinate transformations, but may be more general. The main object here was to determine the relationship between the number of derivatives of the descriptors, particularly time derivatives, and the number of constraints which result in the canonical formalism. First of all, assuming the Lagrangian only contains first time derivatives, one obtains again the condition that the relation between momenta and time derivatives cannot be inverted. There exist I null vectors for the matrix relationship and, as a result, I primary constraints. It follows that although the momenta involves $\dot{y}_A$, their transformation under (19.14) contains only P time derivatives of the descriptors. Therefore, while the generator of a canonical transformation may contain $P+1$ spatial derivatives, it will contain only P time derivatives of the descriptors. Since the generator of the canonical transformation is an integral over the spacelike surface $x^0 = $ constant,

$$ C = \int \mathcal{C} d^3 x, $$

all the spatial derivatives of the descriptors ξ^i can be removed by an integration by parts, so that one can write (the index (p) indicates the number of time derivatives of ξ^i)

$$ \mathcal{C} = \sum_{i=1}^{I} \sum_{p=0}^{P} {}^{p} A_i \xi^{ip}. $$

Assuming that (19.14) is an invariant transfomation group, it follows that the change in the functional form of the Hamiltonian is a linear combination of the primary constraints. That change is also given by the total time derivative of the generating functional which now contains time derivatives of the descriptors up to the $P + 1$th order. That is,

$$ \delta' H = \int \delta w^i g_i = \frac{\partial C}{\partial t} + [C, H] $$

where the $g_i = 0$ are the primary constraints. The change in $\dot{y}_A$ as a result of the transformation is the same as that which results from the change in the Hamiltonian. One concludes that δw^i contains a term linear in the $P + 1$th time derivative of ξ^i. Therefore, one finds that ${}^{P} A_i = 0$ are the primary constraints and the remaining ${}^{P} A_i = 0$ are the secondary constraints which may also contain a combination of primary constraints. So there are $P + 1$ levels of constraints with I constraints at each level. That means that for the coordinate transformations, where $P = 1$ and $I = 4$, there are two levels and eight constraints in total. Unfortunately, this argument does not give us an explicit answer for the Poisson brackets between constraints for at each step the answer is arbitrary up to a linear combination of the primary constraints.

19.6 Lagrangian Formalism

Around this time, Feynman [21] and Schwinger [22] had begun to formulate quantum field theory through the action using the configuration and velocity field functions as the dynamical variables. Therefore, Bergmann and Schiller undertook to investigate how the results obtained in the canonical formalism appear in the Lagrangian formalism. They undertook to study more general transformations than the coordinate transformations. Therefore, they asked for the changes in the Lagrangian due to transformations which could involve point transformations on configuration space or on the velocity space as well as the coordinate transformations. Thus, they considered a transformation $\bar{\delta} y_A = f_A$ where f_A may depend on first derivatives of the field variables as well as the y_A. These are not necessarily invariant transformations, but, if solutions are mapped into solutions, the Lagrangian may change by the addition of a divergence as well as in form. Thus, they arrive at (19.7). By restricting the transformation so that the functional form of the Lagrangian does not change by the appearance of second time derivatives in the Lagrangian, one arrives at conditions on the f_A, part of which read

$$\partial^{A\cdot} Q^0 \partial^{A\cdot} f_B \pi^B = 0, \pi^B = \partial^{A\cdot} L, \qquad \partial^{A\cdot} \equiv \frac{\partial}{\partial \dot{y}_A}. \qquad (19.15)$$

Here, π^A is understood to be a function of $y_A, \cdots, \dot{y}_A$. This condition restricts the transformations to the canonical transformations of the Hamiltonian formalism. If the definition of π^A can be inverted to eliminate $\dot{y}_A$, the generating density can be defined so that

$$f_A = \partial_A(\pi^B f_B - Q^0), \qquad \delta_A = \frac{\partial}{\partial \pi^A}. \qquad (19.16)$$

With appropriate normalization of f_A and Q^ρ relative to a quasi-inverse [20], the above equation remains valid even in the singular case.

The above considerations are true whether or not the transfomation leaves the functional form of the Lagrangian unchanged. If one requires covariance for the theory, the results of Section 3 are recovered. When one expands the identity (19.8a), one finds that the term with third time derivatives will vanish only if $u_{A\mu} = u_{A\mu}{}^0$ satisfies

$$u_{A\mu} \Lambda^{AB} = 0, \qquad \Lambda^{AB} \equiv \partial^{A\cdot} \partial^{B\cdot} L. \qquad (19.17)$$

Now, if we ask for the change in π^A as a function of $\dot{y}_A$ we find

$$\delta \pi^A = \Lambda^A{}_B \delta \dot{y}_B, \qquad (19.18)$$

so that $\delta \pi^A = 0$ for $\delta \dot{y}_A = \lambda^\mu u_{A\mu}$. This exhibits the extent of the indeterminacy of $\dot{y}_A$ for given π^A. Equation (19.18) also tells us that the momenta satisfy four primary constraints. (In general, the number of primary constraints is equal to the number of independent arbitrary functions in the invariance group.) Note, though, that if the definition of the momenta is substituted, these constraints vanish identically.

We recognize that the generating density defined in (19.16) is the zero component of (19.7). Thus, the generating function is the zero component of a differential conservation relation when the field equations are satisfied. The conserved quantity is the generating functional (Σ is the surface $x^0 = $ constant):

$$C = \int_{\Sigma} (Q^0 \partial^{A0} L \bar{\delta} y_A(x)) d^3 x. \tag{19.19}$$

In general, Q^ρ can be written

$$Q^\rho = Q_\mu{}^\rho \xi^\mu + Q_\mu{}^{\rho\sigma} \xi^\mu{}_{,\sigma}. \tag{19.20}$$

A term with the second derivative of ξ has been omitted for simplicity. It's inclusion exhibits the lack of uniqueness of the resulting superpotential through the addition of the divergence of a quantity antisymmetric in three indices. The argument is easily extended to include this term, but it complicates the description without fixing uniqueness.

Substitute (19.20) into (19.10) with $U^{[\rho\sigma]} = U_\mu{}^{[\rho\sigma]} \xi^\mu$ to obtain

$$u_{A\mu}{}^\rho L^A = -U_\mu{}^{[\rho\sigma]}{}_{,\sigma} + t_\mu{}^\rho, \tag{19.21a}$$

$$U_\mu{}^{[\rho\sigma]} = Q_\mu{}^{\rho\sigma} \partial^{A\rho} L u_{A\mu}{}^\sigma, \tag{19.21b}$$

$$t_\mu{}^\rho = Q_\mu{}^\rho + \partial^{A\rho} L y_{A,\mu}. \tag{19.21c}$$

With these identifications we find that the generating functional can be written

$$C = \int_{\Sigma} \{-U_\mu{}^{[0\sigma]} \xi^\mu{}_{,\sigma} (Q_\mu{}^0 + \partial^{A0} L y_{A,\mu}) \xi^\mu)\} d^3 x,$$
$$= \int_{\Sigma} \{-U_\mu{}^{[00]} \dot{\xi} + (U_\mu{}^{[0s]}{}_{,s} + t_\mu{}^0) \xi^\mu d^3 x. \tag{19.22}$$

Note that although $U_\mu{}^{[00]}$ is identically zero when expressed in terms of $y_A, \cdots y_A$, when the momenta, π^A are introduced, these become the primary constraints. The remaining terms are those field equations lacking second time derivatives, hence the secondary constraints. Thus the generating functional is a linear combination of the primary and secondary constraints. With appropriate boundary conditions, C is a constant of the motion. In that case, when one takes the time derivative of C, no additional constraints arise. This argument can be generalized as with Anderson and Bergmann. Again one finds that the number of primary and secondary constraints is the number of descriptors times the number of time derivatives plus one which appear in the invariant transformation law.

It is interesting that in all these arguments one starts from the fact that the Hamiltonian itself is not zero and one finds that the Hamiltonian is arbitrary up to a linear combination of the primary constraints. However, the existence of secondary constraints in the generating functional expresses the fact that the Hamiltonian also involves the secondary constraints with arbitrary coefficients. However, if the Lagrangian differs from a scalar density by a divergence, as in the case of general relativity, then the resulting Hamiltonian differs from zero also by a divergence.

In general relativity, the Hamiltonian density found is related to $t_\mu{}^0$ and that expression differs from the secondary constraints by a divergence of the superpotential (see (19.21a) with $\rho = 0$).

19.7 Quantization

In the transition to quantum theory, Peter wanted to keep as close as possible to the procedure in standard quantum field theory while recognizing that the ingredient of a fixed background spacetime would be missing. The formalism, however, depends on the existence of an invariant transfomation group. This transformation group is generated by constraints on the phase space. Therefore, the first objective was to try to preserve the algebra of the constraints in terms of the dynamical variables as operators on a Hilbert space rather than as functions or functionals on a phase space. Thus, the constraints are to keep their role as generators of invariant transformations. So, one begins by assuming that the basic Poisson brackets for the variables y_A and π^A go over to the commutation relations

$$[y_A(\mathbf{x}), \ \pi^{\mathbf{A}}(\mathbf{x'})] \ = \ \mathbf{i\hbar}\delta^3(\mathbf{xx'}),$$

where $\mathbf{x}$ is a point on the hypersurface $x^0 = $ constant. Then one hoped to find a factor ordering of the constraints such that the algebra of constraints found by Bergmann and Anderson could be carried over to the quantum algebra. That proved to be difficult and later Jim Anderson proved that such a factor ordering did not exist if the constraints were put in a formally Hermitian order [23]. On the other hand, Artie Komar has argued that the constraints should not become Hermitian operators [24]. Nonetheless, the constraints, however they may carry over as operators, are the generators of the invariant transformations and as such their algebra should be carried over to the quantum theory if at all possible. In the Ashtekar formalism, this problem appears to be resolved [25].

Observable quantities must be invariant under the transformation group. In classical theory, we can establish a frame for observers and give a particular solution in a chosen frame. We know how to construct measurable quantities, the observables, relative to a given frame with a given solution. In a quantum theory of gravity, the dynamical variables become operators and are no longer attached to a particular solution or frame, but are representative of all solutions consistent with a particular Hilbert space. That is, there is no prior frame to which to attach the operators. Therefore, the quantum operators themselves must be constructed out of invariants. So *observables* were defined to be those functionals whose commutators with the constraints vanish modulo the constraints themselves. The search for observables began in the classical theory where the algebra was known and the problem of factor ordering is not a problem. With this definition, the constraints and the Hamiltonian are observables, the constraints being trivial zero observables. But while the Hamiltonian differs from a constraint only by a divergence, the non-zero Hamiltonian is also an observable and should be related to global energy. Dirac also recognized this fact in his definition of H_{main} [26]. However, no other observables were found. But, we know that in gravity there are four independent degrees of freedom per space point apart from the constraints and the Hamiltonian. There should be observables associated with these degrees of freedom.

As observables, the constraints and the Hamiltonian were assumed to become Hermitian operators. However, the notion of Hermiticity depends on the definition of the

Hilbert space and the measure defined on that space. There are problems even if one could find a space of functionals with a measure so that a norm is defined. Not only do the basic variables (y_A, π^A) have c-number commutation relations, but so do the constraints and their canonical conjugates. Such quantities presumably have continuous spectra. But the constraints are to yield only zero when acting on a physically meaningful state vector. So one solution was that the Hilbert space itself should consist only of functionals for which the constraints are trivial operators. That is, all functionals in the Hilbert space, are annihilated by the constraints. Quantities conjugate to the constraints do not act on this space.

All of the above remarks are valid equally for canonical quantization and the Lagrangian procedure discussed by Bergmann and Schiller [20]. But, there are differences. The Lagrangian is constructed out of operators and the variations of the operators are also operators. First of all, the variations cannot be moved freely to the right side of each expression. Secondly, the operator Lagrangian is already a two-index object. If the variations were fully arbitrary, the operator field equations and conservation laws would become four -index objects. Therefore, they restricted the variations to the invariant transformations which depend on c-number descriptors. In this way they were able to recover the principal results of the canonical formalism. In addition, they were able to show how to derive the appropriate commutation relations for the dynamical variables and their time derivatives. However, at that time it was not clear whether the approach from the Lagrangian would lead to a simpler structure than the canonical formalism. As far as I know, this particular use of Lagrangian quantization has not been pursued.

19.8 Conclusion

At this point, one had a good understanding of the classical theory of a general non-linear field theory whose equations of motion are derived from an action. When there is a function group as the invariance group, there are strong and weak conservation laws. The existence of the strong laws on the one hand leads to limitations on the motion of the sources of the field and on the other to constraints whether one introduces the canonical formalism or through the Lagrangian formalism. In either case, the constraints generate the invariant transformations.

One of Peter's original goals has not been realized nor is there any sense in which one thinks it may be satisfied. That is, to be able to recover quantum equations for the sources of the gravitational field in the manner of EIH. We were able to write down a phase space expression for the superpotentials [14], but there seems to be no way to express its action locally so that something like the Schrodinger equation results. On the other hand, the conservation laws are intrinsic to the quantum structure as well as the classical formalism. Therefore, quantum gravity will impose restrictions on interacting matter fields which are its sources. However, how that will appear is still unclear and may be understood only in terms of a fully unified quantum field theory. Looking for restrictions on localized matter does not seem meaningful in quantum field theory.

While a quantum theory of gravity was not completed, the main outlines of what should be kept in the transition from classical theory to quantum theory was discussed in detail. In particular, it is the constraint algebra which should be kept because the constraints are the generators of the invariant transformations. And it is the symmetries which are important in giving meaning to the quantum states. Unfortunately, without knowing something about the topology and geometry of the Hilbert space, one can only postulate that the dynamical operators should be Hermitian.

In this connection, it is worth noting that later Komar [24] suggested that the constraints need not be Hermitian. Indeed, in quantum electrodynamics only the self-dual part of the constraints need be taken as zero operators on the Hilbert space. This is essentially the way the constraints are applied in the Ashtekar formalism [25] where some progress on these questions has taken place.

Apart from laying of the groundwork for a quantum theory of gravity with the construction of the Hamiltonian and study of the constraint algebra, the most important result of this period is the recognition that the observables must commute with the constraints and therefore with the Hamiltonian. This has the strange result that the observables appear to be frozen in time. As a result, one began to think of general relativity as an *already parametrized* theory and one sought, without success, to construct a time variable from within the geometrical structures themselves.

With the more geometrical formulation of the Hamiltonian by Dirac in 1959 [26], the lapse and the shift replace $g^{0\mu}$ and are clearly identified as arbitrary elements which are not dynamical. Essentially, these quantities are the conjugates of the primary constraints. In this structure, the Hamiltonian is constructed from the secondary constraints, the scalar constraint which is quadratic in the momenta and the three-dimensional vector constraints which are linear in the momenta. The latter generate the diffeomorphisms in the space-like surface $x^0 = $ constant while the former generates the evolution off the surface. The observables, of course, must commute with all of these constraints.

While we are now 50 years later still without a quantum theory of gravity, the fundamental ideas developed in those early years by Peter Bergmann and his students continue to form the basis of current work to the extent that one doesn't mention those ideas explicitly any longer. Today, not only is work continuing on the effort to quantize the original Einstein theory, but Einstein's dream of a unified theory is being pursued in string theory and super-gravity.

References

[1] P.G. Bergmann, Phys. Rev. **75**, 680 (1949).
[2] P.G. Bergmann and J. Brunnings, Rev. Mod. Phys. **21**, (1949).
[3] P.A.M. Dirac, Canad. J. Math. **2**, 129 (1950).
[4] P.A.M. Dirac, Canad. J. Math **3**, 1 (1951).
[5] F.A.E. Pirani and A. Schild, Phys. Rev. **79** (1950).
[6] F.A.E. Pirani, private communication.
[7] P.A.M. Dirac, Proc. Roy. Soc. A **246**, 326 (1958).

 [8] P.A.M. Dirac, Proc. Roy. Soc. A **246**, 333 (1958).
 [9] A. Einstein, L. Infeld, and B. Hoffmann, Ann.of Math. **39**, 65 (1938).
[10] A. Einstein and L. Infeld, Ann.of Math. **41**, 455 (1940).
[11] A. Einstein and L. Infeld, Canad. J. Math. **1**,209 (1949).
[12] P.G. Bergmann, R. Penfield, R. Schiller, and H. Zatzkis, Phys. Rev. **80**, 81 (1950).
[13] F.A.E. Pirani, A. Schild, and R. Skinner, Phys. Rev. **87**, 452 (1952).
[14] J.N. Goldberg, Phys. Rev. **89**, 263 (1953).
[15] T. Damour and N. Deruelle, Phys. Lett. **87A**, 81 (1981).
[16] B.R. Iyer and C.M. Will, Phys. Rev. D **52**, 6882 (1995).
[17] R. Penfield, Phys. Rev. **84**, 737 (1951).
[18] R. Penfield, unpublished.
[19] P.G. Bergmann and J.L. Anderson, Phys. Rev. **83**, 1018 (1951).
[20] P.G. Bergmann and R. Schiller, Phys. Rev. **89**, 4 (1953).
[21] R.P. Feynmann, Rev. Mod. Phys. **20**, 367 (1948).
[22] J. Schwinger, Phys. Rev. **82**, 914 (1951).
[23] J.L. Anderson, *Q-Number Coordinate Transformations and the Ordering Problem in General Relativity* in Eastern Theoretical Conference, ed, M.E. Rose (Gordon and Breach, New York, 1963).
[24] A. Komar, Phys. Rev. D**19**, 2908 (1979)
[25] A. Ashtekar, Lectures on Non-perturbative Canonical Gravity, notes by R. Tate (World Scientific, Singapore, 1991).
[26] P.A.M. Dirac, Phys. Rev. **114**, 924 (1959).

A Biased and Personal Description of GR at Syracuse University, 1951–1961

E. T. Newman

Department of Physics and Astronomy, University of Pittsburgh, Pittsburgh, PA 15260, U.S.A.;
newman@pitt.edu

Summary. In mid-century, General Relativity was largely in the doldrums. Though at the time I was completely unaware of it, there were perhaps only four or five active groups around the world working in GR; Hamburg (Jordan), London (Bondi), Princeton (Wheeler), Warsaw (Infeld) and Syracuse (Bergmann). I had the privilege and pleasure of being a member of the Syracuse group working under Peter G. Bergmann. I would like to describe some of the things that took place there, who were the active participants, who we interacted with, what was accomplished and finally conjecture what role we played in the revitalization of relativity in the late 1950s and early 1960s.

20.1 Preliminaries

As a preliminary remark I want to say that I am not an historian of science — I am (I think) a working mathematical physicist — and I do not know the modus operandi of historians. I once read a history book — the biography of Erwin Schrodinger — and, from it, I thought that historians were only interested in the well-known licentious behavior of physicists. But our good friend John Norton quickly and definitely informed me that I had been misinformed. So I came to this meeting with a talk prepared about general relativity (GR) at Syracuse University in the years 1951–61; completely leaving out the rich details that Norton thought inappropriate. But then listening to all the talks in the first three conference days, I had the realization that I still did not understand how historians of science understood history. Every talk in the first three days dealt with Relativity *before* 1950. So on the night before my talk, thinking that history ended in 1950, I redid my notes so that I could describe to you GR at Syracuse University before 1950. That did present a problem, since there was not much GR done at Syracuse University before 1950 — but I managed to make up some interesting facts. Then to my dismay, this morning, just before my talk, I heard two lectures describing events that took place in the second half of the 1900s. And again I realized that I was wrong. In desperation, in the last 15 minutes, I went back to an earlier and more noble draft of my talk. Unfortunately, by now, it is slightly schizophrenic and disorganized.

But levity aside, the fact is that I really did not know what would be of interest to professional historians. What appears in the following report is my guess — and with it a hope that I have touched on some topics of interest.

20.2 Subject

I start with the premise that general relativity (GR) in the late 1940s in the U.S. was in the doldrums[1]. This view arose as my personal observation made as an undergraduate student at New York University. At the time, nobody seemed either interested in or knowledgeable about GR nor did I find it easy to get information about potential graduate schools that had GR programs. In fact the only school I did find was Syracuse University — which eventually became my academic home for five years. (Jean Eisenstaedt (1) recently pointed out to me a paper of his that gives a much more complete and objective description of this stagnation of the field during this period.)

There are several points that I would like to try to convey to you. The main one being that in the early 1950s, we saw the beginnings of a reawakening of the field — in retrospect one can see that it occurred almost simultaneously in the several different schools (to be discussed below) — but in my inexperienced eyes I only saw it occurring at Syracuse University. The emphasis in my talk will be on the GR group, at Syracuse, under Peter G. Bergmann during the years (1950–61). I will describe the personnel there, a bit of their subsequent careers and some of the external interactions. The second (closely associated) point to be made was the remarkable number and high level of the collaborations that developed during this renaissance, between the different schools from around the world — and the total absence (as far as I could see) of professional jealousies or conflicts. I believe that the very high level of scientific activity in the different groups with the subsequent interactions between groups played a critical role in the rebirth of interest in GR. A prime example of this, (I emphasize and describe in more detail later), is that in the brief period, 1960–62, essentially, the entire theory of gravitational radiation was developed by the strong interaction of many workers from Syracuse, London, Hamburg and Warsaw via personal contacts and word of mouth communication. The published material came afterwards with, to the best of my knowledge, complete attributions and acknowledgments. Though there is no way to prove or document it, I am quite convinced that the high quality of the science came, at least partially, from this free exchange of ideas.

Though Bergmann was deeply involved in many different research projects (e.g., quantum gravity and the search for observables, gravitational radiation and statistical mechanics; see below) the main emphasis here will be on the development of radiation theory. The quantum gravity aspects of the Syracuse program are better known and have been already reported on (2).

20.3 Personal Calendar; for Perspective

Since I am describing my own observations, personal experiences and biases, it seemed to me that, for perspective, my personal calender might be of some interest and value.

1. I graduated as a physics major from New York University in the spring of 1951 and went to Syracuse in Sept. 1951 as a grad student to work with Peter Bergmann. {Anecdote; a few weeks after I arrived in Syracuse I saw that a well-known left-wing journalist, I.F. Stone was giving a public talk. I went to the talk but with considerable trepidation since I had come from a fairly left-wing family background and the time was at the peak of the Joe McCarthy witch-hunt period. In my mild state of paranoia I actually had my collar turned up so that I would not be recognized. As I sat there, to my joy almost the entire Syracuse relativity group openly walked in, talking and laughing happily. From that moment on I felt at home.}

2. Since it was in the midst of the Korean war and I was in danger of being drafted, I stopped for a Master's degree (in 1955) before continuing on to the Ph.D. (1956).

3. I joined the faculty of the Physics Dept., University of Pittsburgh 1956 and remained there until the present — with many leaves of absence.

4. I spent six months in Europe (1958); visiting Copenhagen, Hamburg, London, Dublin, Liverpool. This was my first exposure to the European Relativity Community and was, for me, an eye opener and of the greatest importance in my scientific development.

5. I made yearly return visits to Syracuse, 1956–60.

6. In 1961 I returned for the year to Syracuse. For me this was again a period of great importance.

7. Over the subsequent years I have retained a close relationship with the Syracuse relativity group.

20.4 Participants

Peter G. Bergmann, the leader of the group, though born in Germany, received much of his training with Philip Frank in Prague before coming to the USA as Einstein's assistant. After a brief period on the faculty at Lehigh University, he joined the Syracuse faculty in 1947, remaining there until his retirement in 1982. Though he had many graduate students and post-doctoral fellows over the years, I will list and comment on only those I knew in the 1950s. I will give the names and approximate dates the participants were at Syracuse and then give a few of the salient items of their subsequent careers. Unless stated otherwise, everyone remained as well-known researchers in GR.

The list is divided into two periods; the early years (1951–56) when I was in residence at Syracuse and then the later years (1959–61) when I was a frequent visitor.

I. early years;

a. Josh Goldberg; (grad. student, 1947–52); Wright-Patterson AFB, Syracuse Univ.

b. Jim Anderson; (grad. student, 48–53?), Univ. of Maryland, Stevens Inst. of Tech.

c. Ralph Schiller; (grad. student, 47–53?), Stevens Inst. of Tech. (Switched fields to Biophysics)

d. Ezra T. Newman; (grad.student, 51–56), Univ. of Pittsburgh

e. Irwin Goldberg; (grad.student, 52–57??), Drexel Univ. (He appears to have dropped out completely. I could not find any mention of him any place.)

f. Al Janis; (grad.student; 53–57), Univ. of Pittsburgh

g. Rainer Sachs; (grad.stud; 54–58?), Kings College, London, Berkeley (after an extremely distinguished career in GR, he switched to Biophysics)

h. Jeff Winicour; (grad.stud; 59–64?), Wright-Patterson AFB, Univ. of Pittsburgh

i. Arthur Komar; (post-doc., 58...?), Syr. Univ., Yeshiva Univ.

II. Later Years ˜1959–61 (*=returned, after many years, to Syracuse for a long term visit);

a. Roger Penrose; (Cambridge) — Birkbeck College, London, Oxford Univ., (Rouse Ball Prof of Mathematics)

b. Ivor Robinson; (Cambridge) — Univ of Texas

c. Andrzej Trautman; (Univ. of Warsaw); Univ. of Warsaw

d. Engelbert Schucking; (Hamburg); Univ of Texas, NYU

e. Melvin Schwarz; – Queens College, NY

f. *ET Newman

g. *R. Schiller

h. *A. Komar

i. *R. Sachs returned to Syracuse for short visits in 1961

j. Juergen Ehlers; (Hamburg) Univ. of Texas, Max Planck Inst. Munich, Einstein Institute, Golm, (Director Emeritus)

The intent of this list is to show how the influence of the Syracuse GR Group — the former graduate students, the post-doctoral fellows and long term visitors — spread through the US and Europe. There is no suggested implication that it had a greater or lesser influence than any of the other groups that will be mentioned shortly.

20.5 Major Research Interests of the Syracuse Group

The major research interest of the Syracuse GR group — especially so in the earlier years (1949–58), though the interest continued for many more years — was in Canonical Quantum Gravity. My belief is that the resurgence of world-wide interest in quantum gravity was largely due to the Syracuse group. Someplace in the middle, considerable overlap developed with the work of Dirac and the Princeton group. One of the main contributions made during this period was to the theory of pathological Lagrangians, their related constrained Hamiltonian systems and the search for observables. I myself worked in this area until 1956. The story of this research direction was reported by Josh Goldberg (2) in (1998). I will make no further mention of it other than to say that it was scientific questions concerning quantum gravity that led to the interest in gravitational radiation.

A second research topic of considerable interest to the Bergmann group was the Foundations of Statistical Mechanics. Though there was a great deal of personal interaction between the students doing GR and those doing Statistical Mechanics, it is not clear to me if there was any real scientific spin-off from these interactions.

A third topic of interest was the Theory of Gravitational Radiation & the Asymptotic Behavior of the Gravitational Field. This began with a paper by Bergmann and Sachs in (1958). The balance of this report is devoted to this subject. We will discuss in detail the interactions between members of the different groups and the specific technical ideas that grew from individuals and how they spread.

Remark 1. I point out that Wright-Patterson Air Force Base provided financial support for the Syracuse and King's College groups (among several other relativity groups) from the mid 1950s to the early 1970s — during a most productive period. A question often asked is why did they do so. Though I was not privy to any internal Air Force information, once, when I spent a three month period working at the base, a full-time base-scientist remarked to me that they hoped to be able to understand and perhaps develop anti-gravity devices. It does seem likely that this idea played some role in their financial support. I have always hoped and believed that someone there understood that fundamental science should be supported and was valuable in its own right. I never saw any pressure from them to develop anti-gravity ideas.

20.6 Parallel Developments

At the same time that the Syracuse group was developing, unknown to me was the parallel growth of several other groups. The main ones, from my perspective, were the Princeton Group under John Wheeler — their overlap with the Syracuse group was in the field of quantum gravity — and the Kings College, London, group under Hermann Bondi with the overlap being in gravitational radiation. Other groups playing basic roles were the Univ. of Hamburg group under P. Jordan and the Univ. of Warsaw group under L. Infeld.

Though there is not, in any sense, a unique way to organize the associations, I will give a *rough* grouping of the main players with their early close associations;

I. Bergmann, Goldberg, Sachs, Newman,

II. Bondi, Pirani, Trautman, Penrose, Robinson.

Referring now only to the theory of asymptotically flat space-times and gravitational radiation, I will briefly describe (mainly from my memory) how these main players interacted with each other and what were the scientific/technical ideas that were developed. [In the appendix, I have included a technical glossary of the terms used. A star (*) near a technical term will indicate that its definition can be found in the appendix.] I find it impossible to know precisely who had what idea first — the publication information of (some of) the main papers [also included in the appendix] are not at all a good indicator of when the ideas were developed or even who was the first to propose an idea.

One of the first papers to seriously approach the issue of gravitational radiation was that of Goldberg in 1955 (following discussions with Bergmann) where it was shown, via the EIH approximation for equations of motion, that there was a radiation reaction term in the force law.

Probably the first major idea, for the direct study of gravitational radiation was that of Felix Pirani, namely, to use the degenerate structure (whose existence was first pointed out by E. Cartan in 1922 and rediscovered by the Russian mathematician Petrov) of the principal null vectors* (pnv) of the Weyl tensor* for the discussion of radiation. This material was eventually referred to as the Petrov–Pirani–Penrose Classification* of the Weyl tensor. (From an historical point of view it is perhaps of interest to note that Cayley, studying the algebraic classification of 6×6 matrices had already found essentially the same classification.) Closely associated with Pirani's work was that of his close collaborator Trautman, who found (by generalizing Sommerfeld's work on radiation in Maxwell Theory) the asymptotic fall-off properties of the Weyl tensor for outgoing radiation.

Bondi made probably the major contribution with his realization that the most appropriate way to study gravitational radiation was to introduce null coordinate systems, i.e., systems where one of the coordinates formed a family of null surfaces. He then applied this idea to obtain the asymptotic solution to the Einstein equations with the assumption of axial symmetry and analyticity in 1/r near infinity. The profound result from this work was the proof of the existence of gravitational radiation and the resulting mass loss from the gravitating system—the Bondi mass-loss formula. It has recently been suggested[2] that at about the same time Trautman had obtained the same result.

Sachs, after his initial work with Bergmann, on the asymptotic behavior of linearized multipole fields, moved to Hamburg and, interacting with Ehlers and Schucking, began his generalizations of Bondi's work. First dropping the axial symmetry condition and giving it a covariant formulation, he gave a very general form of the asymptotic radiation metrics. Included there was a geometric formulation of the remaining coordinate freedom (first observed by Bondi) now known as the Bondi–Metzner–Sachs group, the asymptotic symmetry group of a radiation space-time. He also stressed the geometric meaning and importance of several of the relevant variables of the theory, namely the optical parameters;* (the shear, the divergence and the twist of a null congruence) as well as giving their dynamical equations, the optical equations.*

A major result that was developed in this period was the so-called Peeling Theorem*; a powerful detailed description of the algebraic properties of the asymptotic Weyl tensor which associated the different coefficients of powers of (1/r) with the Petrov–Pirani–Penrose Classification.* Though, I believe, much of this insight must have come from Sachs, I do not know exactly who first gave its precise formulation.

Following Ivor Robinson, who first pointed out the importance of studying shear-free null geodesic congruences, Goldberg and Sachs, interacting in London with Bondi's group, proved the beautiful Goldberg–Sachs Theorem.* This theorem shows, for vacuum metrics with degenerate principal null vectors, that the degenerate vectors were tangent to null geodesics and had vanishing shear.* Closely related and in the same period, Robinson and Trautman, interacting in London, integrated the vacuum Einstein equations for spaces with a degenerate principal null vector that had nonvanishing divergence but had vanishing twist. These two works stimulated a great deal of activity over the years.

Starting in a different direction, Penrose (1961) reintroduced[3] and greatly extended the use of spinor algebra and calculus into GR. At first many thought it was just another formal method for stating the Einstein equations but with little practical use. However in 1962 Penrose (using spinors) with Newman (working with the tetrad calculus) developed the spin-coefficient formalism which turned out to be a very powerful tool for the study of GR.[4] They obtained all the results of Bondi, Sachs and Goldberg in a much simpler fashion, the Peeling Theorem was almost obvious and they were able to drop the Bondi–Sachs assumption of analyticity for the asymptotic behavior of the radiation fields. Many further details and applications for radiation theory arose from this formalism. Soon after this Penrose extended these ideas with his introduction of Null Infinity and the conformal compactification of space-time. Many of the concepts and ideas arising from the theory of asymptotically flat spaces, the use of spinors or tetrad calculus, the use of conformal techniques, etc. are still, 40 odd years later, in active use.

In 1961 Bergmann invited many of these players (Robinson, Trautman, Schucking, Newman, Penrose) to Syracuse where many of these ideas were extensively discussed and developed.

The point of this brief history was to highlight the remarkable scientific developments coming from so many places and people, that occurred in such a brief period of time. It seems clear that these results played a major role in the revitalization of GR in the second half of the 20th century. I find it difficult to believe that all the effort and money devoted to the detection of gravitational radiation would have been expended without, for example, the Bondi mass-loss theorem.

20.7 Postlude

I look back on the years (1951–61) as one of the most exciting scientific and personal times of my life. I felt close to virtually all the participants and even up to the present, I keep in close contact with many or most. A sad fact for relativity is that two of the very best, Ray Sachs and Felix Pirani, with no explanation to the community, simply dropped out of relativity at the height of their intellectual powers. I have asked them both for explanations—with no satisfactory answers. On the other hand they both seem to be perfectly content. Pirani has been writing very successful children's books, popular science books and even a play. Sachs has made a completely new scientific career in Radiobiology, Computational Biology and Mathematical Biology. Most of the others have had excellent productive careers in Mathematical Relativity —with Penrose being probably the world's most eminent or famed relativist. {Anecdote: To illustrate my high regard for both Penrose and Sachs, I wish to recount a slightly, for me, embarrassing tale of many years ago. I had submitted a paper to Journal of Math. Phys. (on what is now known as the Kerr–Newman metric) which came back with some excellent referee critical comments that I completely accepted and agreed with. A short time late, while talking with Penrose, I commented to him about this excellent referee report; telling him that it was such a well written report that only one person in the world could have done it, namely Ray Sachs. Penrose, rather sheepishly, replied "maybe there was someone else who could have done it".}

Final Comment: Though I can not think of a single effect or equation or metric in relativity that I would call or refer to as the "Bergmann ....", nor is there a single paper of his that, I could say created a revolution in thought, nevertheless I believe that he was among the clearest and deepest thinkers in relativity. He played a key role in developing the directions the field took — from Quantum Gravity to Radiation Theory — through his publications, his university courses, his gentle but strong influence on his students during the long talks and walks, his conference reports and lectures. His influence in keeping the field alive was inestimable. In addition, he was one of the kindest, most intellectually honest and honorable scientists I have known — and he emphasized to his students the importance of these attributes. And he was loved by his students, post-docs and colleagues — and they carried on the traditions and love for physics and GR in particular, that he had imparted to them.

20.8 Apologia

I want to emphasize that the story I have told here, regarding gravitational radiation and asymptotically flat spaces, is largely from memory — dipping on occasions — into books and papers for references and some memory help from a few friends. If I have errors of fact, I do apologize and hope that I can correct them. However the judgments as to the most important and influential scientific contributions were mine — though perhaps easily argued with. I know that I have left out many friends and colleagues whose work did play a significant role in the research directions described here — some whose direct influence I could easily feel and see were E. Schucking, A. Janis, T. Unti — others, whose influence was there, but further afield from me, were A. Schild and R. Geroch. Probably I have slighted and perhaps even hurt colleagues; but they should know it was done unwittingly — they can blame an aging memory.

References

[1] Jean Eisenstaedt, *Low Water Mark of General Relativity, 1925–1955*, Proceedings of the 1986 Osgood Hill Conference, Einstein Studies Vol.I, D. Howard and J. Sachel Editor, Birkhäuser Boston, 1986.
[2] Josh Goldberg, report to the History of GR Conference, Notre Dame, 1998, to be published.

20.9 Appendix

20.9.1 Technical Glossary

1. The Weyl Tensor; defined from the Curvature Tensor by

$$C_{abcd} = R_{abcd} - \frac{1}{2}(g_{ac}R_{bd} - g_{ad}R_{bc} + g_{bd}R_{ac} - g_{bc}R_{ad})$$
$$- \frac{R}{6}(g_{ad}g_{bc} - g_{ac}g_{bd}).$$

2. Principal Null Vector (pnv) of Weyl Tensor; Four independent algebraic solutions for l_a;

$$l_{[c}C_{a]ef[b}l_{d]}l^e l^f = 0$$
$$g^{ab}l_a l_b = 0$$

sometimes degeneracies (two or more coinciding) => algebraically special.

3. Algebraic Classification; Petrov–Pirani–Penrose Classification

$$Type\ \mathrm{I} \text{or} general = [1, 1, 1, 1],$$
$$Type\mathrm{II} = [2, 1, 1],$$
$$Type\mathrm{D} = [2, 2],$$
$$Type\mathrm{III} = [3, 1],$$
$$Type\mathrm{IV} \text{ or Null} = [4].$$

4. Optical Parameters; for a null geodesic field l_a; $l^a \nabla_a l^b = 0$,

$$divergence = \rho = -\frac{1}{2}\nabla_a l^a,$$
$$shear = |\sigma| = \frac{1}{\sqrt{2}}(\nabla_{(a}l_{b)}\nabla^a l^b - \frac{1}{2}(\nabla_a l^a)^2)^{\frac{1}{2}},$$
$$twist = \nabla_{[a}l_{b]}\nabla^a l^b.$$

5. Optical equations with affine parameter r,

$$\partial_r \rho = \rho^2 + \sigma\bar{\sigma} + Ricci \text{ component},$$
$$\partial_r \sigma = 2\rho\sigma + Weyl \text{ component}.$$

6. Goldberg–Sachs Theorem; Degenerate pnv l^a, $G_{ab} = 0$; ⟺

$$l^a \nabla_a l^b = 0; \text{ i.e.} null geodesic,$$
$$\sigma = 0; \text{ shearfree}.$$

7. Peeling Theorem; Algebraic Properties of Asymptotic Weyl Tensor, $G_{ab} = 0$.

$$C_{aefb} = r^{-1}C_{aefb}^{IV} + r^{-2}C_{aefb}^{II} + r^{-3}C_{aefb}^{II} + r^{-4}C_{aefb}^{I} + O(r^{-5}).$$

8. Bondi mass-loss; the mass term in C_{aefb}^{II} is a monotonically decreasing function in time.

20.9.2 Major contributions to radiation theory

1. J. Goldberg, PR 99, 1873-83 (1955)
2. F. Pirani, (1956), Bull. Acad. Polo. Sci, III, **5**, p. 143 (Introduced Petrov Classification of Algebra of the Weyl tensor.)
3. R. Sachs, P.G. Bergmann; 1958, Phys. Rev. 112, p. 674 (linear theory, definition of multipoles)
4. A. Trautman, King's College Notes; Lectures on General Relativity, 1958, eventually revised and published in "Lectures on General Relativity" Vol.1, Prentice Hall, 1965 (Sommerfeld radiation conditions applied to GR, very influential set of notes.) and recently republished as a "golden oldie" in the GRG Journal.
5. R. Penrose, Ann. Phys., 1960, 10, p. 171. (Major exposition of Spinor Calculus and GR)
6. R. Sachs, Proc. Roy. Soc., 1961, 264, p. 309 (Introduced optical parameters, shear, divergence, twist, asymptotic structure of curvature tensor)
7. F. Pirani, 1961, King's College Notes; published in "Lectures on General Relativity", Vol.1, Prentice Hall, 1965 (Introduce the Petrov Classification of Weyl tensors)
8. J. Goldberg, R. Sachs, (1962), Acta Physica Polonica, Vol. XII, p12 (The Goldberg–Sachs Theorem; Princ. Null Vectors of Algebraically Special Metrics)
9. H. Bondi, M. van der Burg, A. Metzner, 1962, Proc. Roy. Soc. 269, p.21, (Introduction of null coordinates, asymptotic solutions of Einstein equations, mass loss Theorem, BMS group)
10. R. Sachs, Proc. Roy. Soc., 1962, 270, p. 103 (Generalized Bondi work, elucidated the BMS group)
11. E. Newman, R. Penrose, JMP, 1962, 3, p.566 (systematic use of tetrad calculus and spinor analysis, Goldberg–Sachs Theorem, Peeling)
12. E. Newman and T. Unti, 1962, JMP, 3, p. 892 (Asymptotic Integration of the Einstein Eqs, the BMS group)
13. I. Robinson and A. Trautman, 1962, Proc. Roy. Soc. 265 p.463, (Integrated most of the Einstein Eqs. for twist-free algebraically special metrics)
14. R. Penrose, 1963, Phys.Rev. Lttrs, 10, p. 66, (Introduced Null Infinity and Conformal Compactification of Space-Time)

Interaction Chart for the early days of Radiation Theory — much simplified
(Joint Publications)
1. Bergmann–Sachs
2. Goldberg–Sachs
3. Robinson–Trautman
4. Newman-Penrose
5. Bergmann–Schucking–Robinson
(Institutional Interactions)
1. Bergmann, Sachs, Goldberg, Newman
2. Schucking, Ehlers, Sachs
3. Bondi, Pirani, Sachs

Notes

[1] doldrums;.a. A period of stagnation or slump. b. A period of depression or unhappy listlessness.

[2] It was P. Chrusciel, a former student of Trautman, who made this suggestion. Its status is, however, not at all clear. Trautman, who is extremely modest, has not entered into the debate. The issue, which is certainly not a contentious one, is often resolved by different authors by just referring to the result as "the Bondi–Trautman Mass Loss Theorem" with no discussion.

[3] For an early discussion of spinors in GR see, for example, W. L. Bade and H. Jehle, Rev. Mod. Phys. **25**, 714, (1953)

[4] In 1981 it received the Citation Index Award for being one of the most cited papers in GR.